Bill Bryson's opening lines were:
'I come from Des Moines. Somebody had to.'
This is what followed:

The Lost Continent
A road trip around the puzzle that is small-town America introduces the world to the adjective 'Brysonesque'.
> *'A very funny performance, littered with wonderful lines and memorable images'* LITERARY REVIEW

Neither Here nor There
Europe never seemed funny until Bill Bryson looked at it.
> *'Hugely funny (not snigger-snigger funny but great-big-belly-laugh-till-you-cry funny)'* DAILY TELEGRAPH

Made in America
A compelling ride along the Route 66 of American language and popular culture.
> *'A tremendously sassy work, full of zip, pizzazz and all those other great American qualities'* INDEPENDENT ON SUNDAY

Notes from a Small Island
A paean to Bryson's beloved Britain captures the very essence of the original 'green and pleasant land'.
> *'Not a book that should be read in public, for fear of emitting loud snorts'* THE TIMES

A Walk in the Woods
Bryson's punishing (by his standards) hike across the celebrated Appalachian Trail, the longest footpath in the world.
> *'This is a seriously funny book'* SUNDAY TIMES

Notes from a Big Country
Bryson brings his inimitable wit to bear on that strangest of phenomena – the American way of life.
> 'Not only hilarious but also insightful and informative'
> INDEPENDENT ON SUNDAY

Down Under
An extraordinary journey to the heart of another big country – Australia.
> 'Bryson is the perfect travelling companion . . . When it comes to travel's peculiars the man still has no peers' THE TIMES

A Short History of Nearly Everything
Travels through time and space to explain the world, the universe and everything.
> 'Truly impressive . . . It's hard to imagine a better rough guide to science' GUARDIAN

The Life and Times of the Thunderbolt Kid
Quintessential Bryson – a funny, moving and perceptive journey through his childhood.
> 'He can capture the flavour of the past with the lightest of touches' SUNDAY TELEGRAPH

At Home
On a tour of his own house, Bill Bryson gives us an instructive and entertaining history of the way we live.
> 'A work of constant delight and discovery . . . don't leave home without it' SUNDAY TELEGRAPH

One Summer
Bryson travels back in time to a forgotten summer, when America came of age, took centre stage and changed the world for ever.
> 'Has history ever been so enjoyable?' MAIL ON SUNDAY

Winner of the Aventis Prize for Science Writing and the Descartes Science Communication Prize

Praise for *A Short History of Nearly Everything*

'A truly remarkable achievement . . . a bravura performance, which will give much pleasure' *Nature*

'This most enjoyable of books . . . A travelogue of science, with a witty, engaging and well-informed guide'
The Times

'Hugely readable and never obfuscating' *Sunday Times*

'Mr Bryson has a natural gift for clear and vivid expression. I doubt that a better book for the layman about the findings of modern science has been written'
Sunday Telegraph

'Truly impressive . . . It's hard to imagine a better rough guide to science' *Guardian*

'A fascinating idea, and I can't think of many writers, other than Bryson, who would do it this well. It's the sort of book I would have devoured as a teenager. It might well turn unsuspecting young readers into scientists'
Evening Standard

'The very book I have been looking for most of my life . . . Bryson wears his knowledge with aplomb'
Daily Mail

'Lucid, thoughtful and, above all, entertaining' *Scotsman*

www.billbryson.co.uk

www.penguin.co.uk

'Possibly the best scientific primer ever published'
Economist

'It is one of the book's great achievements that Bryson is able to weave a satisfying universal narrative without sparing the reader one whit of scientific ignorance or doubt . . . It represents a wonderful education, and all schools would be better places if it were the core science reader on the curriculum'   *Times Literary Supplement*

'The travel writer gives us a guide to "time, space, the world, the universe and everything". Bryson promises to make geology, chemistry and even particle physics fun and understandable. Move over Stephen Hawking'
FHM

'Genuinely readable and useful . . . Nobody who reads it will ever look at the world around them in the same way again'   *Daily Express*

'Bill Bryson has an unmatched gift for explaining the most difficult subjects in the clearest possible way. If, like me, your brain tends to go numb when faced with terms like plate tectonics, genome, relativity theory, big bang and particle physics, then it is more than likely that [this book] is the cure you have always been looking for'
Mail on Sunday

'One of the most impressive aspects is the breadth of its coverage . . . The huge number of readers who are likely to engage with this book will enjoy themselves while painlessly imbibing a lot of good science . . . Sheer brilliance'   *Times Higher Educational Supplement*

The Road to Little Dribbling
Two decades after *Notes from a Small Island*, Bill Bryson takes a new amble around Britain, to rediscover the beautiful, eccentric and endearing country he calls home.
  '*Clever, witty, entertaining*' INDEPENDENT ON SUNDAY

The Body
Bill Bryson at his very best, this is a must-read owner's manual for everybody.
  '*Readable and useful ... not just comprehensive, but quirky.*'
  THE TIMES

Also by Bill Bryson

The Lost Continent
Mother Tongue
Troublesome Words
Neither Here Nor There
Made in America
Notes from a Small Island
A Walk in the Woods
Notes from a Big Country
(published in the USA as *I'm a Stranger Here Myself*)
Down Under
(published in the USA as *In a Sunburned Country*)
African Diary
A Short History of Nearly Everything
The Life and Times of the Thunderbolt Kid
Shakespeare (Eminent Lives series)
Bryson's Dictionary for Writers and Editors
Icons of England
At Home
One Summer: America 1927
The Road to Little Dribbling
The Body

A SHORT HISTORY OF NEARLY EVERYTHING 2.0

Fully revised and updated

BILL BRYSON

doubleday

TRANSWORLD PUBLISHERS

UK | USA | Canada | Ireland | Australia
India | New Zealand | South Africa

Transworld is part of the Penguin Random House group of companies
whose addresses can be found at global.penguinrandomhouse.com.

Penguin Random House UK, One Embassy Gardens,
8 Viaduct Gardens, London SW11 7BW

penguin.co.uk

First published in Great Britain 2003 by Doubleday
an imprint of Transworld Publishers

This revised and updated edition published 2025

003

Copyright © Bill Bryson 2003, 2025
Text illustrations © Neil Gower 2003, 2025

The moral right of the author has been asserted.

Every effort has been made to obtain the necessary permissions with
reference to copyright material, both illustrative and quoted. We apologize
for any omissions in this respect and will be pleased to make the
appropriate acknowledgements in any future edition.

Penguin Random House values and supports copyright. Copyright fuels creativity,
encourages diverse voices, promotes freedom of expression and supports a vibrant culture.
Thank you for purchasing an authorized edition of this book and for respecting intellectual
property laws by not reproducing, scanning or distributing any part of it by any means
without permission. You are supporting authors and enabling Penguin Random House
to continue to publish books for everyone. No part of this book may be used or reproduced
in any manner for the purpose of training artificial intelligence technologies or systems.
In accordance with Article 4(3) of the DSM Directive 2019/790, Penguin Random House
expressly reserves this work from the text and data mining exception.

Typeset in 11.5/13pt Bembo Book MT Pro by Six Red Marbles UK, Thetford, Norfolk
Printed and bound in Great Britain by Clays Ltd, Elcograf S.p.A.

The authorized representative in the EEA is Penguin Random House Ireland,
Morrison Chambers, 32 Nassau Street, Dublin D02 YH68.

A CIP catalogue record for this book is available from the British Library.

ISBN:
9781529941050 (cased)
9781529966817 (tpb)

Penguin Random House is committed to a sustainable future
for our business, our readers and our planet. This book is made
from Forest Stewardship Council® certified paper.

To Jonny, Xanthe and Chris, and in memory of Jon Davidson.

CONTENTS

Introduction		1
I:	Lost in the Cosmos	
1	How to Build a Universe	11
2	Welcome to the Solar System	22
3	The Reverend Evans's Universe	33
II:	The Size of the Earth	
4	The Measure of Things	49
5	The Stone-Breakers	71
6	Science Red in Tooth and Claw	89
7	Elemental Matters	109
III:	A New Age Dawns	
8	Einstein's Universe	131
9	The Mighty Atom	152
10	Getting the Lead Out	169
11	Muster Mark's Quarks	181
12	The Earth Moves	198

IV:	Dangerous Planet	
13	Bang!	215
14	The Fire Below	236
15	Dangerous Beauty	254

V:	Life Itself	
16	Lonely Planet	269
17	Into the Troposphere	286
18	The Bounding Main	303
19	The Rise of Life	326
20	Small World	341
21	Life Goes On	361
22	Goodbye to All That	375
23	The Richness of Being	392
24	Cells	419
25	Darwin's Singular Notion	432
26	The Stuff of Life	449

VI:	The Road to Us	
27	Ice Time	471
28	The Mysterious Biped	486
29	The Restless Ape	505
30	Goodbye	517

Acknowledgements	529
Bibliography	533
Notes	549
Index	619

The physicist Leo Szilard once announced to his friend Hans Bethe that he was thinking of keeping a diary: 'I don't intend to publish. I am merely going to record the facts for the information of God.' 'Don't you think God knows the facts?' Bethe asked. 'Yes,' said Szilard. 'He knows the facts, but He does not know this version of the facts.'

Hans Christian von Baeyer, Taming the Atom

INTRODUCTION

Welcome. And congratulations. I am delighted that you could make it. Getting here wasn't easy, I know. In fact, I suspect it was a little tougher than you realize.

To begin with, for you to be here now trillions of drifting atoms had somehow to assemble in an intricate and curiously obliging manner to create you. It's an arrangement so specialized and particular that it has never been tried before and will only exist this once. For the next many years (we hope) these tiny particles will uncomplainingly engage in all the billions of deft, cooperative efforts necessary to keep you intact and let you experience the supremely agreeable but generally under-appreciated state known as existence.

Why atoms take this trouble is a bit of a puzzle. Being you is not a gratifying experience at the atomic level. For all their devoted attention, your atoms don't actually care about you – indeed, don't even know that you are there. They don't even know that *they* are there. They are mindless particles, after all, and not even themselves alive. (It is a slightly arresting notion that if you were to pick yourself apart with tweezers, one atom at a time, you would produce a mound of fine atomic dust, none of which had ever been alive but all of which had once

INTRODUCTION

been you.) Yet somehow for the period of your existence they will answer to a single rigid impulse: to keep you you.

The bad news is that atoms are fickle and their time of devotion is fleeting – fleeting indeed. Even a long human life adds up to only about 700,000 hours. And when that modest milestone flashes into view, or at some other point thereabouts, for reasons unknown your atoms will close you down, then silently disassemble and go off to be other things. And that's it for you.

Still, you may rejoice that it happens at all. Generally speaking, in the universe it doesn't, so far as we can tell. This is decidedly odd because the atoms that so liberally and congenially flock together to form living things on Earth are exactly the same atoms that decline to do it elsewhere. Whatever else it may be, at the level of chemistry life is fantastically mundane. All that is required to make you, or any other living thing, is some carbon, hydrogen, oxygen and nitrogen, a little calcium, a dash of sulphur, a light dusting of other very ordinary elements – nothing you wouldn't find in any ordinary pharmacy – and that's all you need. The only thing special about the atoms that make you is that they make you. That is, of course, the miracle of life.

Without atoms there would be no water or air or rocks, no stars and planets, no distant gassy clouds or swirling nebulae or any of the other things that make the universe so agreeably material. Atoms are so numerous and necessary that we easily overlook that they needn't actually exist at all. There is no law that requires the universe to fill itself with small particles of matter or to produce light and gravity and the other properties on which our existence hinges. There needn't actually *be* a universe. For a very long time there wasn't. There were no atoms and no universe for them to float about in. There was nothing – nothing at all anywhere.

So thank goodness for atoms. But the fact that you have atoms and that they assemble in such a willing manner is only part of what got you here. To be here now, alive in the twenty-first

INTRODUCTION

century and smart enough to know it, you also had to be the beneficiary of an extraordinary string of biological good fortune. Survival on Earth is a surprisingly tricky business. Of the billions and billions of species of living things that have existed since the dawn of time, most – 99.99 per cent, it has been suggested – are no longer around. Life on Earth, you see, is not only brief but dismayingly tenuous. It is a curious feature of our existence that we come from a planet that is very good at promoting life but even better at extinguishing it.

The average species on Earth lasts for only between 1 million and 4 million years, so if you wish to be around for billions of years, you must be as fickle as the atoms that made you. You must be prepared to change everything about yourself – shape, size, colour, species affiliation, everything – and to do so repeatedly. That's much easier said than done, because the process of change is random. To get from 'protoplasmal primordial atomic globule' (as Gilbert and Sullivan put it) to sentient upright modern human has required you to mutate new traits over and over in a precisely timely manner for an exceedingly long while. So at various periods over the last 3.8 billion years you have abhorred oxygen and then doted on it, grown fins and limbs and jaunty sails, laid eggs, flicked the air with a forked tongue, been sleek, been furry, lived underground, lived in trees, been as big as a deer and as small as a mouse, and a million things more. The tiniest deviation from any of these evolutionary imperatives and you might now be licking algae from cave walls or lolling walrus-like on some stony shore or disgorging air through a blowhole in the top of your head before diving 60 feet for a mouthful of delicious sandworms.

Not only have you been lucky enough to be attached since time immemorial to a favoured evolutionary line, but you have also been extremely – make that miraculously – fortunate in your personal ancestry. Consider the fact that for 3.8 billion years, a period of time older than the Earth's mountains and rivers and oceans, every one of your forebears on both sides has

INTRODUCTION

been attractive enough to find a mate, healthy enough to reproduce, and sufficiently blessed by fate and circumstances to live long enough to do so. Not one of your pertinent ancestors was squashed, devoured, drowned, starved, stuck fast, untimely wounded or otherwise deflected from its life's quest of delivering a tiny charge of genetic material to the right partner at the right moment to perpetuate the only possible sequence of hereditary combinations that could result – eventually, astoundingly, and all too briefly – in you.

This is a book about how it happened – in particular, how we went from there being nothing at all to there being something, and then how a little of that something turned into us, and also some of what happened in between and since. That's rather a lot to cover, of course, which is why the book is called *A Short History of Nearly Everything*, even though it isn't really. It couldn't be. But with luck by the time we finish it may feel as if it is.

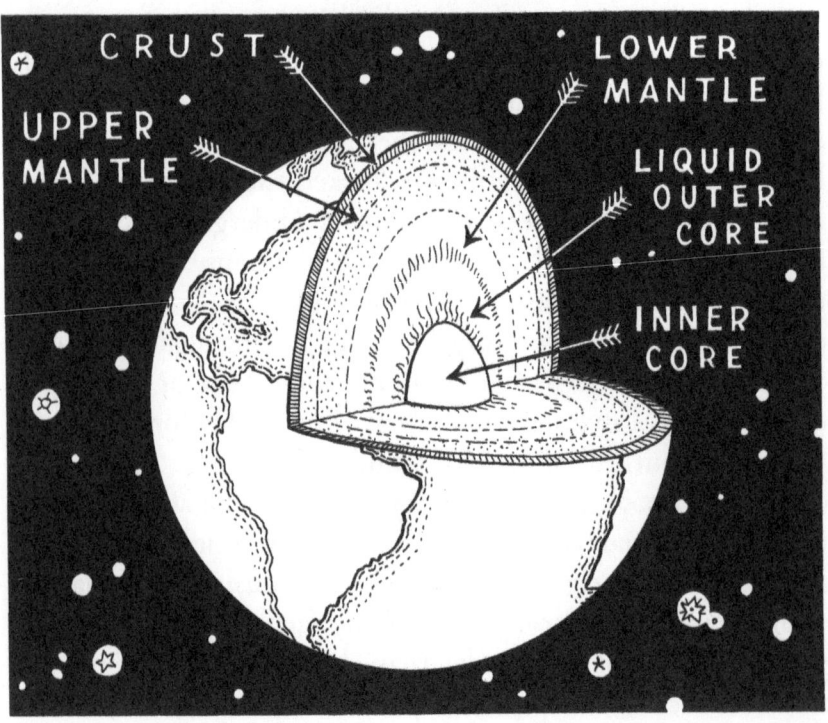

INTRODUCTION

My own starting point, for what it is worth, was a school science book that I had when I was in elementary school. The book was a standard-issue 1950s schoolbook – battered, unloved, grimly hefty – but near the front it had an illustration that just captivated me: a cutaway diagram showing the Earth's interior as it would look if you cut into the planet with a large knife and carefully withdrew a wedge representing about a quarter of its bulk.

It's hard to believe that there was ever a time when I had not seen such an illustration before, but evidently I had not, for I clearly remember being transfixed. I suspect, in honesty, that my initial interest was based on a private image of streams of unsuspecting eastbound motorists in the American plains states plunging over the edge of a sudden 4,000-mile-high cliff running between Central America and the North Pole, but gradually my attention did turn in a more scholarly manner to the scientific import of the drawing and the realization that the Earth consisted of discrete layers, ending in the centre with a glowing sphere of iron and nickel, which was as hot as the surface of the Sun, according to the caption, and I remember thinking with real wonder: 'How do they *know* that?'

I didn't doubt the correctness of the information for an instant – I still tend to trust the pronouncements of scientists in the way I trust those of surgeons, plumbers and other possessors of arcane and privileged knowledge – but I couldn't for the life of me conceive how any human mind could work out what spaces thousands of miles below us, that no eye had ever seen and no X-ray could penetrate, could look like and be made of. To me that was just a miracle. That has been my position with science ever since. Excited, I took the book home that night and opened it before dinner – an action that I expect prompted my mother to feel my forehead and ask if I was all right – and, starting with the first page, I read.

And here's the thing. It wasn't exciting at all. It wasn't actually altogether comprehensible. Above all, it didn't answer any

INTRODUCTION

of the questions that the illustration stirred up in a normal enquiring mind: how did we end up with a Sun in the middle of our planet and how do they know how hot it is? And if it is burning away down there, why isn't the ground under our feet hot to the touch? And why isn't the rest of the interior melting – or is it? And when the core at last burns itself out, will some of the Earth slump into the void, leaving a giant sinkhole on the surface? And how do you *know* this? *How did you figure it out?*

But the author was strangely silent on such details – indeed, silent on everything but anticlines, synclines, axial faults and the like. It was as if he wanted to keep the good stuff secret by making all of it soberly unfathomable. As the years passed, I began to suspect that this was not altogether a private impulse. There seemed to be a mystifying universal conspiracy among textbook authors to make certain that the material they dealt with never strayed into the realm of the frankly exciting.

I now know that there is a happy abundance of science writers who pen the most lucid and thrilling prose – Timothy Ferris, Richard Fortey and Tim Flannery are three that jump out from a single station of the alphabet (and that's not even to mention the late but god-like Richard Feynman) – but, sadly, none of them wrote any textbook I ever used.

All mine were written by men (it was always men) who held the interesting notion that everything became clear when expressed as a formula and the amusingly deluded belief that the children of America would appreciate having chapters end with a section of questions they could mull over in their own time. So I grew up convinced that science was supremely dull, but suspecting that it needn't be, and not really thinking about it at all if I could help it. This, too, became my position for a long time.

Then, much later, when I was well into adulthood, I was on a long flight across the Pacific, staring idly out the window at

INTRODUCTION

moonlit ocean, when it occurred to me with a certain uncomfortable forcefulness that I didn't know the first thing about the only planet I was ever going to live on. I had no idea, for example, why the oceans were salty but the Great Lakes weren't. Didn't have the faintest idea. I didn't know if the oceans were growing more salty with time or less, and whether ocean salinity levels was something I should be concerned about or not. (I am very pleased to tell you that until the late 1970s scientists didn't know the answers to these questions either. They just didn't talk about it very audibly.)

And ocean salinity, of course, represented only the merest sliver of my ignorance. I didn't know what a proton was, or a protein, didn't know a quark from a quasar, didn't understand how geologists could look at a layer of rock on a canyon wall and tell you how old it was – didn't know anything, really. I became gripped by a quiet, unwonted but insistent urge to know a little about these matters and to understand above all how people figured them out. That to me remained the greatest of all amazements – how scientists work things out. How does anybody know how much the Earth weighs or how old its rocks are or what really is way down there in the centre? How can they know how and when the universe started and what it was like when it did?

How do they know what goes on inside an atom? And how, come to that – or perhaps above all, on reflection – can scientists so often seem to know nearly everything but then still not be able to predict an earthquake or even tell us whether we should take an umbrella with us to the races next Wednesday?

So I decided that I would devote a portion of my life – nearly five years altogether, as it now turns out – to reading books and journals and finding saintly, patient experts prepared to answer a lot of outstandingly uninformed questions. The idea was to see if it isn't possible to understand and appreciate – marvel at, enjoy even – the wonder and accomplishments of science at a

INTRODUCTION

level that isn't too technical or demanding, but isn't entirely superficial either.

That was my idea and my hope, and that is what the book that follows is intended to do. Anyway, we have a great deal of ground to cover and much less than 700,000 hours in which to do it, so let's begin.

I

LOST IN THE COSMOS

They're all in the same plane. They're all going around in the same direction ... It's perfect, you know. It's gorgeous. It's almost uncanny.
                      Astronomer Geoffrey Marcy describing the solar system

1

HOW TO BUILD A UNIVERSE

No matter how hard you try you will never be able to grasp just how tiny, how spatially unassuming, is a proton. It is just way too small.

A proton is an infinitesimal part of an atom, which is itself of course an insubstantial thing. Protons are so small that a little dib of ink like the dot on this 'i' can hold something in the region of 10,000,000,000,000,000,000,000 of them. So protons are exceedingly microscopic, to say the very least.

Now imagine if you can (and of course you can't) shrinking one of those protons down to a billionth of its normal size into a space so small that it would make a proton look enormous. Now pack into that tiny, tiny space about an ounce of matter. Excellent. You are ready to start a universe. I'm assuming of course that you wish to build an inflationary universe. If you'd prefer instead to build a more old-fashioned, standard Big Bang universe, you'll need additional materials. In fact, you will need to gather up everything there is – every last mote and particle of matter between here and the edge of creation – and squeeze it into a spot so infinitesimally compact that it has no dimensions at all. It is known as a singularity.

In either case, get ready for a really big bang. Naturally, you will wish to retire to a safe place to observe the spectacle.

Unfortunately, there is nowhere to retire to because outside the singularity there is no *where*. When the universe begins to expand, it won't be spreading out to fill a larger emptiness. The only space that exists is the space it creates as it goes.

It is natural but wrong to visualize the singularity as a kind of pregnant dot hanging in a dark, boundless void. But there is no space, no darkness. The singularity has no around around it. There is no space for it to occupy, no place for it to be. We can't even ask how long it has been there – whether it has just lately popped into being, like a good idea, or whether it has been there for ever, quietly awaiting the right moment. Time doesn't exist. There is no past for it to emerge from.

And so, from nothing, our universe begins.

In a single blinding pulse, a moment of glory much too swift and expansive for any form of words, the singularity assumes heavenly dimensions, space beyond conception. The first lively second (a second that many cosmologists will devote careers to shaving into ever-finer wafers) produces gravity and the other forces that govern physics. In less than a minute the universe is a million billion miles across and growing fast. There is a lot of heat now, 10 billion degrees of it, enough to begin the nuclear reactions that create the lighter elements – principally hydrogen and helium, with a dash (about one atom in a hundred million) of lithium. In three minutes, 98 per cent of all the matter there is or will ever be has been produced. We have a universe. It is a place of the most wondrous and gratifying possibility, and beautiful, too. And it was all done in about the time it takes to make a sandwich.

Until quite recently cosmologists argued over whether that moment of creation was 10 billion years ago or twice that or something in between, but in 2003, using data from a NASA spacecraft with the uncompromisingly nerdy name of Wilkinson Microwave Anisotropy Probe (WMAP), the age of the universe was fixed at a reassuringly precise 13.77 billion years, with a range of uncertainty of just 0.4 per cent.

There is of course a great deal we don't know, and much of what we think we know we haven't known, or thought we've known, for long. Even the notion of the Big Bang is quite a recent one. The idea had been kicking around since the 1920s when Georges Lemaître, a Belgian priest-scholar, first tentatively proposed it, but it didn't really become an active notion in astrophysics until the mid-1960s, when two young radio astronomers made an extraordinary and inadvertent discovery.

Their names were Arno Penzias and Robert Wilson. In 1964, they were trying to make use of a large communications antenna owned by Bell Laboratories at Holmdel, New Jersey, but they were troubled by a persistent background noise – a steady, steamy hiss that made any experimental work impossible. The noise was unrelenting and unfocused. It came from every point in the sky, day and night, through every season. For a year the young astronomers did everything they could think of to track down and eliminate the noise. They tested every electrical system. They rebuilt instruments, checked circuits, wiggled wires, dusted plugs. They climbed into the dish and placed duct tape over every seam and rivet. They climbed back into the dish with brooms and scrubbing brushes and carefully swept it clean of what they referred to in a later paper as 'white dielectric material', or what is known more commonly as bird shit. Nothing they tried worked.

Unknown to them, just 50 kilometres away at Princeton University a team of scientists led by Robert Dicke was working on how to find the very thing they were trying so diligently to get rid of. The Princeton researchers were pursuing an idea that had been suggested in the 1940s by the Russian-born astrophysicist George Gamow: that if you looked deep enough into space you should find some cosmic background radiation left over from the Big Bang. Gamow calculated that by the time it had crossed the vastness of the cosmos the radiation would reach Earth in the form of microwaves. In a more recent paper he had even suggested an instrument that might do the job:

the Bell antenna at Holmdel. Unfortunately, neither Penzias and Wilson, nor any of the Princeton team, had read Gamow's paper.

The noise that Penzias and Wilson were hearing was, of course, the noise that Gamow had postulated. They had found the edge of the universe, or at least the visible part of it, 90 billion trillion miles away. They were 'seeing' the first photons – the most ancient light in the universe – though time and distance had converted them to microwaves, just as Gamow had predicted. In his book *The Inflationary Universe*, Alan Guth provides an analogy that helps to put this finding in perspective. If you think of peering into the depths of the universe as like looking down from the hundredth floor of the Empire State Building (with the hundredth floor representing now and street level representing the moment of the Big Bang), at the time of Wilson and Penzias's discovery the most distant galaxies anyone had ever detected were on about the sixtieth floor and the most distant things – quasars – were on about the twentieth. Penzias and Wilson's finding pushed our acquaintance with the visible universe to within half an inch of the lobby floor.

Still unaware of what caused the noise, Wilson and Penzias phoned Dicke at Princeton and described their problem to him in the hope that he might suggest a solution. Dicke realized at once what the two young men had found. 'Well, boys, we've just been scooped,' he told his colleagues as he hung up the phone. Soon afterwards the *Astrophysical Journal* published two articles: one by Penzias and Wilson describing their experience with the hiss, the other by Dicke's team explaining its nature. Although Penzias and Wilson had not been looking for cosmic background radiation, didn't know what it was when they had found it, and hadn't described or interpreted its character in any paper, they received the 1978 Nobel Prize in Physics. The Princeton researchers got only sympathy. According to Dennis Overbye in *Lonely Hearts of the Cosmos*, neither Penzias nor

Wilson altogether understood the significance of what they had found until they read about it in the *New York Times*.

Incidentally, disturbance from cosmic background radiation is something we have all experienced. Tune your television to any channel it doesn't receive and about 1 per cent of the dancing static you see is accounted for by this ancient remnant of the Big Bang. The next time you complain that there is nothing on, remember that you can always watch the birth of the universe.

Although everyone calls it the Big Bang, many books caution us not to think of it as an explosion in the conventional sense. It was, rather, a vast, sudden expansion on a whopping scale. So what caused it?

One notion is that perhaps the singularity was the relic of an earlier, collapsed universe – that ours is just one of an eternal cycle of expanding and collapsing universes, like the bladder on an oxygen machine. Others attribute the Big Bang to what they call 'a false vacuum' or 'a scalar field' or 'vacuum energy' – some quality or thing, at any rate, that introduced a measure of instability into the nothingness that was. It seems impossible that you could get something from nothing, but the fact that once there was nothing and now there is a universe is evident proof that you can. It may be that our universe is merely part of many larger universes, some in different dimensions, and that Big Bangs are going on all the time all over the place. Or it may be that space and time had some other forms altogether before the Big Bang – forms too alien for us to imagine – and that the Big Bang represents some sort of transition phase, where the universe went from a form we can't understand to one we almost can. As Dr Andrei Linde, a cosmologist at Stanford, once noted, 'These are very close to religious questions.'

The Big Bang theory isn't about the bang itself but about what happened after the bang. Not long after, mind you. By doing a lot of maths and watching carefully what goes on in particle accelerators, scientists believe they can look back to 10^{-43} seconds after the moment of creation, when the universe was still so small that you would have needed a microscope to find it. We mustn't swoon over every extraordinary number that comes before us, but it is perhaps worth latching on to one from time to time just to be reminded of their ungraspable and amazing breadth. Thus 10^{-43} is 0.0000000000000000000000 00000000000000000001, or one-ten-million-trillion-trillion-trillionths of a second.*

Most of what we know, or believe we know, about the early moments of the universe is thanks to an idea called inflation theory first propounded in 1979 by a junior particle physicist then at Stanford, now at MIT, named Alan Guth. He was thirty-two years old and, by his own admission, had never done anything much before. He would probably never have had his great theory except that he happened to attend a lecture on the Big Bang given by none other than Robert Dicke. The lecture inspired Guth to take an interest in cosmology, and in particular in the birth of the universe.

---

* A word on scientific notation. Since very large numbers are cumbersome to write and nearly impossible to read, scientists use a shorthand involving powers (or multiples) of ten in which, for instance, 10,000,000,000 is written 10^{10} and 6,500,000 becomes 6.5×10^6. The principle is based very simply on multiples of ten: 10×10 (or 100) becomes 10^2; $10 \times 10 \times 10$ (or 1,000) is 10^3; and so on, obviously and indefinitely. The little superscript number signifies the number of zeroes following the larger principal number. Negative notations provide essentially a mirror image, with the superscript number indicating the number of spaces to the right of the decimal point (so 10^{-4} means 0.0001). Though I salute the principle, it remains an amazement to me that anyone seeing '1.4×10^9 km³' would see at once that that signifies 1.4 billion cubic kilometres, and no less a wonder that they would choose the former over the latter in print (especially in a book designed for the general reader, where the example was found). On the assumption that many readers are as unmathematical as I am, I will use notations sparingly, though they are occasionally unavoidable, not least in a chapter dealing with things on a cosmic scale.

HOW TO BUILD A UNIVERSE

The eventual result was the inflation theory, which holds that a fraction of a moment after the dawn of creation, the universe underwent a sudden dramatic expansion. It inflated – in effect ran away with itself, doubling in size every 10^{-34} seconds. The whole episode may have lasted no more than 10^{-30} seconds – that's one-million-million-million-million-millionths of a second – but it changed the universe from something you could hold in your hand to something at least 10,000,000,000,000,000,000,000,000 times bigger. Inflation theory explains the ripples and eddies that make our universe possible. Without it, there would be no clumps of matter and thus no stars, just drifting gas and everlasting darkness. According to Guth's theory, at one ten-millionth of a trillionth of a trillionth of a trillionth of a second, gravity emerged. After another ludicrously brief interval it was joined by electromagnetism and the strong and weak nuclear forces – the stuff of physics. These were followed an instant later by shoals of elementary particles – the stuff of stuff. From nothing at all, suddenly there were swarms of photons, protons, electrons, neutrons and much else – between 10^{79} and 10^{89} of each, according to the standard Big Bang theory.

Such quantities are of course ungraspable. It is enough to know that in a single instant we were endowed with a universe that was vast – at least 100 billion light years across, according to the theory, but possibly any size up to infinite – and perfectly arrayed for the creation of stars, galaxies and other complex systems.*

---

* A question that naturally arises is how the universe could expand faster than the speed of light if, as we all know, nothing moves faster than the speed of light. The simple answer is that the law applies only to objects moving *within* space, and in the case of universal expansion it is space itself that is on the move, not the objects within it. Imagine a loaf of raisin bread rising in a warm oven. As it expands, the raisins embedded in it move further apart, but they are not travelling *through* the dough, but rather just moving along with it. If the raisins were moving independently, they could not go faster than the speed of light, but the loaf/universe has no such constraints.

What is extraordinary from our point of view is how well it turned out for us. If the universe had formed just a tiny bit differently – if gravity were fractionally stronger or weaker, if the expansion had proceeded just a little more slowly or swiftly – then there might never have been stable elements to make you and me and the ground we stand on. Had gravity been a trifle stronger, the universe itself might have collapsed like a badly erected tent without precisely the right values to give it the necessary dimensions and density and component parts. Had it been weaker, however, nothing would have coalesced. The universe would have remained for ever a dull, scattered void.

This is one reason why some experts believe that there may have been many other big bangs, perhaps trillions and trillions of them, spread through the mighty span of eternity, and that the reason we exist in this particular one is that this is one that we *could* exist in. As Edward P. Tryon of Columbia University once put it: 'In answer to the question of why it happened, I offer the modest proposal that our Universe is simply one of those things which happen from time to time.' To which adds Guth: 'Although the creation of a universe might be very unlikely, Tryon emphasized that no one had counted the failed attempts.'

Martin Rees (formally Baron Rees of Ludlow), Britain's former Astronomer Royal, has suggested that there may be many universes, possibly an infinite number, each with different attributes, in different combinations, and that we simply live in one that combines things in the way that allows us to exist. He has made an analogy with a very large clothing store: 'If there is a large stock of clothing, you're not surprised to find a suit that fits. If there are many universes, each governed by a differing set of numbers, there will be one where there is a particular set of numbers suitable to life. We are in that one.'

Rees notes that six numbers in particular govern our universe, and that if any of these values were changed even very slightly things could not be as they are. For example, for the universe to exist as it does requires that hydrogen be converted to helium in a precise but comparatively stately manner – specifically, in a way that converts seven one-thousandths of its mass to energy. Lower that value very slightly – from 0.07 per cent to 0.06 per cent, say – and no transformation could take place: the universe would consist of hydrogen and nothing else. Raise the value very slightly – to 0.08 per cent – and bonding would be so wildly prolific that the hydrogen would long since have been exhausted. In either case, with the slightest tweaking of the numbers the universe as we know and need it would not be here.

I should say that everything is just right *so far*. In the long term, gravity may turn out to be a little too strong; one day it may halt the expansion of the universe and bring it collapsing in upon itself, until it crushes itself down into another singularity, possibly to start the whole process over again. On the other hand, it may be too weak, in which case the universe will keep racing away for ever until everything is so far apart that there is no chance of material interactions, so that the universe becomes a place that is very roomy, but inert and dead. The third option is that gravity is perfectly pitched – 'critical density' is the cosmologists' term for it – and that it will hold the universe together at just the right dimensions to allow things to go on indefinitely. Cosmologists, in their lighter moments, sometimes call this the 'Goldilocks effect' – that everything is just right. (For the record, these three possible universes are known respectively as closed, open and flat.)

Now, the question that has occurred to all of us at some point is: what would happen if you travelled out to the edge of the universe and, as it were, put your head through the curtains? Where would your head *be* if it were no longer in the universe? What would you find beyond? The answer, disappointingly, is that you can never get to the edge of the universe. That's not because it would take too long to get there – though of course it would – but because even if you travelled outward and outward in a straight line, indefinitely and pugnaciously, you would never arrive at an outer boundary. Instead, you would come back to where you began (at which point, presumably, you would rather lose heart in the exercise and give up). The reason for this is that the universe bends, in a way we can't adequately imagine, in conformance with Einstein's theory of relativity (which we will get to in due course). For the moment it is enough to know that we are not adrift in some large, ever-expanding bubble. Rather, space curves, in a way that allows it to be boundless but finite. Space cannot even properly be said to be expanding because, as the physicist and Nobel laureate Steven Weinberg long ago noted, 'solar systems and galaxies are not expanding, and space itself is not expanding.' Rather, the galaxies are rushing apart. It is all something of a challenge to intuition. Or, as the biologist J. B. S. Haldane once famously observed: 'The universe is not only queerer than we suppose; it is queerer than we *can* suppose.'

The analogy that is usually given for explaining the curvature of space is to try to imagine someone from a universe of flat surfaces, who had never seen a sphere, being brought to Earth. No matter how far he roamed across the planet's surface, he would never find an edge. He might eventually return to the spot where he had started, and would of course be utterly confounded to explain how that had happened. Well, we are in the same position in space as our puzzled flatlander, only we are flummoxed by a higher dimension.

Just as there is no place where you can find the edge of the universe, so there is no place where you can stand at the centre and say: 'This is where it all began. This is the centremost point of it all.' We are *all* at the centre of it all. Actually, we don't know that for sure; we can't prove it mathematically. Scientists just assume that we can't really be the centre of the universe – think what that would imply – but that the phenomenon must be the same for all observers in all places. Still, we don't actually know.

For us, the universe goes only as far as light has travelled in the billions of years since the universe was formed. This visible universe – the universe we know and can talk about – is estimated to be about 94 billion light years across. But according to most theories the universe at large – the meta-universe, as it is sometimes called – is vastly roomier still. According to Rees, the number of light years to the edge of this larger, unseen universe would be written not 'with ten zeroes, not even with a hundred, but with millions'. In short, there's more space than you can imagine already without going to the trouble of trying to envision some additional beyond.

For a long time the Big Bang theory had one gaping hole that troubled a lot of people – namely, that it couldn't begin to explain how we got here. Although 98 per cent of all the matter that exists was created with the Big Bang, that matter consisted exclusively of light gases: the helium, hydrogen and lithium that we mentioned earlier. Not one particle of the heavy stuff so vital to our own being – carbon, nitrogen, oxygen and all the rest – emerged from the gaseous brew of creation. But – and here's the troubling point – to forge these heavy elements, you need the kind of heat and energy thrown off by a Big Bang. Yet there has been only one Big Bang and it didn't produce them. So where did they come from? Interestingly, the man who found the answer to that question was a cosmologist who heartily despised the Big Bang as a theory and coined the term Big Bang facetiously, as a way of mocking it.

We'll get to him shortly, but before we turn to the question of how we got here, it might be worth taking a few minutes to consider just where exactly 'here' is.

2

WELCOME TO THE SOLAR SYSTEM

Astronomers these days can do the most amazing things. If someone struck a match on the Moon, they could spot the flare. From the tiniest throbs and wobbles of distant stars they can infer the size and character and even potential habitability of planets much too remote to be seen – planets so distant that it would take us half a million years in a spaceship to get there. With their radio telescopes they can capture wisps of radiation so preposterously faint that the total amount of energy collected from outside the solar system by all of them together since collecting began (in 1951) is 'less than the energy of a single snowflake striking the ground', in the words of Carl Sagan. That was in 1980, so the total has clearly gone up but the amount of accumulated energy is still infinitesimal.

In short, there isn't a great deal that goes on in the universe that astronomers can't find when they have a mind to. Which is why it is all the more remarkable to reflect that until 1978 no one had ever noticed that Pluto has a moon. In the summer of that year, a young astronomer named James Christy at the Lowell Observatory in Flagstaff, Arizona, was making a routine examination of photographic images of Pluto when he saw that there was something there – something blurry and uncertain but definitely other than Pluto. Consulting a colleague named Robert Harrington, he concluded that what he was looking at was a moon. And it wasn't just any moon. Relative to the planet, it was the biggest moon in the solar system.

This was actually something of a blow to Pluto's status as a planet, which had never been terribly robust anyway. Since previously the space occupied by the moon and the space occupied by Pluto were thought to be one and the same, it meant that Pluto was much smaller than anyone had supposed – smaller even than Mercury. Indeed, several moons in the solar system, including our own, are larger.

Now, a natural question is why it took so long for anyone to find a moon in our own solar system. In fact, it may surprise you to hear, new moons are being found all the time. In 2023 alone, the International Astronomical Union recognized twelve new moons around Jupiter, bringing its total to ninety-five, nearly three times the number known when this book was first published. Only a few weeks after that announcement, Saturn's moon count was increased by a whopping eighty-three, bringing its total to 145, and in early 2025 the total was increased again, by 128, to a truly spectacular 274, or more moons than all the rest of the planets in the solar system put together. Discoveries have been coming so thick and fast, in fact, that new, smaller discoveries are no longer being given names from mythology, as they were traditionally, but just dry catalogue designations like S/2021 J3 and S/2017 J7.

The reason these moons were not found earlier is that they are mostly small – only a mile or two across – and often orbiting far out from their host planet, and sometimes orbiting in the opposite direction from their fellow moons. Many of the new moons are shaped like potatoes and are really just big rocks. Indeed, the only universally accepted definition of a moon at present is that it is a solid object orbiting a planet, irrespective of its size. By that definition there are almost certainly hundreds, possibly thousands, more to be discovered.

Indeed, Pluto, we now know, hasn't just one moon but *five*:

Christy's discovery of 1978, called Charon, and four others (Nix, Styx, Kerberos and Hydra), which were found between 2005 and 2012.

As for Pluto itself, it's been a tough few years. Its reign as our most distant planet didn't even last a single human lifetime. It was discovered, rather miraculously, in 1930 by a young man from Kansas named Clyde Tombaugh who had no formal training in astronomy and was looking for something else anyway. Tombaugh had been hired by the Lowell Observatory in Arizona to try to find a massive, gassy planet on the edge of the solar system that the observatory's wealthy founder, Percival Lowell, was certain was out there.

Lowell is best remembered now for his long-influential belief that Mars was covered with canals built by industrious Martians for purposes of conveying water from polar regions to the dry but productive lands nearer the equator. Lowell's other abiding conviction was that there existed, somewhere out beyond Neptune, an undiscovered ninth planet, dubbed Planet X. Lowell based this belief on irregularities he detected in the orbits of Uranus and Neptune, and spent years obsessively but unsuccessfully searching for it.

After Lowell's death, it fell to Tombaugh to continue the quest. Although Tombaugh had never attended college and his knowledge of the universe came largely from reading *Popular Astronomy* magazine, he was extraordinarily diligent, and after a year's patient searching he somehow spotted Pluto, a faint point of light in a glittery firmament. It was a miraculous find, and what made it all the more striking was that the observations on which Lowell had predicted the existence of a planet beyond Neptune proved to be comprehensively erroneous. Tombaugh could see at once that the new planet was nothing like the massive gasball Lowell had postulated, but any reservations he or anyone else had about the character of the new planet were soon swept aside in the delirium that attended almost any big news story in that easily excited age. This was the first American-discovered planet, and no one was going to be distracted by the thought that it was really just a distant icy dot. It was named Pluto at least partly because the first two letters made a monogram from Lowell's initials. Lowell was posthumously hailed everywhere as a genius of the first order and Tombaugh was largely forgotten, except among planetary astronomers, who tend to revere him.

So the solar system had a ninth planet, but it was quite an odd and mysterious one, in large part because it was so distant and anomalously small. Many astronomers thought it not a planet at all, but merely the largest object so far found in a zone of galactic debris known as the Kuiper Belt. The Kuiper Belt was actually theorized by an astronomer named F. C. Leonard in 1930, but the name honours Gerard Kuiper, a Dutch native working in America, who expanded the idea.

It is certainly true that Pluto doesn't act much like the other planets. Whereas the others orbit on more or less the same plane, Pluto's orbital path is tipped out of alignment at an angle of 17 degrees, like the brim of a hat tilted rakishly on someone's head. Its orbit is so irregular that for substantial periods on each of its lonely circuits around the Sun it is closer to us than Neptune is. For most of the 1980s and 1990s, Neptune was in fact the solar system's most far-flung planet. Only on 11 February 1999 did Pluto return to the outside lane, there to remain for the next 228 years.

So if Pluto really was a planet, it was certainly an odd one. It would mean that our planetary system consisted of four rocky inner planets, four gassy outer giants, and a tiny, solitary iceball.

Matters came to a head of sorts in the summer of 2005 when astronomers from the Palomar Observatory in California announced the discovery of a Kuiper Belt object even bigger than Pluto, following an eccentric orbit that takes it up to three times further out from the Sun than Pluto – so far out that a single orbit of the Sun takes it 560 years. The discoverers informally named it Xena, after the warrior princess in a popular

1990s television show, but eventually — and, let's be frank, mercifully — that was discarded for the more seemly classical name Eris, the Greek goddess of strife.

The following year, at an unusually lively meeting in Prague, the International Astronomical Union decreed that Eris was not a proper planet at all, but fell into a new category to be called dwarf planets. Even more controversially, the same meeting likewise demoted Pluto (along with Ceres, which lives in the asteroid belt between Mars and Jupiter and never had planetary status anyway). Interestingly, it was only then that astronomers got round to creating a formal definition of what constitutes a planet. They decided it must meet three criteria: it must orbit the Sun, it must be massive and spherical, and it must exert enough gravity to absorb or scatter any rocky debris that passes nearby. Pluto failed on this last count.

Ironically, just under a decade after its demotion, NASA's New Horizon spacecraft flew past Pluto and discovered it was way more interesting than previously supposed. Far from being an inert ball of ice, Pluto had active glaciers, mountains as big as those in the Rockies, probably an ocean of liquid water beneath the surface, and signs of volcanic activity. It was, in short, one of the liveliest places in the solar system.

But it's still not a planet.

So now we have a solar system with eight planets, five dwarf planets (Pluto, Ceres, Makemake, Haumea and Eris), at least 400 moons, more than 1.3 million asteroids, and a little under 4,000 comets. But, as we have already seen, many of those numbers change nearly as fast as they can be recorded. As for Pluto, in the 76 years it was a planet it didn't even manage one full orbit of the Sun.

The central point in all this is that we live in a solar system that in many ways we still hardly know. Compared with the vastness of the universe at large, our solar system may seem comparatively compact and neighbourly, but in fact it is roomy beyond our capacity to imagine.

WELCOME TO THE SOLAR SYSTEM

Let us pretend, for purposes of edification and entertainment, that we are about to go on a journey by rocketship. We won't go terribly far – just to the edge of our own solar system – but we need to get a fix on how big a place space is and what a small part of it we occupy. Now the bad news, I'm afraid, is that we won't be home for supper. Even at the speed of light it would take seven hours to reach Pluto. But of course we can't travel at anything like that speed. We'll have to go at the speed of a spaceship, and these are rather more lumbering. The Voyager 1 spacecraft, which launched from Earth in 1977 and used a rare alignment of Jupiter, Saturn, Uranus and Neptune as a series of celestial slingshots to fling it onward at ever greater speeds, has travelled 15 billion miles in half a century and is increasing that distance by nearly a million miles every day, but it is still nowhere near the edge of the solar system.*

So get ready for a long trip. As we proceed, you will quickly notice that space is extremely well named and rather dismayingly uneventful. Our solar system may be the liveliest thing for trillions of miles, but all the visible stuff in it – the Sun, the planets and their moons, the tumbling rocks of the asteroid belt, the 4,500 or so known comets and all the other miscellaneous drifting detritus – fills less than a trillionth of the available space. You also quickly realize that none of the maps you have ever seen of the solar system was drawn remotely to scale. Most schoolroom charts show the planets coming one after the other at neighbourly intervals – the outer giants actually cast shadows over each other in many illustrations – but this is a necessary deceit to get them all on the same piece of paper. Neptune in reality isn't just a little bit beyond Jupiter, it's *way* beyond Jupiter – five

---

* Voyager 1 and its sister craft Voyager 2 were designed to last for only about five years, but as of this writing were still sending back usable data to Earth – an achievement all the more remarkable when you consider that their onboard computers 'have less memory than the key fob that opens your car door', according to Linda Spilker, a planetary scientist at the NASA Jet Propulsion Laboratory.

times further from Jupiter than Jupiter is from us, so far out that it receives only 3 per cent as much sunlight as Jupiter.

On a model of the solar system to scale, with the Earth reduced to about the diameter of a pea, Jupiter would be over 300 metres away and Pluto would be 2.5 kilometres distant (and about the size of a bacterium, so you wouldn't be able to see it anyway). On the same scale, Proxima Centauri, our nearest star, would be 16,000 kilometres away. Even if you shrank down everything so that Jupiter was as small as the full stop at the end of this sentence and Pluto was no bigger than a molecule, Pluto would still be over 10 metres away. So the solar system is really quite enormous. By the time we reach Pluto, we have come so far that the Sun – our dear, warm, skin-tanning, life-giving Sun – has shrunk to the size of a pinhead. It is little more than a bright star. In such a lonely void you can begin to understand how even quite significant objects have escaped attention.

Now, the other thing you will notice as we speed past Pluto is that we are speeding past Pluto. If you check your itinerary, you will see that this is a trip to the edge of our solar system, and I'm afraid we're not there yet. Pluto may be the last object traditionally marked on schoolroom charts, but the solar system doesn't end there. In fact, it isn't even close to ending there. We won't get to the solar system's edge until we have passed through the Oort Cloud* – a vast and, it must be said, entirely hypothesized realm of drifting comets that is thought to encircle our solar system like a giant soap bubble. Because no one has ever seen the Oort Cloud, we can't be exactly certain where it begins and ends, but the best estimates are that it will take us about 300 years to reach it and perhaps 30,000 to pass out the other side. (Other sources think it could take more like 6,000 years. Whatever the exact figure, it will take generations.)

---

* Properly called the Öpik–Oort Cloud, it is named for the Estonian astronomer Ernst Öpik, who hypothesized its existence in 1932, and for the Dutch astronomer Jan Oort, who refined the calculations eighteen years later.

WELCOME TO THE SOLAR SYSTEM

Of course we have no prospect of such a journey. Based on what we know now and can reasonably imagine, there is absolutely no prospect that any human being will ever visit the edge of our own solar system – ever. It is just too far. The basic unit of measure in the solar system is the Astronomical Unit, or AU, representing the distance from the Sun to the Earth (and defined, very precisely, as 149,597,870.7 kilometres). Pluto is about 40 AUs from us, the heart of the Oort Cloud about 50,000. In a word, it is remote.

But let's pretend again, for convenience, that we have actually made it to the Oort Cloud. The first thing you might notice is how very peaceful it is out here. We're a long way from anywhere now – so far from our own Sun that it's not even the brightest star in the sky. It is a remarkable thought that that distant tiny twinkle has enough gravity to hold all these comets in orbit. It's not a very strong bond, so the comets drift in a stately manner, moving at only about 220 miles an hour. From time to time one of these lonely comets is nudged out of its normal orbit by some slight gravitational perturbation. Sometimes they are ejected into the emptiness of space, never to be seen again, but sometimes they fall into a long orbit around the Sun. About three or four of these a year, known as long-period comets, pass through the inner solar system. Just occasionally these stray visitors smack into something solid, like Earth. That's why we've come out here now – because the comet we have come to see has just begun a long fall towards the centre of the solar system. It is headed for, of all places, Manson, Iowa. It is going to take a long time to get there – 3 or 4 million years at least – so we'll leave it for now, and return to it much later in the story.

So that's your solar system. And what else is out there, beyond the solar system? Well, nothing much and a great deal, depending on how you look at it.

In the short term, it's nothing much. The most perfect vacuum ever created by humans is not as empty as the emptiness of interstellar space. And there is a great deal of this nothingness until you get to the next bit of something. Our nearest neighbour in the cosmos, Proxima Centauri, which is part of the three-star cluster known as Alpha Centauri, is 4.3 light years away, almost nothing in galactic terms, but still 100 million times further than a trip to the Moon. To reach it by spaceship would take perhaps 50,000 years, and even if you made the trip you still wouldn't be anywhere except at a lonely clutch of stars in the middle of a vast nowhere. To reach the next landmark of consequence, Sirius, would involve another 4.6 light years of travel. And so it would go if you tried to star-hop your way across the cosmos. Just reaching the centre of our own galaxy would take far longer than we have existed as beings.

Space, let me repeat, is enormous. The average distance between stars out there is over 30 million million kilometres. Even at speeds approaching those of light, these are fantastically challenging distances for any travelling individual. Of course, it is *possible* that alien beings travel billions of miles to amuse themselves by planting crop circles in Wiltshire or frightening the daylights out of some poor guy in a pickup truck on a lonely road in Arizona (they must have teenagers, after all), but it does seem unlikely.

Still, statistically the probability that there are other thinking beings out there is good. Nobody knows how many stars there are in the Milky Way – estimates range from 100 billion or so to perhaps 400 billion – and the Milky Way is just one of unknowable billions of galaxies in the observable universe (some estimates now put the number as high as 2 trillion). In the 1960s, a professor at Cornell named Frank Drake, intrigued by the massive numbers, worked out a famous equation designed to calculate the chances of advanced life existing in the cosmos, based on a series of diminishing probabilities.

Under Drake's equation you divide the number of stars in a selected portion of the universe by the number of stars that are likely to have planetary systems; divide that by the number of planetary systems that could theoretically support life; divide that by the number on which life, having arisen, advances to a state of intelligence; and so on. At each such division, the number shrinks colossally – yet even with the most conservative inputs the number of advanced civilizations just in the Milky Way always works out to be somewhere in the millions.

What an interesting and exciting thought. We may be only one of millions of advanced civilizations just in our own galaxy. Unfortunately, space being spacious, the average distance between any two of these civilizations is reckoned to be at least 200 light years, which is a great deal more than merely saying it makes it sound. It means, for a start, that even if these beings know we are here and are somehow able to see us in their telescopes, they're watching light that left Earth 200 years ago. So they're not seeing you and me. They're watching people who don't know what an atom is, or a gene, or even a railway train. Any message we receive from these observers is likely to begin by congratulating us on the handsomeness of our horses and our mastery of whale oil. Two hundred light years is a distance so far beyond us as to be, well, just beyond us. As the British astronomer Martin Rees has pointed out, 'Even if a life-bearing planet were nearby, it would in all probability be at a different stage of evolution. And the life there could be based on something else altogether – methane rather than water, say – which would make it entirely unlike our own.'

Such is likely to be the case with a planet called K2-18b, which in April 2025 was reported to contain an abundance of dimethyl sulfide, a molecule that occurs on Earth only as a by-product of living organisms, thus making K2-18b the clearest candidate yet for life elsewhere in the universe. There is no telling at this stage, however, how complex that life may be, if it exists at all, and in any case K2-18b is 720 trillion miles away, so well beyond any practical contact or close examination. Dimethyl sulfide, for what it is worth, smells horrible.

So even if we are not literally alone, in all practical terms we are. Carl Sagan calculated the number of probable planets in the universe at as many as 10 billion trillion – a number vastly beyond imagining. But what is equally beyond imagining is the amount of space through which they are lightly scattered.

'If we were randomly inserted into the universe,' Sagan wrote, 'the chances that you would be on or near a planet would be less than one in a billion trillion trillion.' (That's 10^{33}, or 1 followed by thirty-three zeroes.)

Sagan died in 1996 at the cruelly early age of sixty-two, so those calculations will have changed, but his melancholy conclusion has not. 'Worlds are precious,' he wrote.

3

THE REVEREND EVANS'S UNIVERSE

For years and years, when the skies were clear and the Moon not too bright, the Reverend Robert Evans, a quiet and cheerful man, would lug a bulky telescope on to the back sun-deck of his home in the Blue Mountains of Australia, about 80 kilometres west of Sydney, and do something extraordinary. He would look deep into the past and find dying stars.

Looking into the past was, of course, the easy part. Glance at the night sky and what you see is history and lots of it – not the stars as they are now, but as they were when their light left them. For all we know, the North Star,* our faithful companion, might actually have burned out last January or in 1854 or at any time since the early eighteenth century and news of it just hasn't reached us yet. The best we can say – can ever say – is that it was still burning on this date 323 years ago. Stars die all the time. What Bob Evans could do better than anyone else who ever tried was spot these moments of celestial farewell. He was, in short, the world's greatest hunter of supernovae.

A supernova occurs when a giant star, one much bigger than

---

* The distance to the North Star, or Polaris to use its official name, was only definitively calculated in 2012, putting it considerably closer to Earth than previously thought.

our own Sun, collapses and then spectacularly explodes, releasing in an instant the energy of 100 billion suns, burning for a time more brightly than all the stars in its galaxy. 'It's like a trillion hydrogen bombs going off at once,' Evans told me when we first met at his home in 2001. A semi-retired minister in the Uniting Church of Australia, Evans was the most modest and in many ways most exceptional of the people I met for this book.

Despite the violence of their dying explosions, most supernovae are so unimaginably distant that their light reaches us as no more than the faintest twinkle. For the month or so that they are visible, all that distinguishes them from the other stars in the sky is that they occupy a point of space that wasn't filled before. It is these anomalous, very occasional pricks in the crowded dome of the night sky that the Reverend Evans had an uncanny – indeed, quite unparalleled – knack for spotting.

To understand what a feat this was, imagine a standard dining-room table covered in a black tablecloth and throwing a handful of salt across it. The scattered grains can be thought of as a galaxy. Now imagine 1,500 more tables like the first one – enough to make a single line 2 miles long – each with a random array of salt across it. Now add one grain of salt to any table and let Bob Evans walk among them. At a glance he would spot it. That grain of salt was the supernova.

Evans's was a talent so exceptional that Oliver Sacks, in *An Anthropologist on Mars*, devoted a passage to him in a chapter on autistic savants – quickly adding that 'there is no suggestion that he is autistic'. Evans, who never met Sacks, laughed at the suggestion that he might be either autistic or a savant, but was powerless to explain quite where his talent came from.

'I just seem to have a knack for memorizing star fields,' he told me, with a frankly apologetic look, when I visited him and his wife, Elaine, in their picture-book bungalow on a tranquil edge of the village of Hazelbrook, out where Sydney finally ends and the boundless Australian bush begins. 'I'm not particularly good at other things,' he added. 'I don't remember names well.'

'Or where he's put things,' called Elaine from the kitchen. He nodded frankly again and grinned, then asked me if I'd like to see his telescope. I had imagined that Evans would have a proper observatory in his back yard – a scaled-down version of a Mount Wilson or Palomar, with a sliding domed roof and a mechanized chair that would be a pleasure to manoeuvre. In fact, he led me not outside but to a crowded storeroom off the kitchen where he kept his books and papers and where his telescope – a white cylinder about the size and shape of a household hot-water tank – rested in a home-made, swivelling plywood mount. When he wished to observe, he would carry them, in two trips, to the small sun-deck off the kitchen. Between the overhang of the roof and the feathery tops of eucalyptus trees growing up from the slope below, he had only a letterbox view of the sky, but he assured me it was more than good enough for his purposes.

The term supernova was coined in the 1930s by a memorably odd astrophysicist named Fritz Zwicky. Born in Bulgaria and raised in Switzerland, Zwicky came to the California Institute of Technology in the 1920s and there at once distinguished himself by his abrasive personality and erratic talents. He didn't seem to be outstandingly bright, and many of his colleagues considered him little more than 'an irritating buffoon'. A fitness fanatic, he would often drop to the floor of the Caltech dining hall or some other public area and do one-armed push-ups to demonstrate his virility to anyone who seemed inclined to doubt it. He was notoriously aggressive, his manner eventually becoming so intimidating that his closest collaborator, a gentle man named Walter Baade, refused to be left alone with him. Among other things, Zwicky accused Baade, who was

German, of being a Nazi, which he was not. On at least one occasion Zwicky threatened to kill Baade, who worked up the hill at the Mount Wilson Observatory, if he saw him on the Caltech campus.

But Zwicky was also capable of insights of the most startling brilliance. In the early 1930s he turned his attention to a question that had long troubled astronomers: the appearance in the sky of occasional unexplained points of light, new stars. Improbably, he wondered if the neutron — the subatomic particle that had just been discovered in England by James Chadwick, and was thus both novel and rather fashionable — might be at the heart of things. It occurred to him that if a star collapsed to the sort of densities found in the core of atoms, the result would be an unimaginably compacted core. Atoms would literally be crushed together, their electrons forced into the nucleus, forming neutrons. You would have a neutron star. Imagine a million really weighty cannonballs squeezed down to the size of a marble and — well, you're still not even close. The core of a neutron star is so dense that a single spoonful of matter from it would weigh more than 500 billion kilograms. A spoonful! But there was more. Zwicky realized that after the collapse of such a star there would be a huge amount of energy left over — enough to make the biggest bang in the universe. He called these resultant explosions supernovae. They would be — they are — the biggest events in creation.

On 15 January 1934 the journal *Physical Review* published a very concise abstract of a presentation that had been conducted by Zwicky and Baade the previous month at Stanford University. Despite its extreme brevity — one paragraph of twenty-four lines — the abstract contained an enormous amount of new science: it provided the first reference to supernovae and to neutron stars; convincingly explained their method of formation; correctly calculated the scale of their explosiveness; and, as a kind of concluding bonus, connected supernova explosions to the production of a mysterious new phenomenon called cosmic rays,* which had recently been found swarming through the universe. These ideas were revolutionary, to say the least. The existence of neutron stars wouldn't be confirmed for thirty-four years, and cosmic rays, though long since confirmed, still remain little understood. Altogether, the abstract was, in the words of the astrophysicist and Nobel laureate Kip S. Thorne, 'one of the most prescient documents in the history of physics and astronomy'.

Interestingly, Zwicky had almost no understanding of why any of this would happen. According to Thorne, 'he did not understand the laws of physics well enough to be able to substantiate his ideas.' Zwicky's talent was for big ideas. Others – Baade mostly – were left to do the mathematical sweeping up.

Zwicky was also the first to recognize that there wasn't nearly enough visible mass in the universe to hold galaxies together, and that there must be some other gravitational influence – what we now call dark matter. One thing he failed to see was that if a neutron star shrank enough it would become so dense that even light couldn't escape its immense gravitational pull. You would have a black hole. Unfortunately, Zwicky was held in such disdain by most of his colleagues that his ideas attracted almost no notice. When, five years later, the great Robert Oppenheimer turned his attention to neutron stars in a landmark paper, he made not a single reference to any of Zwicky's work, even though Zwicky had been working for years on the same problem in an office just down the corridor. Zwicky's deductions concerning dark matter wouldn't attract serious

---

* Cosmic rays, first posited in 1925 by Robert Millikan of the University of Chicago, aren't rays at all but charged particles, moving at nearly the speed of light. The name has stuck even though it is wrong (and that doesn't happen often in science). To what extent cosmic rays influence life and climate on Earth is still debated.

attention for nearly four decades. We can only assume that he did a lot of push-ups in this period.

Surprisingly little of the universe is visible to us when we incline our heads to the sky. Only about 6,000 stars are visible to the naked eye from Earth, and only about 2,000 can be seen from any one spot. With binoculars the number of stars you can see from a single location rises to about 50,000, and with a small 2-inch telescope it leaps to 300,000. With a 16-inch telescope, such as Evans used, you begin to count not in stars but in galaxies. From his deck, Evans told me, he estimated he could see between 50,000 and 100,000 galaxies, each containing tens of billions of stars. These were of course respectable numbers, but even with so much to take in, supernovae are extremely rare. A star can burn for billions of years, but it dies just once and quickly, and only a few dying stars explode. Most expire quietly, like a camp fire at dawn. In a typical galaxy, consisting of 100 billion stars, a supernova will occur on average once every 200 or 300 years. Looking for a supernova, therefore, was a little like standing on the observation platform of the Empire State Building with a telescope and searching windows around Manhattan in the hope of finding, let us say, someone lighting a twenty-first birthday cake.

So when a hopeful and softly spoken minister got in touch to ask if they had any usable field charts for hunting supernovae, the astronomical community thought he was out of his mind. At the time Evans had a 10-inch telescope – a very respectable size for amateur star-gazing, but hardly the sort of thing with which to do serious cosmology – and he was proposing to find one of the universe's rarer phenomena. In the whole of astronomical history before Evans started looking in 1980, fewer than sixty supernovae had been found.

Evans, however, had certain advantages. Most observers, like most people generally, are in the northern hemisphere, so he had a lot of sky largely to himself, especially at first. He also had speed and his uncanny memory. Large telescopes are cumbersome things, and much of their operational time is consumed in being manoeuvred into position. Evans could swing his little 16-inch telescope around like a tail-gunner in a dogfight, spending no more than a couple of seconds on any particular point in the sky. In consequence, he could observe perhaps 400 galaxies in an evening while a large professional telescope would be lucky to do fifty or sixty.

Looking for supernovae is mostly a matter of not finding them. From 1980 to 1996 he averaged two discoveries a year – not a huge pay-off for hundreds of nights of peering and peering. Once he found three in fifteen days, but another time he went three years without finding any at all.

'There is actually a certain value in not finding anything,' he said. 'It helps cosmologists to work out the rate at which galaxies are evolving. It's one of those rare areas where the absence of evidence *is* evidence.'

On a table beside the telescope were stacks of photos and papers relevant to his pursuits, and he showed me some of them. If you have ever looked through popular astronomical publications, and at some time you must have, you will know that they are generally full of richly luminous colour photos of distant nebulae and the like – fairy-lit clouds of celestial light of the most delicate and moving splendour. Evans's working images were nothing like that. They were just blurry black-and-white photos with little points of haloed brightness. One he showed me depicted a swarm of stars in which lurked a trifling flare that I had to put close to my face to discern. This, Evans told me, was a star in a constellation called Fornax from a galaxy known to astronomy as NGC1365. (NGC stands for New General Catalogue, where these things are recorded. Once it was a heavy book on someone's desk in Dublin; today, needless to say, it's a database.) For 60 million years, the light from this star's spectacular demise travelled through space until one night in August 2001 it arrived at Earth in the form of a puff of radiance, the tiniest brightening, in the night sky.

It was, of course, Robert Evans on his eucalypt-scented hillside who spotted it.

'There's something satisfying, I think,' Evans said, 'about the idea of light travelling for millions of years through space and *just* at the right moment as it reaches Earth someone looks at the right bit of sky and sees it. It just seems right that an event of that magnitude should be witnessed.'

Supernovae do much more than simply impart a sense of wonder. They come in several types (one of them discovered by Evans), and of these, one in particular, known as the Ia supernova, is important to astronomy because these supernovae always explode in the same way, with the same critical mass. For this reason they can be used as 'standard candles' – benchmarks by which to measure the brightness (and hence relative distance) of other stars, and thus to measure the expansion rate of the universe. Supernovae are also an important source of gravitational waves, which were first detected by astrophysicists in 2015, to much jubilation, and which will be discussed a bit further on.

In 1987 Saul Perlmutter at the Lawrence Berkeley Laboratory in California, needing more Ia supernovae than visual sightings were providing, set out to find a more systematic method of searching for them. Perlmutter devised a nifty system using sophisticated computers and charge-coupled devices – in essence, really good digital cameras. It automated supernova-hunting. Telescopes could now take thousands of pictures and let a computer detect the telltale bright spots that marked a supernova explosion. In five years, with the new technique, Perlmutter and his colleagues at Berkeley found forty-two supernovae. 'With CCDs you can aim a telescope at the sky and go watch television,' Evans said with a touch of dismay. 'It took all the romance out of it.'

I asked him if he was tempted to adopt the new technology.

'Oh, no,' he said, 'I enjoy my way too much. Besides' – he gave a nod at the photo of his latest supernova and smiled – 'I can still beat them sometimes.'

Robert Evans died on 8 November 2022, aged eighty-five. He had discovered forty-seven supernovae, a number that is unlikely ever to be surpassed.

The question that naturally occurs is: what would it be like if a star exploded nearby? Our nearest stellar neighbour, as we have seen, is Alpha Centauri, 4.3 light years away. I had imagined that if there were an explosion there we would have 4.3 years to watch the light of this magnificent event spreading across the sky, as if tipped from a giant can. What would it be like if we had four years and four months to watch an inescapable doom advancing towards us, knowing that when it finally arrived it would blow the skin right off our bones? Would people still go to work? Would farmers plant crops? Would anyone deliver them to the shops?

Weeks later, back in the town in New Hampshire where I then lived, I put these questions to John Thorstensen, an astronomer at Dartmouth College. 'Oh no,' he said, laughing. 'The news of such an event travels out at the speed of light, but so does the destructiveness, so you'd learn about it and die from it in the same instant. But don't worry, because it's not going to happen.'

For the blast of a supernova explosion to kill you, he explained, you would have to be 'ridiculously close' – probably within ten light years or so. 'The danger would be various types of radiation – cosmic rays and so on.' These would produce fabulous auroras, shimmering curtains of spooky light that would fill the whole sky. This would not be a good thing. Anything potent enough to put on such a show could well blow away the magnetosphere, the magnetic zone high above the Earth that normally protects us from ultraviolet rays and other cosmic assaults. Without the magnetosphere anyone unfortunate enough to step into sunlight would pretty quickly take on the appearance of, let us say, an overcooked pizza.

The reason we can be reasonably confident that such an

event won't happen in our corner of the galaxy, Thorstensen said, is that it takes a particular kind of star to make a supernova in the first place. A candidate star must be ten to twenty times as massive as our own Sun, and 'we don't have anything of the requisite size that's that close. The universe is a mercifully big place.' The nearest likely candidate, he added, is Betelgeuse, whose various sputterings have for years suggested that something interestingly unstable is going on there. But Betelgeuse is 642.5 light years away – distant enough not to be a worry.

Only half a dozen times in recorded history have supernovae been close enough to be visible to the naked eye. One was a blast in 1054 that created the Crab Nebula. Another, in 1604, made a star bright enough to be seen during the day for over three weeks. The most recent was in 1987, when a supernova flared in a zone of the cosmos known as the Large Magellanic Cloud, but that was only barely visible and only in the southern hemisphere – and it was a comfortably safe 169,000 light years away.

Supernovae are significant to us in one other decidedly central way. Without them we wouldn't be here. You will recall the cosmological conundrum with which we ended the first chapter – that the Big Bang created lots of light gases but no heavy elements. Those came later, but for a very long time nobody could figure out *how* they came later. The problem was that you needed something really hot – hotter even than the middle of the hottest stars – to forge carbon and iron and the other elements without which we would be distressingly immaterial. Supernovae provided the explanation, and it was an English cosmologist almost as singular in manner as Fritz Zwicky who worked it out.

He was a Yorkshireman named Fred Hoyle. Hoyle, who died in 2001, was described in an obituary in *Nature* as a 'cosmologist and controversialist', and both of those he most certainly was. He was, according to *Nature*'s obituary, 'embroiled in controversy for most of his life' and 'put his name to much rubbish'. He claimed, for instance, and without evidence, that the Natural History Museum's treasured fossil of an archaeopteryx was a forgery along the lines of the Piltdown hoax, causing much exasperation to the museum's palaeontologists, who had to spend days fielding phone calls from journalists all over the world. He also believed that the Earth was seeded from space not only by life but also by many of its diseases, such as influenza and bubonic plague, and suggested at one point that humans evolved projecting noses with the nostrils underneath as a way of keeping cosmic pathogens from falling into them.

It was he who coined the term Big Bang, for a radio broadcast in 1949. He pointed out that nothing in our understanding of physics could account for why everything, gathered to a point, would suddenly and dramatically begin to expand. Hoyle favoured a steady-state theory in which the universe was constantly expanding and continually creating new matter as it went. Hoyle also realized that if stars imploded they would liberate huge amounts of heat – 100 million degrees or more, enough to begin to generate the heavier elements in a process known as nucleosynthesis. In 1957, working with others, Hoyle showed how the heavier elements were formed in supernova explosions. For this work, W. A. Fowler, one of his collaborators, received a Nobel Prize. Hoyle, shamefully, did not.

According to Hoyle's theory, an exploding star would generate enough heat to create all the new elements and spray them into the cosmos where they would form gaseous clouds – the interstellar medium, as it is known – that could eventually coalesce into new solar systems. With the new theories it became possible at last to construct plausible scenarios for how we got here. What we now think we know is this:

About 4.6 billion years ago, a great swirl of gas and dust some 24 billion kilometres across accumulated in space where we are now and began to aggregate. Virtually all of it – 99.9 per cent of the mass of the solar system – went to make the Sun. Out of the floating material that was left over, two microscopic grains floated close enough together to be joined by electrostatic forces. This was the moment of conception for our planet. All over the inchoate solar system, the same was happening. Colliding dust grains formed larger and larger clumps. Eventually the clumps grew large enough to be called planetesimals. As these endlessly bumped and collided, they fractured or split or recombined in endless random permutations, but in every encounter there was a winner, and some of the winners grew big enough to dominate the orbit around which they travelled.

It all happened remarkably quickly. To grow from a tiny cluster of grains to a baby planet some hundreds of kilometres across is thought to have taken only a few tens of thousands of years. In just 200 million years, possibly less, the Earth was essentially formed, though still molten and subject to constant bombardment from all the debris that remained floating about.

At this point, about 4.4 billion years ago, an object the size of Mars crashed into the Earth, blowing out enough material to form a companion sphere, the Moon. Within weeks, it is thought, the flung material had reassembled itself into a single clump, and within a year it had formed into the spherical rock that companions us yet. Most of the lunar material, it is thought, came from the Earth's crust, not its core, which is why the Moon has so little iron while we have a lot. The theory, incidentally, is almost always presented as a recent one, but in fact it was first proposed in the 1940s by Reginald Daly of Harvard. The only recent thing about it is people paying any attention to it.

When the Earth was only about a third of its eventual size, it was probably already beginning to form an atmosphere, mostly of carbon dioxide, nitrogen, methane and sulphur. Hardly the sort of stuff that we would associate with life, and yet from this noxious stew life formed. Carbon dioxide is a powerful greenhouse gas. This was a good thing, because the Sun was significantly dimmer back then. Had we not had the benefit of a greenhouse effect, the Earth might well have frozen over permanently, and life might never have got a toehold. But somehow life did.

For the next 500 million years the young Earth continued to be pelted relentlessly by comets, meteorites and other galactic debris, which brought water to fill the oceans and the components necessary for the successful formation of life. It was a singularly hostile environment, and yet somehow life got going. Some tiny bag of chemicals twitched and became animate. We were on our way.

Four billion years later, people began to wonder how it had all happened. And it is there that our story next takes us.

II

THE SIZE OF THE EARTH

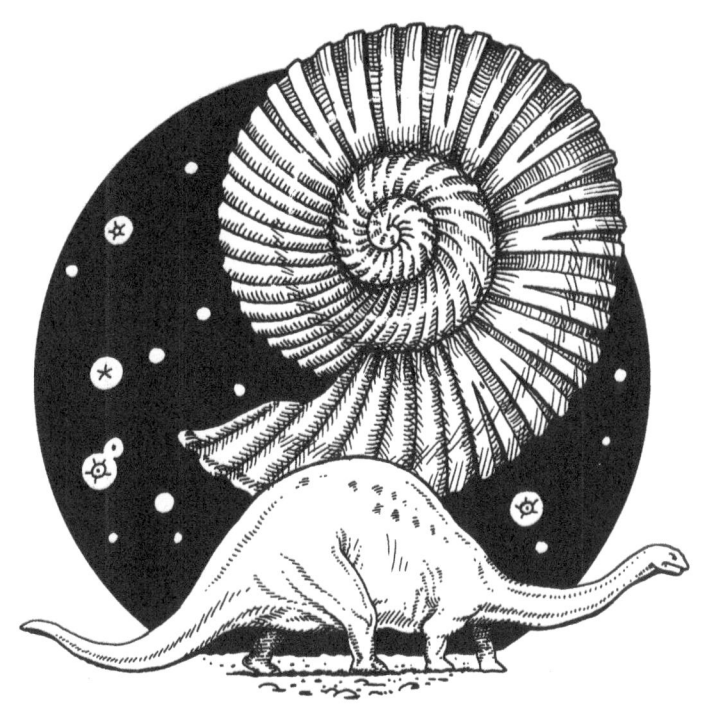

4

THE MEASURE OF THINGS

If you had to select the least convivial scientific field trip of all time, you could certainly do worse than the French Royal Academy of Science's Peruvian expedition of 1735. Led by a hydrologist named Pierre Bouguer and a soldier-mathematician named Charles Marie de La Condamine, it was a party of scientists and adventurers who travelled to Peru with the purpose of triangulating distances through the Andes.

At the time people had lately become infected with a powerful desire to understand the Earth – to determine how old it was, and how massive, where it hung in space, and how it had come to be. The French party's goal was to help settle the question of the circumference of the planet by measuring the length of one degree of meridian (or one-360th of the distance around the planet) along a line reaching from Yarouqui, near Quito, to just beyond Cuenca in what is now Ecuador, a distance of about 320 kilometres.*

---

* Triangulation, their chosen method, was a popular technique based on the geometric fact that if you know the length of one side of a triangle and the angles of two corners, you can work out all its other dimensions without leaving your chair. Suppose, by way of example, that you and I decided we wished to know how far it is to the Moon. Using triangulation, the first thing we must do is put

THE SIZE OF THE EARTH

Almost at once things began to go wrong, sometimes spectacularly so. In Quito, the visitors somehow provoked the locals and were chased out of town by a mob armed with stones. Soon after, the expedition's doctor was murdered in a misunderstanding over a woman. The botanist became deranged. Others died of fevers and falls. The third most senior member of the party, a man named Jean Godin, ran off with a thirteen-year-old girl and could not be induced to return.

At one point the group had to suspend work for eight months while La Condamine rode off to Lima to sort out a problem with their permits. Eventually he and Bouguer stopped speaking and refused to work together. Everywhere the dwindling party went it was met with the deepest suspicions from officials who found it difficult to believe that a group of French scientists would travel halfway around the world to measure the world. That made no sense at all. Two and a half centuries later, it still seems a reasonable question. Why didn't the French make their measurements in France and save themselves all the bother and discomfort of their Andean adventure?

The answer lies partly with the fact that eighteenth-century scientists, the French in particular, seldom did things simply

---

some distance between us, so let's say for argument that you stay in Paris and I go to Moscow and we both look at the Moon at the same time. Now, if you imagine a line connecting the three principals of this exercise – that is, you and me and the Moon – it forms a triangle. Measure the length of the baseline between you and me and the angles of our two corners and the rest can be simply calculated. (Because the interior angles of a triangle always add up to 180 degrees, if you know the sum of two of the angles you can instantly calculate the third; and knowing the precise shape of a triangle and the length of one side tells you the lengths of the other sides.) This was in fact the method used by a Greek astronomer, Hipparchus of Nicaea, in 150 BC to work out the Moon's distance from the Earth. At ground level the principles of triangulation are the same, except that the triangles don't reach into space but rather are laid side to side on a map. In measuring a degree of meridian, the surveyors would create a sort of chain of triangles marching across the landscape.

if an absurdly demanding alternative was available, and partly with a practical problem that had first arisen with the English astronomer Edmond Halley many years before – long before Bouguer and La Condamine dreamed of going to South America, much less had a reason for doing so.

Halley was an exceptional figure. In the course of a long and productive career he was a sea captain, a cartographer, a professor of geometry at the University of Oxford, deputy controller of the Royal Mint, Astronomer Royal, and inventor of the deep-sea diving bell. He wrote authoritatively on magnetism, tides and the motions of the planets, and fondly on the effects of opium. He invented the weather map and actuarial table, proposed methods for working out the age of the Earth and its distance from the Sun, and even devised a practical method for keeping fish fresh out of season. The one thing he didn't do was discover the comet that bears his name. He merely recognized that the comet he saw in 1682 was the same one that had been seen by others in 1456, 1531 and 1607. It didn't become Halley's comet until 1758, some sixteen years after his death.

For all his achievements, however, Halley's greatest contribution to human knowledge may simply have been to take part in a modest scientific wager with two other worthies of his day: Robert Hooke, who is perhaps best remembered now as the first person to describe a cell, and the great and stately Sir Christopher Wren, who was actually an astronomer first and an architect second, though that is not often generally remembered now. In 1683, Halley, Hooke and Wren were dining in London when the conversation turned to the motions of celestial objects. It was known that planets were inclined to orbit in a particular kind of oval known as an ellipse – 'a very specific and precise curve', to quote Richard Feynman – but it wasn't understood why. Wren generously offered a prize worth 40 shillings (equivalent to a couple of weeks' pay) to whichever of the men could provide a solution.

Hooke, who was well known for taking credit for ideas that

weren't necessarily his own, claimed that he had solved the problem already but declined now to share it on the interesting and inventive grounds that it would rob others of the satisfaction of discovering the answer for themselves. He would instead 'conceal it for some time, that others might know how to value it'. If he thought any more on the matter, he left no evidence of it. Halley, however, became consumed with finding the answer, to the point that the following year he travelled to Cambridge and boldly called upon the university's Lucasian Professor of Mathematics, Isaac Newton, in the hope that he could help.

Newton was a decidedly odd figure – brilliant beyond measure, but solitary, joyless, prickly to the point of paranoia, famously distracted (upon swinging his feet out of bed in the morning he would reportedly sometimes sit for hours, immobilized by the sudden rush of thoughts to his head), and capable of the most riveting strangeness. He built his own laboratory, the first at Cambridge, but then engaged in the most bizarre experiments. Once he inserted a bodkin – a long needle of the sort used for sewing leather – into his eye socket and rubbed it around 'betwixt my eye and the bone as near to [the] backside of my eye as I could' just to see what would happen. What happened, miraculously, was nothing – at least, nothing lasting. On another occasion, he stared at the Sun for as long as he could bear, to determine what effect it would have upon his vision. Again he escaped lasting damage, though he had to spend some days in a darkened room before his eyes forgave him.

Set atop these odd beliefs and quirky traits, however, was the mind of a supreme genius – though even when working in conventional channels he often showed a tendency to peculiarity. As a student, frustrated by the limitations of conventional mathematics, he invented an entirely new form, the calculus, but then told no one about it for twenty-seven years. In like manner, he did work in optics that transformed our understanding of light and laid the foundation for the science of spectroscopy, and again chose not to share the results for three decades.

For all his brilliance, real science accounted for only a part of his interests. At least half his working life was given over to alchemy and wayward religious pursuits. These were not mere dabblings but wholehearted devotions. He was a secret adherent of a dangerously heretical sect called Arianism, whose principal tenet was the belief that there had been no Holy Trinity (slightly ironic, since Newton's college at Cambridge was Trinity). He spent endless hours studying the floor plan of the lost Temple of King Solomon in Jerusalem (teaching himself Hebrew in the process, the better to scan original texts) in the belief that it held mathematical clues to the dates of the second coming of Christ and the end of the world. His attachment to alchemy was no less ardent. In 1936, the economist John Maynard Keynes bought a trunk of Newton's papers at auction and discovered with astonishment that they were overwhelmingly preoccupied not with optics or planetary motions, but with a single-minded quest to turn base metals into precious ones. An analysis of a strand of Newton's hair in the 1970s found it contained mercury – an element of interest to alchemists, hatters and thermometer-makers but almost no one else – at a concentration some forty times the natural level. It is perhaps little wonder that he had trouble remembering to get up in the morning.

Quite what Halley expected to get from him when he made his unannounced visit in August 1684 we can only guess. But thanks to the later account of a Newton confidant, Abraham DeMoivre, we do have a record of one of science's most historic encounters:

> In 1684 Dr Halley came to visit at Cambridge [and] after they had some time together the Dr asked him what he thought the curve would be that would be described by the Planets supposing the force of attraction towards the Sun to be reciprocal to the square of their distance from it.

THE SIZE OF THE EARTH

This was a reference to a piece of mathematics known as the inverse square law, which Halley was convinced lay at the heart of the explanation, though he wasn't sure exactly how.

> Sr Isaac replied immediately that it would be an [ellipse]. The Doctor, struck with joy & amazement, asked him how he knew it. 'Why,' saith he, 'I have calculated it,' whereupon Dr Halley asked him for his calculation without farther delay. Sr Isaac looked among his papers but could not find it.

This was astounding – like someone saying he had found a cure for cancer but couldn't remember where he had put the formula. Pressed by Halley, Newton agreed to redo the calculations and produce a paper. He did as promised, but then did much more. He retired for two years of intensive reflection and scribbling, and at length produced his masterwork: the *Philosophiae Naturalis Principia Mathematica* or *Mathematical Principles of Natural Philosophy*, better known as the *Principia*.

Once in a great while, a few times in history, a human mind produces an observation so acute and unexpected that people can't quite decide which is the more amazing – the fact or the thinking of it. The appearance of the *Principia* was one of those moments. It made Newton instantly famous. For the rest of his life he would be draped with plaudits and honours, becoming, among much else, the first person in Britain knighted for scientific achievement. Even the great German mathematician Gottfried von Leibniz, with whom Newton had a long, bitter fight over priority for the invention of the calculus, thought his contributions to mathematics equal to all the accumulated work that had preceded him. 'Nearer the gods no mortal may approach,' wrote Halley in a sentiment that was endlessly echoed by his contemporaries and by many others since.

Although the *Principia* has been called 'one of the most inaccessible books ever written' (Newton intentionally made it difficult so that he wouldn't be pestered by mathematical 'smatterers', as he called them), it was a beacon to those who could follow it. It not only explained mathematically the orbits of heavenly bodies, but also identified the attractive force that got them moving in the first place – gravity. Suddenly every motion in the universe made sense.

At the *Principia*'s heart were Newton's three laws of motion (which state, very baldly, that a thing moves in the direction in which it is pushed; that it will keep moving in a straight line until some other force acts to slow or deflect it; and that every action has an opposite and equal reaction) and his universal law of gravitation. This states that every object in the universe exerts a tug on every other. It may not seem like it, but as you sit here now you are pulling everything around you – walls, ceiling, lamp, pet cat – towards you with your own little (indeed, very little) gravitational field. And these things are also pulling on you. It was Newton who realized that the pull of any two objects is, to quote Feynman again, 'proportional to the mass of each and varies inversely as the square of the distance between them'. Put another way, if you double the distance between two objects, the attraction between them becomes four times weaker. This can be expressed with the formula

$$F = G \frac{Mm}{r^2}$$

which is of course way beyond anything that most of us could make practical use of, but at least we can appreciate that it is elegantly compact. A couple of brief multiplications, a simple division and, bingo, you know your gravitational position wherever you go. It was the first really universal law of nature ever propounded by a human mind, which is why Newton is everywhere regarded with such profound esteem.

The *Principia*'s production was not without drama. To Halley's horror, just as work was nearing completion Newton and Hooke fell into dispute over the priority for the inverse square law and Newton refused to release the crucial third volume, without which the first two made little sense. Only with some frantic shuttle diplomacy and the most liberal applications of flattery did Halley manage finally to extract the concluding volume from the erratic professor.

Halley's traumas were not yet quite over. The Royal Society had promised to publish the work but now pulled out, citing financial embarrassment. The year before, the society had backed a costly flop called *The History of Fishes*, and suspected that the market for a book on mathematical principles would be less than clamorous. Halley, whose means were not great, paid for the book's publication out of his own pocket. Newton, as was his custom, contributed nothing. To make matters worse, Halley at this time had just accepted a position as the society's clerk, and he was informed that the society could no longer afford to provide him with a promised salary of £50 per annum. He was to be paid instead in copies of *The History of Fishes*.

Newton's laws explained so many things – the slosh and roll of ocean tides, the motions of planets, why cannonballs trace a particular trajectory before thudding back to earth, why we aren't flung into space as the planet spins beneath us at hundreds of kilometres an hour* – that it took a while for all their implications to seep in. But one revelation became almost immediately controversial.

This was the suggestion that the Earth is not quite round. According to Newton's theory, the centrifugal force of the Earth's spin should result in a slight flattening at the poles and a bulging at the equator, which would make the planet slightly oblate. That meant that the length of a degree of meridian wouldn't be the same in Italy as it was in Scotland. Specifically, the length would shorten as you moved away from the poles. This was not good news for

---

* How fast you are spinning depends on where you are. The speed of the Earth's spin varies from something over 1,600 kilometres an hour at the equator to zero at the poles. In London the speed is 998 kilometres an hour.

those people whose measurements of the planet were based on the assumption that it was a perfect sphere, which was everyone.

For half a century people had been trying to work out the size of the Earth, mostly by making very exacting measurements. One of the first such attempts was by an English mathematician named Richard Norwood. As a young man Norwood had travelled to Bermuda with a diving bell modelled on Halley's device, intending to make a fortune scooping pearls from the seabed. The scheme failed because there were no pearls and anyway Norwood's bell didn't work, but Norwood was not one to waste an experience. In the early seventeenth century Bermuda was well known among ships' captains for being hard to locate. The problem was that the ocean was big, Bermuda small and the navigational tools for dealing with this disparity hopelessly inadequate. There wasn't even yet an agreed length for a nautical mile. Over the breadth of an ocean the smallest miscalculations would become magnified so that ships often missed Bermuda-sized targets by dismayingly large margins.

Norwood, whose first love was trigonometry and thus angles, decided to bring a little mathematical rigour to navigation, and to that end he determined to calculate the length of a degree.

Starting with his back against the Tower of London, Norwood spent two devoted years marching 208 miles north to York, repeatedly stretching and measuring a length of chain as he went, all the while making the most meticulous adjustments for the rise and fall of the land and the meanderings of the road. The final step was to measure the angle of the sun at York at the same time of day and on the same day of the year as he had made his first measurement in London. From this, he reasoned he could determine the length of one degree of the Earth's meridian and thus calculate the distance around the whole. It was an almost ludicrously ambitious undertaking – a mistake of the slightest fraction of a degree would throw the whole thing out by miles – but in fact, as Norwood proudly declaimed, he was accurate to 'within a scantling' – or, more precisely, to within about 600 yards. In metric terms, his figure worked out at 110.72 kilometres per degree of arc.

In 1637, Norwood's masterwork of navigation, *The Seaman's Practice*, was published and found an immediate following. It went through seventeen editions and was still in print twenty-five years after his death. Norwood returned to Bermuda with his family, where he became a successful planter and devoted his leisure hours to his first love, trigonometry. He survived there for thirty-eight years and it would be pleasing to report that he passed this span in happiness and adulation. In fact, he didn't. On the crossing from England, his two young sons were placed in a cabin with the Reverend Nathaniel White, and somehow so successfully traumatized the young vicar that he devoted much of the rest of his career to persecuting Norwood in any small way he could think of.

Norwood's two daughters brought their father additional pain by making poor marriages. One of the husbands, possibly incited by the vicar, continually laid small charges against Norwood in court, causing him much exasperation and necessitating repeated trips across Bermuda to defend himself. Finally, in the 1650s witchcraft trials came to Bermuda and Norwood spent his final years in severe unease that his papers on trigonometry, with their arcane symbols, would be taken as communications with the devil and that he would be treated to a dreadful execution. So little is known of Norwood that it may in fact be that he deserved his unhappy declining years. What is certainly true is that he got them.

Meanwhile, the momentum for determining the Earth's circumference passed to France. There, the astronomer Jean Picard devised an impressively complicated method of triangulation involving quadrants, pendulum clocks, zenith sectors and telescopes (for observing the motions of the moons of Jupiter). After two years of trundling and triangulating his way across France, in 1669 he announced a more accurate measure of 110.46 kilometres for one degree of arc. This was a great source of pride for the

French but it was predicated on the assumption that the Earth was a perfect sphere – which Newton now said it was not.

To complicate matters, after Picard's death the father and son team of Giovanni and Jacques Cassini repeated Picard's experiments over a larger area and came up with results that suggested that the Earth was fatter not at the equator but at the poles – that Newton, in other words, was exactly wrong. It was this that prompted the Academy of Sciences to dispatch Bouguer and La Condamine to South America to take new measurements.

They chose the Andes because they needed to measure near the equator, to determine if there really was a difference in sphericity there, and because they reasoned that mountains would give them good sightlines. In fact, the mountains of Peru were so constantly lost in cloud that the team often had to wait weeks for an hour's clear surveying. On top of that, they had selected one of the most nearly impossible terrains on Earth. Peruvians refer to their landscape as *muy accidentado* – 'much accidented' – and this it most certainly is. Not only did the French have to scale some of the world's most challenging mountains – mountains that defeated even their mules – but to reach the mountains they had to ford wild rivers, hack their way through jungles, and cross miles of high, stony desert, nearly all of it uncharted and far from any source of supplies. But Bouguer and La Condamine were nothing if not tenacious, and they stuck to the task for nine and a half long, grim, sun-blistered years.

Shortly before concluding the project, word reached them that a second French team, taking measurements in northern Scandinavia (and facing notable discomforts of their own, from squelching bogs to dangerous ice floes), had found that a degree was in fact longer near the poles, as Newton had promised. The Earth was 43 kilometres stouter when measured equatorially than when measured from top to bottom around the poles.

Bouguer and La Condamine thus had spent nearly a decade working towards a result they didn't wish to find only to learn now that they weren't even the first to find it. Listlessly they completed their survey, which confirmed that the first French team was correct. Then, still not speaking, they returned to the coast and took separate ships home.

Something else conjectured by Newton in the *Principia* was that a plumb line hung near a mountain would incline very slightly towards the mountain, affected by the mountain's gravitational mass as well as by the Earth's. This was more than a curious fact. If you measured the deflection accurately and worked out the mass of the mountain, you could calculate the universal gravitational constant – that is, the basic value of gravity, known as G – and along with it the mass of the Earth.

Bouguer and La Condamine had tried this on Peru's Mount Chimborazo, but had been defeated by both the technical difficulties and their own squabbling, and so the notion lay dormant for another thirty years until resurrected in England by Nevil Maskelyne, the Astronomer Royal. In Dava Sobel's popular book *Longitude* Maskelyne is presented as a ninny and villain for failing to appreciate the brilliance of the clockmaker John Harrison, and this may be so; but we are indebted to him in other ways not mentioned in her book, not least for his successful scheme to weigh the Earth.

Maskelyne realized that the nub of the problem lay with finding a mountain of sufficiently regular shape to judge its mass. At his urging, the Royal Society agreed to engage a reliable figure to tour the British Isles to see if such a mountain could be found. Maskelyne knew just such a person – the astronomer and surveyor Charles Mason. Maskelyne and Mason had become friends eleven years earlier while engaged in a project to measure an astronomical event of great importance: the passage of the planet Venus across the face of the Sun. The tireless Edmond Halley had suggested years before that if you measured one of these passages from selected points on the Earth, you could use the principles of triangulation to work out the distance from the Earth to the Sun, and thence to calibrate the distances to all the other bodies in the solar system.

Unfortunately, transits of Venus, as they are known, are an irregular occurrence. They come in pairs eight years apart, but then are absent for a century or more, and there were none in Halley's lifetime.* But the idea simmered and when the next transit fell due in 1761, nearly two decades after Halley's death, the scientific world was ready – indeed, more ready than it had been for an astronomical event before.

With the instinct for ordeal that characterized the age, scientists set off for more than 100 locations around the globe – to Siberia, China, South Africa, Indonesia and the woods of Wisconsin, among many others. France dispatched thirty-two observers, Britain eighteen more, and still others set out from Sweden, Russia, Italy, Germany, Ireland and elsewhere.

It was history's first cooperative international scientific venture, and almost everywhere it ran into problems. Many observers were waylaid by war, sickness or shipwreck. Others made their destinations but opened their crates to find equipment broken or warped by tropical heat. Once again the French seemed fated to provide the most memorably unlucky participants. Jean Chappe spent months travelling to Siberia by coach, boat and sleigh, nursing his delicate instruments over every perilous bump, only to find the last vital stretch blocked by swollen rivers, the result of unusually heavy spring rains, which the locals were swift to blame on him after they saw him pointing strange instruments at the sky. Chappe managed to escape with his life, but with no useful measurements.

Unluckier still was Guillaume le Gentil, whose experiences are wonderfully summarized by Timothy Ferris in *Coming of Age in the Milky Way*. Le Gentil set off from France a year ahead of time to observe the transit from India, but various setbacks left him still at sea on the day of the transit – just about the worst place to be, since steady measurements were impossible

---

* The most recent transits of Venus were in 2004 and 2012. The next ones will be in 2117 and 2125.

on a pitching ship. Undaunted, Le Gentil continued on to India to await the next transit in 1769. With eight years to prepare, he erected a first-rate viewing station, tested and retested his instruments and had everything in a state of perfect readiness. On the morning of the second transit, 4 June 1769, he awoke to a fine day; but just as Venus began its pass, a cloud slid in front of the Sun and remained there for almost exactly the duration of the transit of three hours, fourteen minutes and seven seconds.

Stoically, Le Gentil packed up his instruments and set off for the nearest port, but en route he contracted dysentery and was laid up for nearly a year. Still weakened, he finally made it on to a ship. It was nearly wrecked in a hurricane off the African coast. When at last he reached home, eleven and a half years after setting off, and having achieved nothing, he discovered that his relatives had had him declared dead in his absence and enthusiastically plundered his estate.

In comparison, the disappointments experienced by Britain's eighteen scattered observers were mild. Mason found himself paired with a young surveyor named Jeremiah Dixon and apparently they got along well, for they formed a lasting partnership. Their instructions were to travel to Sumatra and chart the transit there, but after just one night at sea their ship was attacked by a French frigate. (Although scientists were in an internationally cooperative mood, nations weren't.) Mason and Dixon sent a note to the Royal Society observing that it seemed awfully dangerous on the high seas and wondering if perhaps the whole thing oughtn't to be called off. In reply they received a swift and chilly rebuke, noting that they had already been paid, that the nation and scientific community were counting on them, and that their failure to proceed would result in the irretrievable loss of their reputations. Chastened, they sailed on, but en route word reached them that Sumatra had fallen to the French and so they observed the transit inconclusively from the Cape of Good Hope. On the way home they stopped on the lonely Atlantic outcrop of St Helena, where they met Maskelyne, whose observations had been thwarted by cloud cover. Mason and Maskelyne formed a solid friendship and spent several happy, and possibly even mildly useful, weeks charting tidal flows.

Soon afterwards Maskelyne returned to England, where he became Astronomer Royal, and Mason and Dixon – now evidently more seasoned – set off for four long and often perilous years surveying their way through 244 miles of dangerous American wilderness to settle a boundary dispute between the estates of William Penn and Lord Baltimore and their respective colonies of Pennsylvania and Maryland. The result was the famous Mason–Dixon line, which later took on symbolic importance as the dividing line between the slave and free states. (Although the line was their principal task, they also contributed several astronomical surveys, including one of the century's most accurate measurements of a degree of meridian – an achievement that brought them far more acclaim in England than the settling of a boundary dispute between spoiled aristocrats.)

Back in Europe, Maskelyne and his counterparts in Germany and France were forced to the conclusion that the transit measurements of 1761 were essentially a failure. One of the problems, ironically, was that there were too many observations, which when brought together often proved contradictory and impossible to resolve. The successful charting of a Venusian transit fell instead to a little-known Yorkshire-born sea captain named James Cook, who watched the 1769 transit from a sunny hilltop in Tahiti, and then went on to chart and claim Australia for the British Crown. Upon his return there was now enough information for the French astronomer Joseph Lalande to calculate that the mean distance from the Earth to the Sun was a little over 150 million kilometres. (Two further transits in the nineteenth century allowed astronomers to put the figure at 149.59 million kilometres, where it has remained ever since. The precise distance, we now know, is 149.597870691 million kilometres.) The Earth at last had a position in space.

THE SIZE OF THE EARTH

As for Mason and Dixon, they returned to England as scientific heroes and, for reasons unknown, dissolved their partnership. Considering the frequency with which they turn up at seminal events in eighteenth-century science, remarkably little is known about either man. No likenesses exist and few written references. Of Dixon, the *Dictionary of National Biography* notes intriguingly that he was 'said to have been born in a coal mine', but then leaves it to the reader's imagination to supply a plausible explanatory circumstance, and adds that he died at Durham in 1777. Apart from his name and long association with Mason, nothing more is known.

Mason is only slightly less shadowy. We know that in 1772, at Maskelyne's behest, he accepted the commission to find a suitable mountain for the gravitational deflection experiment, at length reporting back that the mountain they needed was in the central Scottish Highlands, just above Loch Tay, and was called Schiehallion. Nothing, however, would induce him to spend a summer surveying it. He never returned to the field again. His next known movement was in 1786 when, abruptly and mysteriously, he turned up in Philadelphia with his wife and eight children, apparently on the verge of destitution. He had not been back to America since completing his survey there eighteen years earlier and had no known reason for being there, nor any friends or patrons to greet him. A few weeks later he was dead.

With Mason refusing to survey the mountain, the job fell to Maskelyne. So, for four months in the summer of 1774, Maskelyne lived in a tent in a remote Scottish glen and spent his days directing a team of surveyors, who took hundreds of measurements from every possible position. To find the mass of the mountain from all these numbers required a great deal of tedious calculating, for which a mathematician named Charles Hutton was engaged. The surveyors had covered a map with scores of figures, each marking an elevation at some point on or around the mountain. It was essentially just a confusing mass of numbers, but Hutton noticed that if he used a pencil to connect points of equal height, it all became much more orderly. Indeed, one could instantly get a sense of the overall shape and slope of the mountain. He had invented contour lines.

Extrapolating from his Schiehallion measurements, Hutton calculated the mass of the Earth at 5,000 million million tons, from which could reasonably be deduced the masses of all the other major bodies in the solar system, including the Sun. So from this one experiment we learned the masses of the Earth, the Sun, the Moon, the other planets and *their* moons, and got contour lines into the bargain – not bad for a summer's work.

Not everyone was satisfied with the results, however. The shortcoming of the Schiehallion experiment was that it was not possible to get a truly accurate figure without knowing the actual density of the mountain. For convenience, Hutton had assumed that the mountain had the same density as ordinary stone, about 2.5 times that of water, but this was little more than an educated guess.

One improbable-seeming person who turned his mind to the matter was a country parson named John Michell, who resided in the Yorkshire village of Thornhill. Despite his remote and comparatively humble situation, Michell was one of the great scientific thinkers of the eighteenth century, and much esteemed for it. Among a great deal else, he perceived the wave-like nature of earthquakes, conducted much original research into magnetism and gravity, and, quite extraordinarily, envisioned the possibility of black holes 200 years before anyone else – a leap that not even Newton could make. When the German-born musician William Herschel decided his real interest in life was astronomy, it was Michell to whom he turned for instruction in making telescopes, a kindness for which planetary science has been in his debt ever since.*

---

* In 1781 Herschel became the first person in the modern era to discover a planet. He wanted to call it George, after the British monarch, but was overruled. Instead it became Uranus, after the Greek god of the sky.

But of all that Michell accomplished, nothing was more ingenious or had greater impact than a machine he designed and built for measuring the mass of the Earth. Unfortunately he died before he could conduct the experiments, and both the idea and the necessary equipment were passed on to a brilliant but magnificently retiring London scientist named Henry Cavendish.

Cavendish is a book in himself. Born into a life of sumptuous privilege – his grandfathers were Dukes, respectively, of Devonshire and Kent – he was the most gifted English scientist of his age, but also the strangest. He suffered, in the words of one of his few biographers, from shyness to a 'degree bordering on disease'. Any human contact was for him a source of the deepest discomfort.

Once he opened his door to find an Austrian admirer, freshly arrived from Vienna, on the front step. Excitedly, the Austrian began to babble out praise. For a few moments Cavendish received the compliments as if they were blows from a blunt object and then, unable to take any more, fled down the path and out the gate, leaving the front door wide open. It was some hours before he could be coaxed back to the property. Even his housekeeper communicated with him by letter.

Although he did sometimes venture into society – he was particularly devoted to the weekly scientific soirées of the great naturalist Sir Joseph Banks – it was always made clear to the other guests that Cavendish was on no account to be approached or even looked at. Those who sought his views were advised to wander into his vicinity as if by accident and to 'talk as it were into vacancy'. If their remarks were scientifically worthy they might receive a mumbled reply, but more often than not they would hear a peeved squeak (his voice appears to have been high-pitched) and turn to find an actual vacancy and the sight of Cavendish fleeing for a more peaceful corner.

His wealth and solitary inclinations allowed him to turn his house in Clapham into a large laboratory where he could range undisturbed through every corner of the physical sciences – electricity, heat, gravity, gases, anything to do with the composition of matter. The second half of the eighteenth century was a time when people of a scientific bent grew intensely interested in the physical properties of fundamental things – gases and electricity in particular – and began seeing what they could do with them, often with more enthusiasm than sense. In America, Benjamin Franklin famously risked his life by flying a kite in an electrical storm. In France, a chemist named Pilâtre de Rozier tested the flammability of hydrogen by gulping a mouthful and blowing across an open flame, proving at a stroke that hydrogen is indeed explosively combustible and that eyebrows are not necessarily a permanent feature of one's face. Cavendish, for his part, conducted experiments in which he subjected himself to graduated jolts of electrical current, diligently noting the increasing levels of agony until he could keep hold of his quill, and sometimes his consciousness, no longer.

In the course of a long life Cavendish made a string of signal discoveries – among much else, he was the first person to isolate hydrogen and the first to combine hydrogen and oxygen to form water – but almost nothing he did was entirely divorced from strangeness. To the continuing exasperation of his fellow scientists, he often alluded in published work to the results of experiments that he had not told anyone about. In his secretiveness he didn't merely resemble Newton, but actively exceeded him. His experiments with electrical conductivity were a century ahead of their time, but unfortunately remained undiscovered until that century had passed. Indeed, the greater part of what he did wasn't known until the late nineteenth century, when the Cambridge physicist James Clerk Maxwell took on the task of editing Cavendish's papers, by which time credit for his discoveries had nearly always been given to others.

Among much else, and without telling anyone, Cavendish discovered or anticipated the law of the conservation of energy, Ohm's Law, Dalton's Law of Partial Pressures, Richter's Law of

THE SIZE OF THE EARTH

Reciprocal Proportions, Charles's Law of Gases, and the principles of electrical conductivity. That's just some of it. According to the science historian J. G. Crowther, he also foreshadowed 'the work of Kelvin and G. H. Darwin on the effect of tidal friction on slowing the rotation of the earth, and Larmor's discovery, published in 1915, on the effect of local atmospheric cooling . . . the work of Pickering on freezing mixtures, and some of the work of Rooseboom on heterogeneous equilibria'. Finally, he left clues that led directly to the discovery of the group of elements known as the noble gases, some of which are so elusive that the last of them wasn't found until 2002. But our interest here is in Cavendish's last known experiment when, in the late summer of 1797, at the age of sixty-seven, he turned his attention to the crates of equipment that had been left to him – evidently out of simple scientific respect – by John Michell.

When assembled, Michell's apparatus looked like nothing so much as an eighteenth-century version of a Nautilus weight-training machine. It incorporated weights, counterweights, pendulums, shafts and torsion wires. At the heart of the machine were two 350-pound lead balls, which were suspended beside two smaller spheres. The idea was to measure the gravitational deflection of the smaller spheres by the larger ones, which would allow the first measurement of the elusive force known as the gravitational constant, and from which the weight (strictly speaking the mass)* of the Earth could be deduced.

Because gravity holds planets in orbit and makes falling objects land with a bang, we tend to think of it as a powerful

---

* To a physicist, mass and weight are two quite different things. Your mass stays the same wherever you go, but your weight varies depending on how far you are from the centre of some other massive object like a planet. Travel to the Moon and you will be much lighter but no less massive. On Earth, for all practical purposes, mass and weight are the same and so the terms can be treated as synonymous, at least outside the classroom.

force, but it isn't really. It is only powerful in a kind of collective sense, when one massive object, like the Sun, holds on to another massive object, like the Earth. At an elemental level gravity is extraordinarily unrobust. Each time you pick up a book from a table or a coin from the floor you effortlessly overcome the gravitational exertion of an entire planet. What Cavendish was trying to do was measure gravity at this extremely featherweight level.

Delicacy was the keyword. Not a whisper of disturbance could be allowed into the room containing the apparatus, so Cavendish took up a position in an adjoining room and made his observations with a telescope aimed through a peephole. The work was incredibly exacting, involving seventeen delicate, interconnected measurements, which together took nearly a year to complete. When at last he had finished his calculations, Cavendish announced that the Earth weighed a little over 13,000,000,000,000,000,000,000,000 pounds, or 6 billion trillion metric tons, to use the modern measure.*

Today, scientists have at their disposal machines so precise they can detect the weight of a single bacterium and so sensitive that readings can be disturbed by someone yawning 75 feet away, but they have not significantly improved on Cavendish's measurements of 1797. The current best estimate for the Earth's weight is 5.9725 billion trillion tonnes, a difference of only about 1 per cent from Cavendish's finding. Interestingly, all of this merely confirmed estimates made by Newton 110 years before Cavendish without any experimental evidence at all.

At all events, by the late eighteenth century scientists knew very precisely the shape and dimensions of the Earth and its distance from the Sun and planets; and now Cavendish, without even leaving home, had given them its weight. So you might think that determining the age of the Earth would be relatively

---

* A metric ton (in Britain, 'tonne') is 2,204 pounds, compared with 2,000 pounds for an American ton and 2,240 pounds for a British one.

straightforward. After all, the necessary materials were literally at their feet. But no. Human beings would split the atom and invent television, nylon and instant coffee before they could figure out the age of their own planet.

To understand why, we must travel north to Scotland and begin with a brilliant and genial man, of whom few have ever heard, who had just invented a new science called geology.

5

THE STONE-BREAKERS

At just the time that Henry Cavendish was completing his experiments in London, 400 miles away in Edinburgh another kind of concluding moment was about to take place with the death of James Hutton. This was bad news for Hutton, of course, but good news for science as it cleared the way for a man named John Playfair to rewrite Hutton's work without fear of embarrassment.

Hutton was by all accounts a man of the keenest insights and liveliest conversation, a delight in company, and without rival when it came to understanding the mysterious slow processes that shaped the Earth. Unfortunately, it was beyond him to set down his notions in a form that anyone could begin to understand. He was, as one biographer observed with an all but audible sigh, 'almost entirely innocent of rhetorical accomplishments'. Nearly every line he penned was an invitation to slumber. Here he is in his 1795 masterwork, *A Theory of the Earth with Proofs and Illustrations*, discussing ... well, something:

> The world which we inhabit is composed of the materials, not of the earth which was the immediate predecessor of the present, but of the earth which, in ascending from the present, we consider

as the third, and which had preceded the land that was above the surface of the sea, while our present land was yet beneath the water of the ocean.

Yet almost single-handedly, and quite brilliantly, he created the science of geology and transformed our understanding of the Earth.

Hutton was born in 1726 into a prosperous Scottish family, and enjoyed the sort of material comfort that allowed him to pass much of his life in a genially expansive round of light work and intellectual betterment. He studied medicine, but found it not to his liking and turned instead to farming, which he followed in a relaxed and scientific way on the family estate in Berwickshire. Tiring of field and flock, in 1768 he moved to Edinburgh, where he founded a successful business producing sal ammoniac from coal soot, and busied himself with various scientific pursuits. Edinburgh at that time was a centre of intellectual vigour and Hutton luxuriated in its enriching possibilities. He became a leading member of a society called the Oyster Club, where he passed his evenings in the company of men such as the economist Adam Smith, the chemist Joseph Black and the philosopher David Hume, as well as such occasional visiting sparks as Benjamin Franklin and James Watt.

In the tradition of the day, Hutton took an interest in nearly everything, from mineralogy to metaphysics. He conducted experiments with chemicals, investigated methods of coal mining and canal building, toured salt mines, speculated on the mechanisms of heredity, collected fossils and propounded theories on rain, the composition of air and the laws of motion, among much else. But his particular passion was geology.

Among the questions that attracted interest in that fanatically inquisitive age was one that had puzzled people for a very long time – namely, why ancient clam shells and other marine fossils were so often found on mountaintops. How on earth did they get there? Those who thought they had a solution fell into two opposing camps. One group, known as the Neptunists, were convinced that everything on the Earth, including sea shells in improbably lofty places, could be explained by rising and falling sea levels. They believed that mountains, hills and other features were as old as the Earth itself, and were changed only when water washed over them during periods of global flooding.

Opposing them were the Plutonists, who noted that volcanoes and earthquakes, among other enlivening agents, continually changed the face of the planet but clearly owed nothing to wayward seas. The Plutonists also raised awkward questions about where all the water went when it wasn't in flood. If there was enough of it at times to cover the Alps, then where, pray, was it during times of tranquillity, such as now? Their belief was that the Earth was subject to profound internal forces as well as surface ones. However, they couldn't convincingly explain how all those clam shells got up there.

It was while puzzling over these matters that Hutton had a series of exceptional insights. From looking at his own farmland, he could see that soil was created by the erosion of rocks and that particles of this soil were continually washed away and carried off by streams and rivers and redeposited elsewhere. He realized that if such a process were carried to its natural conclusion then the Earth would eventually be worn quite smooth. Yet everywhere around him there were hills. Clearly there had to be some additional process, some form of renewal and uplift, that created new hills and mountains to keep the cycle going. The marine fossils on mountaintops, he decided, had not been deposited during floods but had risen along with the mountains themselves. He also deduced that it was heat within the Earth that created new rocks and continents and thrust up mountain chains. It is not too much to say that geologists wouldn't grasp the full implications of this thought until 200 years later, when finally they adopted the concept of plate tectonics. Above all, what Hutton's theories suggested was that the processes that shaped the Earth required huge amounts of time, far more than anyone had ever dreamed. There were enough insights here to transform utterly our understanding of the planet.

In 1785 Hutton worked his ideas up into a long paper, which was read at consecutive meetings of the Royal Society of Edinburgh. It attracted almost no notice at all. It's not hard to see why. Here, in part, is how he presented it to his audience:

> In the one case, the forming cause is in the body which is separated; for, after the body has been actuated by heat, it is by the reaction of the proper matter of the body, that the chasm which constitutes the vein is formed. In the other case, again, the cause is extrinsic in relation to the body in which the chasm is formed. There has been the most violent fracture and divulsion; but the cause is still to seek; and it appears not in the vein; for it is not every fracture and dislocation of the solid body of our earth, in which minerals, or the proper substances of mineral veins, are found.

Needless to say, almost no one in the audience had the faintest idea what he was talking about. Encouraged by his friends to expand his theory, in the touching hope that he might somehow stumble on to clarity in a more expansive format, Hutton spent the next ten years preparing his magnum opus, which was published in two volumes in 1795. Together the two books ran to nearly 1,000 pages and were, remarkably, worse than even his most pessimistic friends had feared. Nearly half the completed work now consisted of quotations from French sources, still in the original French. A third volume was so unenticing that it wasn't published until 1899, more than a century after Hutton's death, and the fourth and concluding volume was never published at all. Hutton's *Theory of the Earth* is a strong candidate for the least-read important book in science (or at least, it would be if there weren't so many others). Even Charles Lyell, the greatest geologist of the following century and a man who read everything, admitted he couldn't get through it.

Luckily, Hutton had a Boswell in the form of John Playfair, a professor of mathematics at the University of Edinburgh and a close friend, who not only could write silken prose but – thanks to many years at Hutton's elbow – actually understood what Hutton was trying to say, most of the time. In 1802, five years after Hutton's death, Playfair produced a simplified exposition of the Huttonian principles entitled *Illustrations of the Huttonian Theory of the Earth*. The book was gratefully received by those who took an active interest in geology, which in 1802 was not a large number. That, however, was about to change. And how.

In the winter of 1807, thirteen like-minded souls in London got together at the Freemasons Tavern at Long Acre, in Covent Garden, to form a dining club to be called the Geological Society. The idea was to meet once a month to swap geological notions over a glass or two of Madeira and a convivial dinner. The price of the meal was set at a deliberately hefty 15 shillings to discourage those whose qualifications were merely cerebral. It soon became apparent, however, that there was a demand for something more properly institutional, with a permanent headquarters where people could gather to share and discuss new findings. In barely a decade membership grew to 400 – still all gentlemen, of course – and the Geological was threatening to eclipse the Royal as the premier scientific society in the country.

The members met twice a month from November until June, when virtually all of them went off to spend the summer doing fieldwork. These weren't people with a pecuniary interest in minerals, you understand, or even academics for the most part, but simply gentlemen with the wealth and time to indulge a hobby at a more or less professional level. By 1830 there were 745 of them, and the world would never see the like again.

It is hard to imagine now, but geology excited the nineteenth century – positively gripped it – in a way that no science ever had before or would again. In 1839, when Roderick Murchison published *The Silurian System*, a plump and ponderous study of a type of rock called greywacke, it was an instant bestseller, racing through four editions, even though it cost 8 guineas a copy and was, in true Huttonian style, unreadable. (As even a Murchison supporter conceded, it had 'a total want of literary attractiveness'.) And when, in 1841, the great Charles Lyell travelled to America to give a series of lectures in Boston, sellout audiences of 3,000 at a time packed into the Lowell Institute to hear his tranquillizing descriptions of marine zeolites and seismic perturbations in Campania.

Throughout the modern, thinking world, but especially in Britain, men of learning ventured into the countryside to do a little 'stone-breaking', as they called it. It was a pursuit taken seriously and they tended to dress with appropriate gravity, in top hats and dark suits, except for the Reverend William Buckland of Oxford, whose habit it was to do his fieldwork in an academic gown.

The field attracted many extraordinary figures, not least the aforementioned Murchison, who spent the first thirty or so years of his life galloping after foxes, converting aeronautically challenged birds into puffs of drifting feathers with buckshot and showing no mental agility whatever beyond that needed to read *The Times* or play a hand of cards. Then he discovered an interest in rocks and became with rather astounding swiftness a titan of geological thinking.

Then there was Dr James Parkinson, who was also an early socialist and author of many provocative pamphlets with titles like 'Revolution without Bloodshed'. In 1794 he was implicated in a faintly lunatic-sounding conspiracy called 'the Pop-gun Plot', in which it was planned to shoot King George III in the neck with a poisoned dart as he sat in his box at the theatre. Parkinson was hauled before the Privy Council for questioning and came within an ace of being dispatched in irons to Australia before the charges against him were quietly dropped. Adopting a more conservative approach to life, he developed an interest in geology and became one of the founding members of the Geological Society and the author of an important geological text, *Organic Remains of a Former World*, which remained in print for half a century. He never caused trouble again. Today, however, we remember him for his landmark study of the affliction then called the 'shaking palsy', but known ever since as Parkinson's disease.

(Parkinson had one other slight claim to fame. In 1785 he became possibly the only person in history to win a natural history museum in a raffle. The museum, in London's Leicester Square, had been founded by Sir Ashton Lever, who had driven himself bankrupt with his unrestrained collecting of natural wonders. Parkinson kept the museum until 1805, when he could no longer support it and the collection was broken up and sold.)

Not quite as remarkable in character but more influential than all the others combined was Charles Lyell. Lyell was born in the year that Hutton died and only 70 miles away, in the village of Kinnordy. Though Scottish by birth, he grew up in the far south of England, in the New Forest of Hampshire, because his mother was convinced that Scots were feckless drunks. As was generally the pattern with nineteenth-century gentlemen scientists, Lyell came from a background of comfortable wealth and intellectual vigour. His father, also named Charles, had the unusual distinction of being a leading authority on the poet Dante and on mosses. (*Orthotricium lyelli*, which most visitors to the English countryside will at some time have brushed up against, is named for him.) From his father Lyell gained an interest in natural history, but it was at Oxford, where he fell under the spell of the Reverend William Buckland – he of the flowing gowns – that the young Lyell began his lifelong devotion to geology. Buckland was a bit of a charming oddity. He had some real achievements, but he is remembered at least as much for his eccentricities. He was particularly noted for a menagerie of wild animals, some large and dangerous, that were allowed to wander through his house and garden, and for his desire to eat his way through every animal in creation. Depending on whim and availability, guests to Buckland's house might be served baked guinea pig, mice in batter, roasted hedgehog or boiled Southeast Asian sea slug. Buckland was able to find merit in them all, except the common garden mole, which he declared disgusting. Almost inevitably, he became the leading authority on coprolites – fossilized faeces – and had a table made entirely out of his collection of specimens.

Even when conducting serious science his manner was generally singular. Once Mrs Buckland found herself being shaken awake in the middle of the night, her husband crying in excitement: 'My dear, I believe that *Cheirotherium*'s footsteps are undoubtedly testudinal.' Together they hurried to the kitchen in their nightclothes. Mrs Buckland made a flour paste, which she spread across the table, while the Reverend Buckland fetched the family tortoise. Plunking it on to the paste, they goaded it forward and discovered to their delight that its footprints did indeed match those of the fossil Buckland had been studying. Charles Darwin thought Buckland a buffoon – that was the word he used – but Lyell appeared to find him inspiring and liked him well enough to go touring with him in Scotland in 1824. It was soon after this trip that Lyell decided to abandon a career in law and devote himself to geology full-time.

Lyell was extremely short-sighted and went through most of his life with a pained squint, which gave him a troubled air. (Eventually he would lose his sight altogether.) His other slight peculiarity was the habit, when distracted by thought, of taking up improbable positions on furniture – lying across two chairs at once or 'resting his head on the seat of a chair, while standing up' (to quote his friend Darwin). Often when lost in thought he would slink so low in a chair that his buttocks would all but touch the floor. Lyell's only real job in life was as professor of geology at King's College in London from 1831 to 1833. It was just at this time that he produced *The Principles of Geology*, published in three volumes between 1830 and 1833, which in many ways consolidated and elaborated upon the thoughts first voiced by Hutton a generation earlier. (Although Lyell never read Hutton in the original, he was a keen student of Playfair's reworked version.)

Between Hutton's day and Lyell's there arose a new geological controversy, which largely superseded, but is often confused with, the old Neptunian–Plutonian dispute. The new battle became an argument between catastrophism and uniformitarianism – unattractive terms for an important and very long-running dispute. Catastrophists, as you might expect from the name, believed that the Earth was shaped by abrupt cataclysmic events – floods, principally, which is why catastrophism and Neptunism are often wrongly bundled together. Catastrophism was particularly comforting to clerics like Buckland because it allowed them to incorporate the biblical flood of Noah into serious scientific discussions. Uniformitarians, by contrast, believed that changes on Earth were gradual and that nearly all Earth processes happened slowly, over immense spans of time. Hutton was much more the father of the notion than Lyell, but it was Lyell most people read, and so he became in most people's minds, then and now, the father of modern geological thought.

Lyell believed that the Earth's shifts were uniform and steady – that everything that had ever happened in the past could be explained by events still going on today. Lyell and his adherents didn't just disdain catastrophism, they detested it. Catastrophists believed that extinctions were part of a series in which animals were repeatedly wiped out and replaced with new sets – a belief that the naturalist T. H. Huxley mockingly likened to 'a succession of rubbers of whist, at the end of which the players upset the table and called for a new pack'. It was too convenient a way to explain the unknown. 'Never was there a dogma more calculated to foster indolence, and to blunt the keen edge of curiosity,' sniffed Lyell.

Lyell's oversights were not inconsiderable. He failed to explain convincingly how mountain ranges were formed and overlooked glaciers as an agent of change. He refused to accept Louis Agassiz's idea of ice ages – 'the refrigeration of the globe', as he dismissively termed it – and was confident that mammals 'would be found in the oldest fossiliferous beds'. He rejected the notion that animals and plants suffered sudden annihilations, and believed that all the principal animal groups – mammals, reptiles, fish and so on – had coexisted since the dawn of time. On all of these he would ultimately be proved wrong.

Yet it would be nearly impossible to overstate Lyell's influence. *The Principles of Geology* went through twelve editions in his lifetime and contained notions that shaped geological thinking far into the twentieth century. Darwin took a first edition with him on the *Beagle* voyage and wrote afterwards that 'the great merit of the *Principles* was that it altered the whole tone of one's mind, and therefore that, when seeing a thing never seen by Lyell, one yet saw it partially through his eyes.' In short, he thought him nearly a god, as did many of his generation. It is a testament to the strength of Lyell's sway that in the 1980s, when geologists had to abandon just a part of his theory to accommodate the impact theory of extinctions, it nearly killed them. But that is another chapter.

Meanwhile, geology had a great deal of sorting out to do, and not all of it went smoothly. From the outset geologists tried to categorize rocks by the periods in which they were laid down, but there were often bitter disagreements about where to put the dividing lines – none more so than a long-running debate that became known as the Great Devonian Controversy. The issue arose when the Reverend Adam Sedgwick of Cambridge claimed for the Cambrian period a layer of rock that Roderick Murchison believed belonged rightly to the Silurian. The dispute raged for years and grew extremely heated. 'De la

Beche is a dirty dog,' Murchison wrote to a friend in a typical outburst. Some sense of the strength of feeling can be gained by glancing through the chapter titles of Martin J. S. Rudwick's excellent and sombre account of the issue, *The Great Devonian Controversy*. These begin innocuously enough with headings such as 'Arenas of Gentlemanly Debate' and 'Unravelling the Greywacke', but then proceed on to 'The Greywacke Defended and Attacked', 'Reproofs and Recriminations', 'The Spread of Ugly Rumours', 'Weaver Recants His Heresy', 'Putting a Provincial in his Place' and (in case there was any doubt that this was war) 'Murchison Opens the Rhineland Campaign'. The fight was finally settled in 1879 with the simple expedient of coming up with a new period, the Ordovician, to be inserted between the Cambrian and Silurian.

Because the British were the most active in the early years of the discipline, British names are predominant in the geological lexicon. *Devonian* is of course from the English county of Devon. *Cambrian* comes from the Roman name for Wales, while *Ordovician* and *Silurian* recall ancient Welsh tribes, the Ordovices and Silures. But with the rise of geological prospecting elsewhere, names began to creep in from all over. *Jurassic* refers to the Jura Mountains on the border of France and Switzerland. *Permian* recalls the former Russian province of Perm in the Ural Mountains. For *Cretaceous* (from the Latin for chalk) we are indebted to a Belgian geologist with the perky name of J. J. d'Omalius d'Halloy.

Originally, geological history was divided into four spans of time: primary, secondary, tertiary and quaternary. The system was too neat to last, and soon geologists were contributing additional divisions while eliminating others. Primary and secondary fell out of use altogether, while quaternary was discarded by some but kept by others. Today only tertiary remains as a common designation everywhere, even though it no longer represents a third period of anything.

Lyell, in his *Principles*, introduced additional units known as epochs or series to cover the period since the age of the dinosaurs, among them Pleistocene ('most recent'), Pliocene ('more recent'), Miocene ('moderately recent') and the rather endearingly vague Oligocene ('but a little recent'). Lyell originally intended to employ '-synchronous' for his endings, giving us such crunchy designations as Meiosynchronous and Pleiosynchronous. The Reverend William Whewell, an influential man, objected on etymological grounds and suggested instead an '-eous' pattern, producing Meioneous, Pleioneous and so on. The '-cene' terminations were thus something of a compromise.

Nowadays, and speaking very generally, geological time is divided first into four great chunks known as eons: Hadean, Archean, Proterozoic and Phanerozoic. These are further divided into four principal eras (though some sources list as many as ten): Precambrian, Palaeozoic (from the Greek meaning 'old life'), Mesozoic ('middle life') and Cenozoic ('recent life'). These eras are further divided into anywhere from a dozen to twenty subgroups, usually called periods though sometimes known as systems. Most of these are also reasonably well known: Cretaceous, Jurassic, Triassic, Silurian, Permian, Devonian and so on.*

Then come Lyell's epochs – the Pleistocene, Miocene, and so on – which apply only to the most recent (but palaeontologically busy) 65 million years; and finally we have a mass of finer subdivisions known as stages or ages. Most of these are named, nearly always awkwardly, after places: *Illinoian*, *Desmoinesian*, *Croixian*, *Kimmeridgian* and so on in like vein. Altogether, according to John McPhee, these number in the 'tens of dozens'. Fortunately, unless you take up geology as a career, you are unlikely ever to hear any of them again.

---

* There will be no testing here, but if you are ever required to memorize them you might wish to remember John Wilford's helpful advice to think of the eras (Precambrian, Palaeozoic, Mesozoic and Cenozoic) as seasons in a year and the periods (Permian, Triassic, Jurassic, etc.) as the months.

Further confusing the matter is that the stages or ages in North America sometimes have different names from the stages in Europe and often only roughly intersect with them in time. Thus the North American Cincinnatian stage mostly corresponds with the Ashgillian stage in Europe, plus a tiny bit of the slightly earlier Caradocian stage. Also, all this changes from textbook to textbook and from person to person, so that some authorities describe seven recent epochs, while others are content with four. In some books, too, you will find the tertiary and quaternary taken out and replaced by periods of different lengths called the Palaeogene and Neogene.

Sometimes, too, you will see the term Phanerozoic used to describe the span encompassing the Cenozoic, Mesozoic and Palaeozoic eras. In recent years, many observers have taken to using the term *Anthropocene* (from the Greek 'new human') to describe the period since the Industrial Revolution, when human activity has consequentially impacted on the Earth's ecosystems, but the term is unofficial. In 2024, the International Commission on Stratigraphy, which rules on geological designations, rejected its adoption on the grounds that there is no precise geological stratum that marks the beginning of such an epoch.

It can all get terribly confusing to non-specialists, but to a geologist these can be matters of passion. 'I have seen grown men glow incandescent with rage over this metaphorical millisecond in life's history,' the British palaeontologist Richard Fortey wrote with regard to a long-running twentieth-century dispute over where the boundary lies between the Cambrian and Ordovician.

At least today we can bring some sophisticated dating techniques to the table. For most of the nineteenth century, geologists could draw on nothing more than the most hopeful guesswork. The frustrating position then was that although they could place the various rocks and fossils in order by age, they had no idea how long any of those ages were. When Buckland speculated on the antiquity of an ichthyosaurus skeleton

THE SIZE OF THE EARTH

he could do no better than suggest that it had lived somewhere between 'ten thousand [and] more than ten thousand times ten thousand' years earlier.

Although there was no reliable way of dating periods, there was no shortage of people willing to try. The best-known early attempt was made in 1650, when Archbishop James Ussher of the Church of Ireland made a careful study of the Bible and other historical sources and concluded, in a hefty tome called *Annals of the Old Testament*, that the Earth had been created at midday on 23 October 4004 BC, an assertion that has amused historians and textbook writers ever since.*

There is a persistent myth, incidentally – and one propounded in many serious books – that Ussher's views dominated scientific beliefs well into the nineteenth century, and that it was Lyell who put everyone straight. Stephen Jay Gould in *Time's Arrow* cites as a typical example this sentence from a popular book of the 1980s: 'Until Lyell published his book, most thinking people accepted the idea that the earth was young.' In fact, no. As Martin J. S. Rudwick puts it, 'No geologist of any nationality whose work was taken seriously by other geologists advocated a timescale confined within the limits of a literalistic exegesis of Genesis.' Even the Reverend Buckland, as pious a soul as the nineteenth century produced, noted that nowhere did the Bible suggest that God made Heaven and Earth on the first day, but merely 'in the beginning'. That beginning, he reasoned, may have lasted 'millions upon millions of years'. Everyone agreed that the Earth was ancient. The question was simply: how ancient?

---

* Although virtually all books find a space for him, there is a striking variability in the details associated with Ussher. Some books say he made his pronouncement in 1650, others in 1654, still others in 1664. Many cite the date of the Earth's reputed beginning as 26 October. At least one book of note spells his name 'Usher'. The matter is interestingly surveyed in Stephen Jay Gould's *Eight Little Piggies*.

THE STONE-BREAKERS

One of the better early ideas at dating the planet came from the ever-reliable Edmond Halley, who in 1715 suggested that if you divided the total amount of salt in the world's seas by the amount added each year, you would get the number of years that the oceans had been in existence, which would give you a rough idea of Earth's age. The logic was appealing, but unfortunately no one knew how much salt was in the sea or by how much it increased each year, which rendered the experiment impracticable.

The first attempt at measurement that could be called remotely scientific was made by the Frenchman Georges-Louis Leclerc, Comte de Buffon, in the 1770s. It had long been known that the Earth radiated appreciable amounts of heat – that was apparent to anyone who went down a coal mine – but there wasn't any way of estimating the rate of dissipation. Buffon's experiment consisted of heating spheres until they glowed white-hot and then estimating the rate of heat loss by touching them (presumably very lightly at first) as they cooled. From this he guessed the Earth's age to be somewhere between 75,000 and 168,000 years old. This was of course a wild underestimate; but it was a radical notion nonetheless, and Buffon found himself threatened with excommunication for expressing it. A practical man, he apologized at once for his thoughtless heresy, then cheerfully repeated the assertions throughout his subsequent writings.

By the middle of the nineteenth century most learned people thought the Earth was at least a few million years old, perhaps even some tens of millions years old, but probably not more than that. So it came as a surprise when in 1859, in *On the Origin of Species*, Charles Darwin announced that the geological processes that created the Weald, an area of southern England stretching across Kent, Surrey and Sussex, had taken, by his calculations, 306,662,400 years to complete. The assertion was remarkable partly for being so arrestingly specific but even more for flying in the face of accepted wisdom about the age of the Earth.* It proved so contentious that Darwin withdrew it from the third edition of the book. The problem at its heart remained, however. Darwin and his geological friends needed the Earth to be old, but no one could come up with a way to make it so.

Unfortunately for Darwin, and for progress, the question came to the attention of the great Lord Kelvin (who, though indubitably great, was then still just plain William Thomson; he wouldn't be elevated to the peerage until 1892, when he was sixty-eight years old and nearing the end of his career, but I shall follow the convention here of using the name retroactively). Kelvin was one of the most extraordinary figures of the nineteenth century – indeed, of any century. The German scientist Hermann von Helmholtz, no intellectual slouch himself, wrote that Kelvin had by far the greatest 'intelligence and lucidity, and mobility of thought' of any man he had ever met. 'I felt quite wooden beside him sometimes,' he added, a bit dejectedly.

The sentiment is understandable, for Kelvin really was a kind of Victorian superman. He was born in 1824 in Belfast, the son of a professor of mathematics at the Royal Academical Institution who soon afterwards transferred to Glasgow. There Kelvin proved himself such a prodigy that he was admitted to Glasgow University at the exceedingly tender age of ten. By the time he had reached his early twenties, he had studied at institutions in London and Paris, graduated from Cambridge (where he won the university's top prizes for rowing and mathematics, and somehow found time to launch a musical society as well), been elected a Fellow of Peterhouse, and written (in French and English) a dozen papers in pure and applied mathematics of such dazzling originality that he had to publish them anonymously for fear of embarrassing his superiors. At the age

---

* Darwin loved an exact number. In a later work, he announced that the number of worms to be found in an average acre of English country soil was 53,767.

of twenty-two he returned to Glasgow to take up a professorship in natural philosophy, a position he would hold for the next fifty-three years.

In the course of a long career (he lived to 1907 and the age of eighty-three), he wrote 661 papers, accumulated sixty-nine patents (from which he grew abundantly wealthy) and gained renown in nearly every branch of the physical sciences. Among much else, he suggested the method that led directly to the invention of refrigeration, devised the scale of absolute temperature that still bears his name, invented the boosting devices that allowed telegrams to be sent across oceans, and made innumerable improvements to shipping and navigation, from the invention of a popular marine compass to the creation of the first depth sounder. And those were merely his practical achievements.

His theoretical work, in electromagnetism, thermodynamics and the wave theory of light, was equally revolutionary.* He had really only one flaw and that was an inability to calculate the correct age of the Earth. The question occupied much of the second half of his career, but he never came anywhere near getting it right. His first effort, in 1862 for an article in a popular magazine called *Macmillan's*, suggested that the Earth was 98

---

* In particular, he elaborated the Second Law of Thermodynamics. A discussion of these laws would be a book in itself, but I offer here this crisp summation by the chemist P. W. Atkins, just to provide a sense of them: 'There are four Laws. The third of them, the Second Law, was recognized first; the first, the Zeroth Law, was formulated last; the First Law was second; the Third Law might not even be a law in the same sense as the others.' In briefest terms, the second law states that a little energy is always wasted. You can't have a perpetual motion device because no matter how efficient, it will always lose energy and eventually run down. The first law says that you can't create energy and the third that you can't reduce temperatures to absolute zero; there will always be some residual warmth. As Dennis Overbye notes, the three principal laws are sometimes expressed jocularly as (1) you can't win, (2) you can't break even, and (3) you can't get out of the game.

million years old, but cautiously allowed that the figure could be as low as 20 million years or as high as 400 million. With remarkable prudence he acknowledged that his calculations could be wrong if 'sources now unknown to us are prepared in the great storehouse of creation' – but it was clear that he thought that unlikely.

With the passage of time Kelvin would become more forthright in his assertions and less correct. He continually revised his estimates downwards, from a maximum of 400 million years to 100 million years, then to 50 million years and finally, in 1897, to a mere 24 million years. Kelvin wasn't being wilful. It was simply that there was nothing in physics that could explain how a body the size of the Sun could burn continuously for more than a few tens of millions of years at most without exhausting its fuel. Therefore it followed that the Sun and its planets were relatively, but inescapably, youthful.

The problem was that nearly all the fossil evidence contradicted this, and suddenly in the nineteenth century there was a *lot* of fossil evidence.

6

SCIENCE RED IN TOOTH AND CLAW

In 1787, someone in New Jersey – exactly who now seems to be forgotten – found an enormous thigh bone sticking out of a stream bank at a place called Woodbury Creek. The bone clearly didn't belong to any species of creature still alive, certainly not in New Jersey. From what little is known now, it is thought to have come from a hadrosaur, a large duck-billed dinosaur. At the time, dinosaurs were unknown. The bone was sent to Dr Caspar Wistar, the nation's leading anatomist, who described it at a meeting of the American Philosophical Society in Philadelphia that autumn. Unfortunately, Wistar failed completely to recognize the bone's significance and merely made a few cautious and uninspired remarks to the effect that it was indeed a whopper. He thus missed the chance, half a century ahead of anyone else, to be the discoverer of dinosaurs. Indeed, the bone excited so little interest that it was put in a storeroom and eventually disappeared altogether. So the first dinosaur bone ever found was also the first to be lost.

That the bone didn't attract greater interest is more than a little puzzling, for its appearance came at a time when America was in a froth of excitement about the remains of large, ancient animals. The cause of this froth was a strange assertion by the great French naturalist the Comte de Buffon – he of the heated spheres from the previous chapter – that living things in the New World were inferior in nearly every way to those of the Old World. America, Buffon wrote in his vast and much-esteemed *Histoire naturelle*, was a land where the water was stagnant, the soil unproductive, and the animals without size or vigour, their constitutions weakened by the 'noxious vapours' that rose from its rotting swamps and sunless forests. In such an environment even the native Indians lacked virility. 'They have no beard or body hair,' Buffon sagely confided, 'and no ardour for the female.' Their reproductive organs were 'small and feeble'.

Buffon's observations found surprisingly eager support among other writers, especially those whose conclusions were not complicated by actual familiarity with the country. A Dutchman named Corneille de Pauw announced in a popular work called *Recherches philosophiques sur les américains* that native American males were not only reproductively unimposing, but 'so lacking in virility that they had milk in their breasts'. Such views enjoyed an improbable durability and could be found repeated or echoed in European texts until near the end of the nineteenth century.

Not surprisingly, such aspersions were indignantly met in America. Thomas Jefferson incorporated a furious (and, unless the context is understood, quite bewildering) rebuttal in his *Notes on the State of Virginia*, and induced his New Hampshire friend General John Sullivan to send twenty soldiers into the northern woods to find a bull moose to present to Buffon as proof of the stature and majesty of American quadrupeds. It took the men two weeks to track down a suitable subject. The moose, when shot, unfortunately lacked the imposing horns that Jefferson had specified, but Sullivan thoughtfully included a rack of antlers from an elk or stag with the suggestion that these be attached instead. Who in France, after all, would know?

Meanwhile, in Philadelphia – Wistar's city – naturalists had begun to assemble the bones of a giant elephant-like creature known at first as 'the great American incognitum' but later identified, not quite correctly, as a mammoth. The first of these bones had been discovered at a place called Big Bone Lick in Kentucky, but soon others were turning up all over. America, it appeared, had once been the home of a truly substantial creature – one that would surely disprove Buffon's foolish Gallic contentions.

In their keenness to demonstrate the incognitum's bulk and ferocity, the American naturalists appear to have become slightly carried away. They overestimated its size by a factor of six and gave it frightening claws, which in fact came from a *Megalonyx*, or giant ground sloth, found nearby. Rather remarkably, they persuaded themselves that the animal had enjoyed 'the agility and ferocity of the tiger', and portrayed it in illustrations as pouncing with feline grace on to prey from boulders. When tusks were discovered, they were forced into the animal's head in any number of inventive ways. One restorer screwed the tusks in upside down, like the fangs of a sabre-toothed cat, which gave it a satisfyingly aggressive aspect. Another arranged the tusks so that they curved backwards, on the engaging theory that the creature had been aquatic and had used them to anchor itself to trees while dozing. The most pertinent consideration about the incognitum, however, was that it appeared to be extinct – a fact that Buffon cheerfully seized upon as proof of its incontestably degenerate nature.

Buffon died in 1788, but the controversy rolled on. In 1795 a selection of bones made their way to Paris, where they were examined by the rising star of palaeontology, the youthful and aristocratic Georges Cuvier. Cuvier was already dazzling people with his genius for taking heaps of disarticulated bones and whipping them into shapely forms. It was said that he could describe the look and nature of an animal from a single tooth or scrap of jaw, and often name the species and genus into the bargain. Realizing that no one in America had thought to write a formal description of the lumbering beast, Cuvier did so, and thus became its official discoverer. He called it a *mastodon* (which means, a touch unexpectedly, 'nipple-teeth').

Inspired by the controversy, in 1796 Cuvier wrote a landmark paper, *Note on the Species of Living and Fossil Elephants*, in which he put forward for the first time a formal theory of extinctions. His belief was that from time to time the Earth experienced global catastrophes in which groups of creatures were wiped out. For religious people, including Cuvier himself, the idea raised uncomfortable implications since it suggested an unaccountable casualness on the part of Providence. To what end would God create species only to wipe them out later? The notion was contrary to the belief in the Great Chain of Being, which held that the world was carefully ordered and that every living thing within it had a place and purpose, and always had and always would. Jefferson for one couldn't abide the thought that whole species would ever be permitted to vanish (or, come to that, to evolve). So when it was put to him that there might be scientific and political value in sending a party to explore the interior of America beyond the Mississippi he leaped at the idea, hoping the intrepid adventurers would find herds of healthy mastodons and other outsized creatures grazing on the bounteous plains. Jefferson's personal secretary and trusted friend Meriwether Lewis was chosen as co-leader, with William Clark, and chief naturalist for the expedition. The person selected to advise him on what to look out for with regard to animals living and deceased was none other than Caspar Wistar.

In the same year – in fact, the same month – that the aristocratic and celebrated Cuvier was propounding his extinction theories in Paris, on the other side of the English Channel a rather more obscure Englishman was having an insight into the value of fossils that would also have lasting ramifications. William Smith was a young supervisor of construction on the Somerset Coal Canal. On the evening of 5 January 1796, he was sitting in a coaching inn in Somerset when he jotted down the notion that would eventually make his reputation.

To interpret rocks, there needs to be some means of correlation, a basis on which you can tell that those carboniferous rocks from Devon are younger than these Cambrian rocks from Wales. Smith's insight was to realize that the answer lay with fossils. At every change in rock strata certain species of fossils disappeared while others carried on into subsequent levels. By noting which species appeared in which strata, you could work out the relative ages of rocks wherever they appeared. Drawing on his knowledge as a surveyor, Smith began at once to make a map of Britain's rock strata, which would be published after many trials in 1815 and would become a cornerstone of modern geology. (The story is comprehensively covered in Simon Winchester's book *The Map that Changed the World*.)

Unfortunately, having had his insight, Smith was curiously uninterested in understanding why rocks were laid down in the way they were. 'I have left off puzzling about the origin of Strata and content myself with knowing that it is so,' he recorded. 'The whys and wherefores cannot come within the Province of a Mineral Surveyor.'

Smith's revelation regarding strata heightened the moral awkwardness concerning extinctions. To begin with, it confirmed that God had wiped out creatures not occasionally but repeatedly. This made Him seem not so much careless as peculiarly hostile. It also made it inconveniently necessary to explain how some species were wiped out while others continued unimpeded into succeeding eons. Clearly, there was more to extinctions than could be accounted for by a single Noachian deluge, as the biblical flood was known. Cuvier resolved the matter to his own satisfaction by suggesting that Genesis applied only to the most recent inundation. God, it appeared, hadn't wished to distract or alarm Moses with news of earlier, irrelevant extinctions.

So, by the early years of the nineteenth century, fossils had taken on a certain inescapable importance, which makes Wistar's failure to see the significance of his dinosaur bone all

THE SIZE OF THE EARTH

the more unfortunate. Suddenly, in any case, bones were turning up all over. Several other opportunities arose for Americans to claim the discovery of dinosaurs, but all were wasted. In 1806 the Lewis and Clark expedition passed through the Hell Creek Formation in Montana, an area where fossil hunters would later literally trip over dinosaur bones, and even examined what was clearly a dinosaur bone embedded in rock, but failed to make anything of it. Other bones and fossilized footprints were found in the Connecticut river valley of New England after a farm boy named Plinus Moody spied ancient tracks on a rock ledge at South Hadley, Massachusetts.

Some of these at least survive — notably the bones of an anchisaurus, which are in the collection of the Peabody Museum at Yale. Found in 1818, they were the first dinosaur bones to be examined and saved, but unfortunately weren't recognized for what they were until 1855. In that same year, 1818, Caspar Wistar died, but he did gain a certain unexpected immortality when a botanist named Thomas Nuttall named a delightful climbing shrub after him. Some botanical purists still insist on spelling it *wistaria*.

By this time, however, palaeontological momentum had moved to England. In 1812, at Lyme Regis on the Dorset coast, an extraordinary child named Mary Anning — aged eleven, twelve or thirteen, depending on whose account you read — found a strange fossilized sea monster, 17 feet long and now known as the ichthyosaurus, embedded in the steep and dangerous cliffs along the English Channel.

It was the start of a remarkable career. Anning would spend the next thirty-five years gathering fossils, which she sold to visitors. (She is commonly held to be the source for the famous tongue-twister 'She sells seashells on the seashore.') She would also find the first plesiosaurus, another marine monster, and one of the first and best pterodactyls. Though none of these was technically a dinosaur, that wasn't terribly relevant at the time since nobody then knew what a dinosaur was. It was enough to realize that the world had once held creatures strikingly unlike anything we might now find.

Anning wasn't merely good at spotting fossils – though she was unrivalled at that – she could extract them with the greatest delicacy and without damage. If you ever have the chance to visit the hall of ancient marine reptiles at the Natural History Museum in London, I urge you to take it, for there is no other way to appreciate the scale and beauty of what this young woman achieved working virtually unaided with the most basic tools in nearly impossible conditions. The plesiosaur alone took her ten years of patient excavation. Although untrained, Anning was also able to provide competent drawings and descriptions for scholars. But even with the advantage of her skills, significant finds were rare and she passed most of her life in considerable poverty.

It would be hard to think of a more overlooked person in the history of palaeontology than Mary Anning, but in fact there was one who came painfully close. His name was Gideon Algernon Mantell and he was a country doctor in Sussex.

Mantell was a lanky assemblage of shortcomings – he was vain, self-absorbed, priggish, neglectful of his family – but never was there a more committed amateur palaeontologist. He was also lucky to have a devoted and observant wife. In 1822, while he was making a house call on a patient in rural Sussex, Mrs Mantell went for a stroll down a nearby lane and in a pile of rubble that had been left to fill potholes she found a curious object – a curved brown stone, about the size of a small walnut. Knowing her husband's interest in fossils, and thinking it might be one, she took it to him. Mantell could see at once it was a fossilized tooth, and after a little study became certain that it was from an animal that was herbivorous, reptilian, extremely large – tens of feet long – and from the Cretaceous period. He was right on all counts; but these were bold conclusions, since nothing like it had been seen before or even imagined.

Aware that his finding would entirely upend what was understood about the past, and urged by his friend the Reverend William Buckland – he of the gowns and experimental appetite – to proceed with caution, Mantell devoted three painstaking years to seeking evidence to support his conclusions. He sent the tooth to Cuvier in Paris for an opinion, but the great Frenchman dismissed it as being from a hippopotamus. (Cuvier later apologized handsomely for this uncharacteristic error.) One day, while doing research at the Hunterian Museum in London, Mantell fell into conversation with a fellow researcher who told him the tooth looked very like those of animals he had been studying, South American iguanas. A hasty comparison confirmed the resemblance. And so Mantell's creature became the iguanodon, after a basking tropical lizard to which it was not in any manner related.

Mantell prepared a paper for delivery to the Royal Society. Unfortunately, it emerged that another dinosaur had been found at a quarry in Oxfordshire and had just been formally described – by the Reverend Buckland, the very man who had urged him not to work in haste. It was the megalosaurus, and the name was actually suggested to Buckland by his friend Dr James Parkinson, the would-be radical and eponym for Parkinson's disease. Buckland, it may be recalled, was foremost a geologist, and he showed it with his work on megalosaurus. In his report, for the *Transactions of the Geological Society of London*, he noted that the creature's teeth were not attached directly to the jawbone, as in lizards, but placed in sockets, in the manner of crocodiles. But, having noticed this much, Buckland failed to realize what it meant: namely, that megalosaurus was an entirely new type of creature. Still, although his report demonstrated little acuity or insight, it was the first published description of a dinosaur – and so it is to Buckland, rather than the far more deserving Mantell, that the credit goes for the discovery of this ancient line of beings.

Unaware that disappointment was going to be a continuing

feature of his life, Mantell continued hunting for fossils – he found another giant, the hylaeosaurus, in 1833 – and purchasing others from quarrymen and farmers until he had probably the largest fossil collection in Britain. Mantell was an excellent doctor and equally gifted bone hunter, but he was unable to support both his talents. As his collecting mania grew, he neglected his medical practice. Soon, fossils filled nearly the whole of his house in Brighton and consumed much of his income. A good deal of the rest went to underwriting the publication of books that few cared to own. *Illustrations of the Geology of Sussex*, published in 1827, sold only fifty copies and left him £300 out of pocket – an uncomfortably substantial sum for the times.

In some desperation Mantell hit on the idea of turning his house into a museum and charging admission, then belatedly realized that such a mercenary act would ruin his standing as a gentleman, not to mention as a scientist – so he allowed people to visit the house for free. They came in their hundreds, week after week, disrupting both his practice and his home life. Eventually he was forced to sell most of his collection to pay off his debts. Soon after, his wife left him, taking two of their four children with her. Remarkably, his troubles were only just beginning.

In the district of Sydenham in south London, at a place called Crystal Palace Park, there stands a strange and largely forgotten sight: the world's first life-sized models of dinosaurs. Once this was one of the most popular attractions in London – in effect, as Richard Fortey once noted, the world's first theme park. Quite a lot about the models is not strictly correct. The iguanodon's thumb has been placed on its nose, as a kind of spike, and it stands on four sturdy legs, making it look like a rather stout and awkwardly overgrown dog. (In life, the iguanodon did not crouch on all fours but was bipedal.) Looking at them now, you would scarcely guess that these odd and lumbering beasts could cause great rancour and bitterness, but they did.

Perhaps nothing in natural history has been at the centre of fiercer and more enduring hatreds than the line of ancient creatures known as dinosaurs.

At the time of the dinosaurs' construction, Sydenham was on the edge of London and its spacious park was considered an ideal place to re-erect the famous Crystal Palace, the glass and cast-iron structure that had been the centrepiece of the Great Exhibition of 1851, and from which the new park naturally took its name. The dinosaurs, built of concrete, were a kind of bonus attraction. On New Year's Eve 1853, a famous dinner for twenty-one prominent scientists was held inside the unfinished iguanodon. Gideon Mantell, the man who had found and identified the iguanodon, was not among them. The person at the head of the table was the greatest star of the young science of palaeontology. His name was Richard Owen and by this time he had already devoted several productive years to making Gideon Mantell's life hell.

Owen had grown up in Lancaster, in the north of England, where he had trained as a doctor. He was a born anatomist, and so devoted to his studies that he sometimes illicitly borrowed limbs, organs and other parts from corpses and took them home for leisurely dissection. Once, while carrying a sack containing the head of a black African sailor that he had just removed, Owen slipped on a wet cobble and watched in horror as the head bounced away from him down the lane and through the open doorway of a cottage, where it came to rest in the front parlour. What the occupants had to say upon finding an unattached head rolling to a halt at their feet can only be imagined. One assumes that they had not formed any terribly advanced conclusions when, an instant later, a fraught-looking young man rushed in, wordlessly retrieved the head and rushed out again.

In 1825, aged just twenty-one, Owen moved to London and soon after was engaged by the Royal College of Surgeons to help organize their extensive, but disordered, collections of medical and anatomical specimens. Most of these had been left to the institution by John Hunter, a distinguished surgeon and tireless collector of medical curiosities, but had never been catalogued or organized, largely because the paperwork explaining the significance of each had gone missing soon after Hunter's death.

Owen swiftly distinguished himself with his powers of organization and deduction. At the same time he showed himself to be a peerless anatomist with instincts for reconstruction almost on a par with the great Cuvier in Paris. He became such an expert on the anatomy of animals that he was granted first refusal on any animal that died at the London Zoological Gardens, and these he would invariably have delivered to his house for examination. Once his wife returned home to find a freshly deceased rhinoceros filling the front hallway. He quickly became a leading expert on all kinds of animals living and extinct – from platypuses, echidnas and other newly discovered marsupials to the hapless dodo and the extinct giant birds called moas that had roamed New Zealand until eaten out of existence by the Maoris. He was the first to describe the archaeopteryx after its discovery in Bavaria in 1861 and the first to write a formal epitaph for the dodo. Altogether he produced some 600 anatomical papers, a prodigious output.

But it was for his work with dinosaurs that Owen is remembered. He coined the term *dinosauria* in 1841. It means 'terrible lizard' and was a curiously inapt name. Dinosaurs, as we now know, weren't all terrible – some were no bigger than rabbits and probably extremely retiring – and the one thing they most emphatically were not was lizards, which are actually of a much older (by 30 million years) lineage. Owen was well aware that the creatures were reptilian and had at his disposal a perfectly good Greek word, *herpeton*, but for some reason chose not to use it. Another, more excusable error (given the paucity of specimens at the time) was his failure to note that dinosaurs constitute not one but two orders of reptiles: the bird-hipped ornithischians and the lizard-hipped saurischians.

Owen was not an attractive person, in appearance or in temperament. A photograph from his late middle years shows him as gaunt and sinister, like the villain in a Victorian melodrama, with long, lank hair and bulging eyes – a face to frighten babies. In manner he was cold and imperious, and he was without scruple in the furtherance of his ambitions. He was the only person Charles Darwin was ever known to hate. Even Owen's son (who soon after killed himself) referred to his father's 'lamentable coldness of heart'.

His undoubted gifts as an anatomist allowed him to get away with the most barefaced dishonesties. In 1857, the naturalist T. H. Huxley was leafing through a new edition of *Churchill's Medical Directory* when he noticed that Owen was listed as Professor of Comparative Anatomy and Physiology at the Government School of Mines, which rather surprised Huxley as that was the position he held. Upon enquiring how *Churchill's* had made such an elemental error, he was told that the information had been provided to them by Dr Owen himself. A fellow naturalist named Hugh Falconer, meanwhile, caught Owen taking credit for one of his discoveries. Others accused him of borrowing specimens, then denying he had done so. Owen even fell into a bitter dispute with the Queen's dentist over the credit for a theory concerning the physiology of teeth.

He did not hesitate to persecute those whom he disliked. Early in his career Owen used his influence at the Zoological Society to blackball a young man named Robert Grant, whose only crime was to have shown promise as a fellow anatomist. Grant was astonished to discover that he was suddenly denied access to the anatomical specimens he needed to conduct his research. Unable to pursue his work, he sank into an understandably dispirited obscurity.

But no one suffered more from Owen's unkindly attentions than the hapless and increasingly tragic Gideon Mantell. After losing his wife, his children, his medical practice and most of his fossil collection, Mantell moved to London. There, in 1841 – the fateful year in which Owen would achieve his greatest glory for naming and identifying the dinosaurs – Mantell was involved in a terrible accident. While crossing Clapham Common in a carriage, he somehow fell from his seat, grew entangled in the reins and was dragged at a gallop over rough ground by the panicked horses. The accident left him bent, crippled and in chronic pain, with a spine damaged beyond repair.

Capitalizing on Mantell's enfeebled state, Owen set about systematically expunging his contributions from the record, renaming species that Mantell had named years before and claiming credit for their discovery for himself. Mantell continued to try to do original research, but Owen used his influence at the Royal Society to ensure that most of his papers were rejected. In 1852, unable to bear any more pain or persecution, Mantell took his own life. His deformed spine was removed and sent to the Royal College of Surgeons where – now here's an irony for you – it was placed in the care of Richard Owen, director of the college's Hunterian Museum.

But the insults had not quite finished. Soon after Mantell's death, an arrestingly uncharitable obituary appeared in the *Literary Gazette*. In it Mantell was characterized as a mediocre anatomist whose modest contributions to palaeontology were limited by a 'want of exact knowledge'. The obituary even removed the discovery of the iguanodon from him and credited it instead to Cuvier and Owen, among others. Though the piece carried no byline, the style was Owen's and no one in the world of the natural sciences doubted the authorship.

By this stage, however, Owen's transgressions were beginning to catch up with him. His undoing began when a committee of the Royal Society – a committee of which he happened to be chairman – decided to award him its highest honour, the Royal Medal, for a paper he had written on an extinct mollusc called the belemnite. 'However,' as Deborah Cadbury notes in her excellent history of the period, *Terrible Lizard*, 'this piece of work was not quite as original as it appeared.' The belemnite, it turned out, had been discovered four years earlier by an amateur naturalist named Chaning Pearce, and the discovery had been fully reported at a meeting of the Geological Society. Owen had been at that meeting, but failed to mention this when he presented a report of his own to the Royal Society – at which, not incidentally, he rechristened the creature *Belemnites owenii* in his own honour. Although Owen was allowed to keep the Royal Medal, the episode left a permanent tarnish on his reputation, even among his few remaining supporters.

Eventually Huxley managed to do to Owen what Owen had done to so many others: he had him voted off the councils of the Zoological and Royal Societies. To round off the retribution, Huxley became the new Hunterian Professor at the Royal College of Surgeons.

Owen would never again do important research, but the latter half of his career was devoted to one unexceptionable pursuit for which we can all be grateful. In 1856 he became head of the natural history section of the British Museum, in which capacity he became the driving force behind the creation of London's Natural History Museum. The grand and beloved gothic heap in South Kensington, opened in 1880, is almost entirely a testament to his vision.

Before Owen, museums were designed primarily for the use and edification of the elite, and even they found it difficult to gain access. In the early days of the British Museum, prospective visitors had to make a written application and undergo a brief interview to determine if they were fit to be admitted at all. They then had to return a second time to pick up a ticket – that is, assuming they had passed the interview – and finally come back a third time to view the museum's treasures. Even then they were whisked through in groups and not allowed to linger. Owen's plan was to welcome everyone, even to the point of encouraging working men to visit in the evening, and to devote most of the museum's space to public displays. He even proposed, very radically, to put informative labels on each display so that people could appreciate what they were viewing. In this, somewhat unexpectedly, he was opposed by T. H. Huxley, who believed that museums should be primarily research institutes. By making the Natural History Museum an institution for everyone, Owen transformed our expectations of what museums are for.

Still, his altruism towards his fellow man generally did not deflect him from more personal rivalries. One of his last official acts was to lobby against a proposal to erect a statue in memory of Charles Darwin. In this he failed. Today, Darwin's statue looks benevolently down on the arriving throngs from the staircase in the entrance hall.

It would be reasonable to suppose that Richard Owen's petty rivalries marked the low point of nineteenth-century palaeontology, but in fact worse was to come, this time from overseas. In America in the closing decades of the century there arose a rivalry even more spectacularly venomous, if not quite as destructive. It was between two strange and ruthless men, Edward Drinker Cope and Othniel Charles Marsh.

They had much in common. Both were spoiled, driven, self-centred, quarrelsome, jealous, mistrustful and ever unhappy. Between them they changed the world of palaeontology. They began as friends and admirers, even naming fossil species after each other, and spent a pleasant week together in 1868. However, something then went wrong between them — nobody is quite sure what — and by the following year they had developed an enmity that would grow into consuming hatred over the next three decades. It is probably safe to say that no two people in the natural sciences have ever despised each other more.

Marsh, the elder of the two by eight years, was a retiring and bookish fellow, with a trim beard and dapper manner, who spent little time in the field and was seldom very good at finding things when he was there. On a visit to the famous dinosaur fields of Como Bluff, Wyoming, he failed to notice the bones that were, in the words of one historian, 'lying everywhere like logs'. But he had the means to buy almost anything he wanted. Although he came from a modest background – his father was a farmer in upstate New York – his uncle was the supremely rich and extraordinarily indulgent financier George Peabody. When Marsh showed an interest in natural history, Peabody had a museum built for him at Yale and provided funds sufficient for him to fill it with almost whatever took his fancy.

Cope was born more directly into privilege – his father was a rich Philadelphia businessman – and was by far the more adventurous of the two. In the summer of 1876 in Montana, while George Armstrong Custer and his troops were being cut down at Little Big Horn, Cope was out hunting for bones nearby. When it was pointed out to him that this was probably not the most prudent time to be taking treasures from Indian lands, Cope thought for a minute and decided to press on anyway. He was having too good a season. At one point he ran into a party of suspicious Crow Indians, but he managed to win them over by repeatedly taking out and replacing his false teeth.

For a decade or so, Marsh and Cope's mutual dislike primarily took the form of quiet sniping, but in 1877 it erupted into grandiose dimensions. In that year a Colorado schoolteacher named Arthur Lakes found bones near Morrison while out hiking with a friend. Recognizing the bones as coming from a 'gigantic saurian', Lakes thoughtfully dispatched some samples to both Marsh and Cope. A delighted Cope sent Lakes $100 for his trouble and asked him not to tell anyone of his discovery, especially Marsh. Confused, Lakes now asked Marsh to pass the bones on to Cope. Marsh did so, but it was an affront that he would never forget.

It also marked the start of a war between the two that became increasingly bitter, underhand and often ridiculous. It sometimes stooped to one team's diggers throwing rocks at the other team's. Cope was caught at one point prising open crates that belonged to Marsh. They insulted each other in print and poured scorn on each other's results. Seldom – perhaps never – has science been driven forward more swiftly and successfully by animosity.

Over the next several years the two men between them increased the number of known dinosaur species in America from nine to almost 150. Nearly every dinosaur that the average person can name – stegosaurus, brontosaurus, diplodocus, triceratops – was found by one or the other of them.* Unfortunately, they worked in such reckless haste that they often failed to note that a new discovery was something already known. Between them they managed to 'discover' a species called *Uintatheres anceps* no fewer than twenty-two times. It took years to sort out some of the classification messes they made. Of the two, Cope's scientific legacy was much the more substantial. In a breathtakingly industrious career, he wrote some 1,400 learned papers and described almost 1,300 new species of fossil (of all types, not just dinosaurs) – more than double Marsh's output in both cases. Cope might have done even more, but unfortunately he went into a rather precipitous descent in his later years. Having inherited a fortune in 1875, he invested unwisely in silver and lost everything. He ended up living in a single room in a Philadelphia boarding house, surrounded by books, papers and bones. Marsh, by contrast, finished his days in a splendid mansion in New Haven. Cope died in 1897, Marsh two years later.

In his final years, Cope developed one other interesting obsession. It became his earnest wish to be declared the type specimen for *Homo sapiens* – that is, to have his bones be the official set for the human race. Normally, the type specimen of a species is the first set of bones found, but since no first set

---

* The notable exception being the *Tyrannosaurus rex*, which was found by Barnum Brown in 1902.

of *Homo sapiens* bones exists there was a vacancy, which Cope desired to fill. It was an odd and vain wish, but no one could think of any grounds to oppose it. To that end, Cope willed his bones to the Wistar Institute, a learned society in Philadelphia endowed by the descendants of the seemingly inescapable Caspar Wistar. Unfortunately, after his bones were prepared and assembled, it was found that they showed signs of incipient syphilis, hardly a feature one would wish to preserve in the type specimen for one's own race. So Cope's petition and his bones were quietly shelved. There is still no type specimen for modern humans.

As for the other players in this drama, Owen died in 1892, a few years before Cope or Marsh. Buckland ended up by losing his mind and finished his days a gibbering wreck in a lunatic asylum in Clapham, not far from where Mantell had suffered his crippling accident. Mantell's twisted spine remained on display at the Hunterian Museum for nearly a century before being mercifully obliterated by a German bomb in the Blitz. What remained of Mantell's collection after his death passed on to his children and much of it was taken to New Zealand by his son Walter, who emigrated there in 1840. Walter became a distinguished Kiwi, eventually attaining the office of Minister of Native Affairs. In 1865 he donated the prime specimens from his father's collection, including the famous iguanodon tooth, to the Colonial Museum (now the Museum of New Zealand) in Wellington, where they have remained ever since.

Of course, dinosaur-hunting didn't end with the deaths of the great nineteenth-century fossil hunters. Indeed, to a surprising extent it had only just begun. In 1898, the year that fell between the deaths of Cope and Marsh, a trove greater by far than anything found before was discovered – noticed, really – at a place called Bone Cabin Quarry, only a few miles from Marsh's prime hunting ground at Como Bluff, Wyoming. There, hundreds and hundreds of fossil bones were to be found weathering out of the hills. They were so numerous, in fact, that someone had built a cabin out of them – hence the name. In just the first two seasons, 100,000 pounds of ancient bones were excavated from the site, and tens of thousands of pounds more came in each of the half dozen years that followed.

The upshot is that by the turn of the twentieth century, palaeontologists had literally tons of old bones to pick over. The problem was that they still didn't have any idea how old any of these bones were. Worse, the agreed ages for the Earth couldn't comfortably support the numbers of aeons and ages and epochs that the past obviously contained. If Earth were really only 20 million years old or so, as the great Lord Kelvin insisted, then whole orders of ancient creatures must have come into being and gone out again practically in the same geological instant. It just made no sense.

Other scientists besides Kelvin turned their minds to the problem and came up with results that only deepened the uncertainty. Samuel Haughton, a respected geologist at Trinity College in Dublin, announced an estimated age for the Earth of 2,300 million years – way beyond anything anybody else was suggesting. When this was drawn to his attention, he recalculated using the same data and put the figure at 153 million years. John Joly, also of Trinity, decided to give Edmond Halley's ocean salts idea a whirl, but his method was based on so many faulty assumptions that he was hopelessly adrift. He calculated that the Earth was 89 million years old – an age that fitted neatly enough with Kelvin's assumptions but unfortunately not with reality.

Such was the confusion that by the close of the nineteenth century, depending on which text you consulted, you could learn that the number of years that stood between us and the dawn of complex life in the Cambrian period was 3 million, 18 million, 600 million, 794 million, or 2.4 billion – or some other number within that range. As late as 1910, one of the most respected estimates, by the American George Becker, put the Earth's age at perhaps as little as 55 million years.

THE SIZE OF THE EARTH

Just when matters seemed most intractably confused, along came another extraordinary figure with a novel approach. He was a bluff and brilliant New Zealand farm boy named Ernest Rutherford, and he produced pretty well irrefutable evidence that the Earth was at least many hundreds of millions of years old, probably rather more.

Remarkably, his evidence was based on alchemy – natural, spontaneous, scientifically credible and wholly non-occult, but alchemy nonetheless. Newton, it turned out, had not been so wrong after all. And exactly how *that* became evident is, of course, another story.

7

ELEMENTAL MATTERS

Chemistry as an earnest and respectable science is often said to date from 1661, when Robert Boyle of Oxford published *The Sceptical Chymist* – the first work to distinguish between chemists and alchemists – but it was a slow and often erratic transition.

Into the eighteenth century scholars could feel oddly comfortable in both camps – like the German Johann Becher, who produced a sober and unexceptionable work on mineralogy called *Physica Subterranea*, but who also was certain that, given the right materials, he could make himself invisible.

Perhaps nothing better typifies the strange and often accidental nature of chemical science in its early days than a discovery made by a German named Hennig Brand in 1669. Brand became convinced that gold could somehow be distilled from human urine. (The similarity of colour seems to have been a factor in his conclusion.) He assembled fifty buckets of human urine, which he kept for months in his cellar. By various recondite processes, he converted the urine first into a noxious paste and then into a translucent waxy substance. None of it yielded gold, of course, but a strange and interesting thing did happen. After a time, the substance began to glow. Moreover, when exposed to air it often spontaneously burst into flame.

The commercial potential for the stuff – which soon became known as phosphorus, from Greek and Latin roots meaning 'light-bearing' – was not lost on eager businesspeople, but the difficulties of manufacture made it too costly to exploit. An ounce of phosphorus retailed for 6 guineas – roughly £1,200 in today's money – or more than gold. At first, soldiers were called on to provide the raw material, but such an arrangement was hardly conducive to industrial-scale production. In the 1750s a Swedish chemist named Karl (or Carl) Scheele devised a way to manufacture phosphorus in bulk without the slop or smell of urine. It was largely because of this mastery of phosphorus that Sweden became, and remains, a leading producer of matches.

Scheele was both an extraordinary and an extraordinarily luckless fellow. A humble pharmacist with little in the way of advanced apparatus, he discovered eight elements – chlorine, fluorine, manganese, barium, molybdenum, tungsten, nitrogen and oxygen – and got credit for none of them. In every case, his finds either were overlooked or made it into publication after someone else had made the same discovery independently. He also discovered many useful compounds, among them ammonia, glycerin and tannic acid, and was the first to see the commercial potential of chlorine as a bleach – all breakthroughs that made other people extremely wealthy.

Scheele's one notable shortcoming was a curious insistence on tasting a little of everything he worked with, including such notoriously disagreeable substances as mercury and hydrocyanic acid (another of his discoveries) – a compound so famously poisonous that 150 years later Erwin Schrödinger chose it as his toxin of choice in a famous thought experiment (see p. 166). Scheele's rashness eventually caught up with him. In 1786, aged just forty-three, he was found dead at his workbench surrounded by an array of toxic chemicals, any one of which could have accounted for the stunned and terminal look on his face.

Were the world just and Swedish-speaking, Scheele would have enjoyed universal acclaim. As it is, the plaudits have tended to go to more celebrated chemists, mostly from the English-speaking world. Scheele discovered oxygen in 1772, but for various heartbreakingly complicated reasons could not get his paper published in a timely manner. Credit went instead to Joseph Priestley, who discovered the same element independently, but latterly, in the summer of 1774. Even more remarkable was Scheele's failure to receive credit for the discovery of chlorine. Nearly all textbooks still attribute chlorine's discovery to Humphry Davy, who did indeed find it, but *thirty-six years* after Scheele.

Although chemistry had come a long way in the century that separated Newton and Boyle from Scheele and Priestley and Henry Cavendish, it still had a long way to go. Right up to the closing years of the eighteenth century (and in Priestley's case a little beyond) scientists everywhere searched for, and sometimes believed they had actually found, things that just weren't there: vitiated airs, dephlogisticated marine acids, phloxes, calxes, terraqueous exhalations and, above all, phlogiston, the substance that was held to be the active agent in combustion. Somewhere in all this, it was thought, there also resided a mysterious *élan vital*, the force that brought inanimate objects to life. No one knew where this ethereal essence lay, but two things seemed probable: that you could enliven it with a jolt of electricity (a notion Mary Shelley exploited to full effect in her novel *Frankenstein*); and that it existed in some substances but not others, which is why we ended up with two branches of chemistry: organic (for those substances that were thought to have it) and inorganic (for those that did not).

Someone of insight was needed to thrust chemistry into the modern age, and it was the French who provided him. His name was Antoine-Laurent Lavoisier. Born in 1743, Lavoisier was a member of the minor nobility (his father had purchased a title for the family). In 1768 he bought a practising share in a deeply despised institution called the Ferme Générale (or General Farm), which collected taxes and fees on behalf of the government. Although Lavoisier himself was by all accounts mild and fair-minded, the company he worked for was neither. For one thing, it did not tax the rich but only the poor, and then often arbitrarily. For Lavoisier, the appeal of the institution was that it provided him with the wealth to follow his principal devotion, science. At his peak, his personal earnings reached 150,000 livres a year – perhaps £12 million in today's money.

Three years after embarking on this lucrative career path he married the fourteen-year-old daughter of one of his bosses. The marriage was a meeting of hearts and minds. Mme Lavoisier had an incisive intellect and soon was working productively alongside her husband. Despite the demands of his job and busy social life, they managed on most days to put in five hours of science – two in the early morning and three in the evening – as well as the whole of Sunday, which they called their *jour de bonheur* (day of happiness). Somehow Lavoisier also found the time to be commissioner of gunpowder, supervise the building of a wall around Paris to deter smugglers, help found the metric system and co-author the handbook *Méthode de Nomenclature Chimique*, which became the bible for agreeing the names of the elements.

As a leading member of the Académie Royale des Sciences, he was also required to take an informed and active interest in whatever was topical – hypnotism, prison reform, the respiration of insects, the water supply of Paris. It was in such a capacity in 1780 that Lavoisier made some dismissive remarks about a new theory of combustion that had been submitted to the academy by a hopeful young scientist. The theory was indeed wrong, but the scientist never forgave him. His name was Jean-Paul Marat.

The one thing Lavoisier never did was discover an element. At a time when it seemed as if almost anybody with a beaker, a flame and some interesting powders could discover something new – and when, not incidentally, some two-thirds of the elements were yet to be found – Lavoisier failed to uncover a single one. It certainly wasn't for want of beakers. Lavoisier had 13,000 of them in what was, to an almost preposterous degree, the finest private laboratory in existence. Instead, he took the discoveries of others and made sense of them. He threw out phlogiston and mephitic airs. He identified oxygen and hydrogen for what they were and gave them both their modern names. In short, he helped to bring rigour, clarity and method to chemistry.

And his fancy equipment did in fact come in very handy. For years, he and Mme Lavoisier occupied themselves with extremely exacting studies requiring the finest measurements. They determined, for instance, that a rusting object doesn't lose weight, as everyone had long assumed, but gains weight – an extraordinary discovery. Somehow, as it rusted the object was attracting elemental particles from the air. It was the first realization that matter can be transformed but not eliminated. If you burned this book now, its matter would be changed to ash and smoke, but the net amount of stuff in the universe would be the same. This became known as the conservation of mass, and it was a revolutionary concept. Unfortunately, it coincided with another type of revolution – the French one – and in this one Lavoisier was entirely on the wrong side.

Not only was he a member of the hated Ferme Générale, but he had enthusiastically built the wall that enclosed Paris – an edifice so loathed that it was the first thing attacked by the rebellious citizens. Capitalizing on this, in 1791 Marat, now a leading voice in the National Assembly, denounced Lavoisier and suggested that it was well past time for his hanging. Soon afterwards the Ferme Générale was shut down. Not long after this Marat was murdered in his bath by an aggrieved young woman named Charlotte Corday, but by this time it was too late for Lavoisier.

In 1793 the Reign of Terror, already intense, ratcheted up to a higher gear. In October Marie Antoinette was sent to the guillotine. The following month, as he and his wife were making tardy plans to slip away to Scotland, Lavoisier was arrested. In May he and thirty-one fellow farmers-general were brought before the Revolutionary Tribunal (in a courtroom presided over by a bust of Marat). Eight were granted acquittals, but Lavoisier and the others were taken directly to the Place de la Révolution (now the Place de la Concorde), site of the busiest of French guillotines. Lavoisier watched his father-in-law beheaded, then stepped up and accepted his fate. Less than three months later, on 27 July, Robespierre himself was dispatched in the same way and in the same place and the Reign of Terror swiftly ended.

A hundred years after his death, a statue of Lavoisier was erected in Paris and much admired until someone pointed out that it looked nothing like him. Under questioning, the sculptor admitted that he had used the head of the mathematician and philosopher the Marquis de Condorcet – apparently he had a spare – in the hope that no one would notice or, having noticed, would care. In the second regard he was correct. The statue of Lavoisier-cum-Condorcet was allowed to remain in place for another half-century until the Second World War, when, one morning, it was quietly taken away and melted down for scrap.

In the early 1800s there arose in England a fashion for inhaling nitrous oxide, or laughing gas, after it was discovered that its use 'was attended by a highly pleasurable thrilling'. For the next half-century it would be the drug of choice for young people. One learned body, the Askesian Society, was for a time devoted to little else. Theatres put on 'laughing gas evenings' where volunteers could refresh themselves with a robust inhalation and then entertain the audience with their comical staggerings. It wasn't until 1846 that anyone got around to finding a practical use for nitrous oxide, as an anaesthetic. Goodness knows how many tens of thousands of people suffered unnecessary agonies under the surgeon's knife because no one had thought of the gas's most obvious practical application.

I mention this to make the point that chemistry, having come so far in the eighteenth century, rather lost its bearings in the first decades of the nineteenth, in much the way that geology would in the early years of the twentieth. Partly it was to do with the limitations of equipment – there were, for instance, no centrifuges until the second half of the century, severely restricting many kinds of experiments – and partly it was social. Chemistry was, generally speaking, a science for business people, for those who worked with coal and potash and dyes, and not for gentlemen, who tended to be drawn to geology, natural history and physics. (This was slightly less true in continental Europe than in Britain, but only slightly.) It is perhaps telling that one of the most important observations of the century, Brownian motion, which established the active nature of molecules, was made not by a chemist but by a Scottish botanist, Robert Brown. (What Brown noticed, in 1827, was that tiny grains of pollen suspended in water remained indefinitely in motion no matter how long he gave them to settle. The cause of this perpetual motion – namely, the actions of invisible molecules – was long a mystery.)

Things might have been worse had it not been for a splendidly improbable character named Count von Rumford, who, despite the grandeur of his title, began life in Woburn, Massachusetts, in 1753 as plain Benjamin Thompson.

Thompson was dashing and ambitious, 'handsome in feature and figure', occasionally courageous and exceedingly bright, but untroubled by anything so inconveniencing as a scruple. At nineteen he married a rich widow fourteen years his senior, but at the outbreak of revolution in the colonies he unwisely sided with the loyalists, for a time spying on their behalf. In the fateful year of 1776, facing arrest 'for lukewarmness in the cause of liberty', he abandoned his wife and child and fled just ahead of a mob of anti-royalists armed with buckets of hot tar, bags of feathers and an earnest desire to adorn him with both.

He decamped first to England and then to Germany, where he served as a military adviser to the government of Bavaria, so impressing the authorities that in 1791 he was named Count von Rumford of the Holy Roman Empire. While in Munich, he also designed and laid out the famous park known as the English Garden. In between these undertakings, he somehow found time to conduct a good deal of solid science. He became the world's foremost authority on thermodynamics and the first to elucidate the principles of the convection of fluids and the circulation of ocean currents. He also invented several useful objects, including a drip coffee-maker, thermal underwear, and a type of range still known as the Rumford fireplace. In 1805, during a sojourn in France, he wooed and married Mme Lavoisier, widow of Antoine-Laurent. The marriage was not a success and they soon parted. Rumford stayed on in France where he died, universally esteemed by all but his former wives, in 1814.

Our purpose in mentioning him here is that in 1799, during a comparatively brief interlude in London, he founded the Royal Institution, yet another of the many learned societies that popped into being all over Britain in the late eighteenth and early nineteenth centuries. For a time it was almost the only institution of standing to actively promote the young science of chemistry, and that was thanks almost entirely to a brilliant young man named Humphry Davy, who was appointed the institution's professor of chemistry shortly after its inception and rapidly gained fame as an outstanding lecturer and productive experimentalist.

Soon after taking up his position, Davy began to bang out new elements one after another – potassium, sodium, magnesium, calcium, strontium, and aluminum or aluminium (depending on which branch of English you favour).* He discovered so many elements not so much because he was serially astute as because he developed an ingenious technique of applying electricity to a molten substance – electrolysis, as it is known. Altogether he discovered a dozen elements, a fifth of the known total of his day. Davy might have done far more, but unfortunately as a young man he developed an abiding attachment to the buoyant pleasures of nitrous oxide. He grew so attached to the gas that he drew on it (literally) three or four times a day. Eventually, in 1829, it is thought to have killed him.

Fortunately, more sober types were at work elsewhere. In 1808, a dour Quaker named John Dalton became the first person to intimate the nature of an atom (progress that will be discussed more completely a little further on) and in 1811 an Italian with the splendidly operatic name of Lorenzo Romano Amadeo Carlo Avogadro, Count of Quarequa and Cerreto, made a discovery that would prove highly significant in the long term – namely, that two equal volumes of gases of any type, if kept at the same pressure and temperature, will contain identical numbers of molecules.

Two things were notable about the appealingly simple Avogadro's Principle, as it became known. First, it provided a basis for more accurately measuring the size and weight of atoms. Using Avogadro's mathematics, chemists were eventually able to work out, for instance, that a typical atom had a diameter of 0.00000001 centimetres, which is very little

---

* The confusion over the aluminum/aluminium spelling arose because of some uncharacteristic indecisiveness on Davy's part. When he first isolated the element in 1808, he called it *alumium*. For some reason he thought better of that and changed it to *aluminum* four years later. Americans dutifully adopted the new term, but many British users disliked *aluminum*, pointing out that it disrupted the *-ium* pattern established by sodium, calcium and strontium, so they added a vowel and syllable. Among his other achievements, Davy also invented the miner's safety lamp.

indeed. And second, almost no one knew about it for almost fifty years.*

Partly this was because Avogadro himself was a retiring fellow – he worked alone, corresponded very little with fellow scientists, published few papers and attended no meetings – but also it was because there were no meetings to attend and few chemical journals in which to publish. This is a fairly extraordinary fact. The Industrial Revolution was driven in large part by developments in chemistry, and yet as an organized science chemistry barely existed for decades. The Chemical Society of London was not founded until 1841 and didn't begin to produce a regular journal until 1848, by which time most learned societies in Britain – Geological, Geographical, Zoological, Horticultural and Linnaean (for naturalists and botanists) – were at least twenty years old and in several cases much more. The rival Institute of Chemistry didn't come into being until 1877, a year after the founding of the American Chemical Society. Because chemistry was so slow to get organized, news of Avogadro's important breakthrough of 1811 didn't begin to become general until the first international chemistry congress, in Karlsruhe, in 1860.

Because chemists worked for so long in isolation, conventions were slow to emerge. Until well into the second half of the century, the formula H_2O_2 might mean water to one

---

* The principle led to the much later adoption of Avogadro's Number, a basic unit of measure in chemistry, which was named for Avogadro long after his death. It is the number of molecules found in 2.016 grams of hydrogen gas (or an equal volume of any other gas). Its value is placed at 6.0221367×10^{23}, which is an enormously large number. Chemistry students have long amused themselves by computing just how large a number it is, so I can report that it is equivalent to the number of popcorn kernels needed to cover the United States to a depth of 9 miles, or cupfuls of water in the Pacific Ocean, or soft-drink cans that would, evenly stacked, cover the Earth to a depth of 200 miles. An equivalent number of American pennies would be enough to make every person on Earth a dollar trillionaire. It is a big number.

chemist but hydrogen peroxide to another. C_2H_4 could signify ethylene or marsh gas. There was hardly a molecule that was uniformly represented everywhere.

Chemists also used a bewildering variety of symbols and abbreviations, often self-invented. Sweden's J. J. Berzelius brought a much-needed measure of order to matters by decreeing that the elements be abbreviated on the basis of their Greek or Latin names, which is why the abbreviation for iron is Fe (from the Latin *ferrum*) and for silver is Ag (from the Latin *argentum*). That so many of the other abbreviations accord with their English names (N for nitrogen, O for oxygen, H for hydrogen and so on) reflects English's latinate nature, not its exalted status. To indicate the number of atoms in a molecule, Berzelius employed a superscript notation, as in H^2O. Later, for no special reason, the fashion became to render the number as subscript: H_2O.

Despite the occasional tidyings-up, chemistry by the second half of the nineteenth century was in something of a mess, which is why everybody was so pleased by the rise to prominence in 1869 of an odd and crazed-looking professor at the University of St Petersburg named Dmitri Ivanovich Mendeleyev.

Mendeleyev (also sometimes spelled Mendeleev or Mendeléef) was born in 1834 at Tobolsk, in the far west of Siberia, into a well-educated, reasonably prosperous and very large family – so large, in fact, that history has lost track of exactly how many Mendeleyevs there were: some sources say Dmitri had thirteen siblings, some say sixteen. All agree, at any rate, that he was the youngest. Luck was not always with the family. When Dmitri was small his father, the headmaster of a local school, went blind and his mother had to go out to work. Clearly an extraordinary woman, she eventually became the manager of a successful glass factory. All went well until 1848, when the factory burned down and the family was reduced to penury. Determined to get her youngest child an education, the indomitable Mrs Mendeleyev hitchhiked with young Dmitri 4,000 miles to St Petersburg – that's equivalent to travelling from London to Equatorial Guinea – and deposited him at the Institute of Pedagogy. Worn out by her efforts, she died soon after.

Mendeleyev dutifully completed his studies and eventually landed a position at the local university. There he was a competent but not terribly outstanding chemist, known more for his wild hair and beard, which he had trimmed just once a year, than for his gifts in the laboratory. However, in 1869, at the age of thirty-five, he began to toy with a way to arrange the elements. At the time, elements were normally grouped in two ways – either by atomic weight (using Avogadro's Principle) or by common properties (whether they were metals or gases, for instance). Mendeleyev's breakthrough was to see that the two could be combined in a single table.

As is often the way in science, the principle had actually been anticipated three years previously by an amateur chemist in England named John Newlands. He suggested that when elements were arranged by weight they appeared to repeat certain properties – in a sense to harmonize – at every eighth place along the scale. Slightly unwisely, for this was an idea whose time had not quite yet come, Newlands called it the Law of Octaves and likened the arrangement to the octaves on a piano keyboard. Perhaps there was something in Newlands' manner of presentation, but the idea was considered fundamentally preposterous and widely mocked. At gatherings, droller members of the audience would sometimes ask him if he could get his elements to play them a little tune. Discouraged, Newlands gave up pushing the idea and soon dropped out of sight altogether.

Mendeleyev used a slightly different approach, placing his elements into groups of seven, but employed fundamentally the same premise. Suddenly the idea seemed brilliant and wondrously perceptive. Because the properties repeated themselves periodically, the invention became known as the Periodic Table. Mendeleyev was said to have been inspired by the card game called solitaire in North America and patience elsewhere, wherein cards are arranged by suit horizontally and by number vertically. Using a broadly similar concept, he arranged the elements in horizontal rows called periods and vertical columns called groups. This instantly showed one set of relationships when read up and down and another when read from side to side. Specifically, the vertical columns group together chemicals that have similar properties, so that copper sits on top of silver and silver sits on top of gold because of their chemical affinities as metals, while helium, neon and argon are in a column made up of gases. (The actual, formal determinant in the ordering is something called their electron valences, and if you want to understand them you will have to enrol in evening classes.) The horizontal rows, meanwhile, arrange the chemicals in ascending order by the number of protons in their nuclei – what is known as their atomic number.

The structure of atoms and the significance of protons will come in a following chapter; for the moment, all that is necessary is to appreciate the organizing principle: hydrogen has just one proton and so it has an atomic number of 1 and comes first on the chart; uranium has 92 protons and so it comes near the end and has an atomic number of 92. In this sense, as Philip Ball has pointed out, chemistry really is just a matter of counting. (Atomic number, incidentally, is not to be confused with atomic weight, which is the number of protons plus the number of neutrons in a given element.)

There was still a great deal that wasn't known or understood. Hydrogen is the most common element in the universe, and yet no one would guess as much for another thirty years. Helium, the second most abundant element, had only been found the year before – its existence hadn't even been suspected before that – and then not on Earth but in the Sun, where it was found with a spectroscope during a solar eclipse, which is why it honours the Greek sun god Helios. It wouldn't be isolated until 1895. Even so, thanks to Mendeleyev's invention, chemistry was now on a firm footing.

For most of us the Periodic Table is a thing of beauty in the abstract, but for chemists it established an immediate orderliness and clarity that can hardly be overstated. 'Without a doubt, the Periodic Table of the Chemical Elements is the most elegant organizational chart ever devised,' wrote Robert E. Krebs in *The History and Use of Our Earth's Chemical Elements* – and you can find similar sentiments in virtually every history of chemistry in print.

Today we have 118 known elements – ninety-four that occur naturally (if not always abundantly) on Earth, plus twenty-four that have been created by humans in labs, particle accelerators, nuclear reactors or even atomic explosions. In Mendeleyev's day just sixty-three elements were known, but part of his cleverness was to realize that the elements as then known didn't make a complete picture. His table predicted, with pleasing accuracy, where new elements would slot in when they were found.

The heaviest element so far discovered is oganesson, synthesized in 2002 in a Russian lab, and named for the Russian nuclear physicist Yuri Oganessian. Its atomic weight is a hefty 118. To call it evanescent would be an exaggeration. As of this writing, just five atoms of oganesson have ever been created.

No one can say how high the numbers might go. Richard Feynman, the celebrated physicist, believed that nothing could exist beyond element 137. Others think the upper limit is in the low 170s. As recently as the 1990s, it was commonly suggested that no elements would ever be found beyond 112.

Naming elements can be a fractious business. In the 1960s, Russian and American scientists both claimed to have found element 104 at about the same time, but it took 30 years to agree on the name rutherfordium for it. Today the final decision for naming lies with the International Union of Pure and Applied Chemistry, which takes the matter so seriously that

it has two advisory boards – The Interdivisional Committee of Terminology, Nomenclature and Symbols and The Chemical Nomenclature and Structure Representation Division (both advised by an assortment of subcommittees and commissions) – to make a basic recommendation, which then goes to a joint working party with the International Union of Pure and Applied Physics for a final determination. All this can take years, so it was big news, at least among chemists, in 2016 when IUPAC and IUPAP announced the names of four elements at once: nihonium (element 113), moscovium (115), tennessine (117) and oganesson (118).

Oganesson was particularly controversial as it is one of only two elements named for a living scientist; the other is seaborgium, after Glenn Seaborg. It took twelve years to agree on its name. A slight irony is that although it takes years to name these new very heavy elements, the elements themselves exist for no more than a few millionths of a second before falling apart.

The nineteenth century held one last important surprise for chemists. It began in 1896, when Henri Becquerel in Paris carelessly left a packet of uranium salts on a wrapped photographic plate in a drawer. When he took the plate out some time later, he was surprised to discover that the salts had burned an impression in it, just as if the plate had been exposed to light. The salts were emitting rays of some sort. Considering the importance of what he had found, Becquerel did a very strange thing: he turned the matter over to a graduate student for investigation. Fortunately the student was a recent émigré from Poland named Marie Curie. Working with her new husband, Pierre, Curie found that certain kinds of rocks poured out constant and extraordinary amounts of energy, yet without diminishing in size or changing in any detectable way. What she and her husband couldn't know – what no one could know until Einstein explained things the following decade – was that the rocks were converting mass into energy in an exceedingly efficient way.

Marie Curie dubbed the effect 'radioactivity'. In the process of their work, the Curies also found two new elements – polonium, which they named after her native country, and radium. In 1903 the Curies and Becquerel were jointly awarded the Nobel Prize in Physics. (Marie Curie would win a second prize, in Chemistry, in 1911, the only person to win in both Chemistry and Physics.)

At McGill University in Montreal the young New Zealand-born Ernest Rutherford became interested in the new radioactive materials. With a colleague named Frederick Soddy he discovered that immense reserves of energy were bound up in these small amounts of matter, and that the radioactive decay of these reserves could account for most of the Earth's warmth. They also discovered that radioactive elements decayed into other elements – that one day you had an atom of uranium, say, and the next you had an atom of lead. This was truly extraordinary. It was alchemy pure and simple; no one had ever imagined that such a thing could happen naturally and spontaneously.

Ever the pragmatist, Rutherford was the first to see that there could be a valuable practical application in this. He noticed that in any sample of radioactive material, it always took the same amount of time for half the sample to decay – the celebrated half-life* – and that this steady, reliable rate of

---

* If you have ever wondered how the atoms determine which 50 per cent will die and which 50 per cent will survive for the next session, the answer is that the half-life is really just a statistical convenience – a kind of actuarial table for elemental things. Imagine you had a sample of material with a half-life of thirty seconds. It isn't that every atom in the sample will exist for exactly thirty seconds or sixty seconds or ninety seconds or some other tidily ordained period. Each atom will in fact survive for an entirely random length of time that has nothing to do with multiples of thirty; it might last until two seconds from now or it might oscillate away for years or decades or centuries to come. No one can say. But what we can say is that for the sample as a whole the rate of disappearance will be such that half the atoms will disappear every thirty seconds. It's an average rate, in other words, and you can apply it to any large sampling. Someone once worked out, for instance, that American dimes have a half-life of about thirty years.

THE SIZE OF THE EARTH

decay could be used as a kind of clock. By calculating backwards from how much radiation a material had now and how swiftly it was decaying, you could work out its age. He tested a piece of pitchblende, the principal ore of uranium, and found it to be 700 million years old – very much older than the age most people were prepared to grant the Earth. In the spring of 1904, Rutherford travelled to London to give a lecture at the Royal Institution, the august organization founded by Count von Rumford only 105 years before, though that powdery and periwigged age now seemed a distant eon compared with the roll-your-sleeves-up robustness of the late Victorians. Rutherford was there to talk about his new disintegration theory of radioactivity, as part of which he brought out his piece of pitchblende. Tactfully – for the ageing Kelvin was present, if not always fully awake – Rutherford noted that Kelvin himself had suggested that the discovery of some other source of heat would throw his calculations out. Rutherford had found that other source. Thanks to radioactivity the Earth could be – and self-evidently was – much older than the 24 million years Kelvin's final calculations allowed. Kelvin beamed at Rutherford's respectful presentation, but was in fact unmoved. He never accepted the revised figures and to his dying day believed his work on the age of the Earth his most astute and important contribution to science – far greater than his work on thermodynamics.

As with most scientific revolutions, Rutherford's new findings were not universally welcomed. John Joly of Dublin strenuously insisted well into the 1930s that the Earth was no more than 89 million years old, and was stopped only then by his own death. Others began to worry that Rutherford had now given them too much time. But even with radiometric dating, as decay measurements became known, it would be decades before we got within a billion years or so of the Earth's actual age. Science was on the right track, but still way out.

Kelvin died in 1907. That year also saw the death of Dmitri Mendeleyev. Like Kelvin, his productive work was far behind him, but his declining years were notably less serene. As he aged, Mendeleyev became increasingly eccentric – he refused to acknowledge the existence of radiation or the electron or anything else much that was new – and difficult. His final decades were spent mostly storming out of labs and lecture halls all across Europe. In 1955, element 101 was named mendelevium in his honour. 'Appropriately,' notes Paul Strathern, 'it is an unstable element.'

Radiation, of course, went on and on, literally and in ways nobody expected. In the early 1900s Pierre Curie began to experience clear signs of radiation sickness – notably dull aches in his bones and chronic feelings of malaise – which doubtless would have progressed unpleasantly. We shall never know for certain because in 1906 he was fatally run over by a carriage while crossing a Paris street. Marie Curie spent the rest of her life working with distinction in the field, helping to found the celebrated Radium Institute of the University of Paris in 1914. Despite her two Nobel Prizes, she was never elected to the Academy of Sciences, in large part because after the death of Pierre she conducted an affair with a married physicist sufficiently indiscreet to scandalize even the French – or at least the old men who ran the academy, which is perhaps another matter.

For a long time it was assumed that anything so miraculously energetic as radioactivity must be beneficial. For years, manufacturers of toothpaste and laxatives put radioactive thorium in their products, and at least until the late 1920s the Glen Springs Hotel in the Finger Lakes region of New York (and doubtless others as well) featured with pride the therapeutic effects of its 'Radio-active mineral springs'. It wasn't banned in consumer products until 1938. By this time it was much too late for Mme Curie, who died of leukaemia in 1934. Radiation, in fact, is so pernicious and long-lasting that even now her papers from the

1890s – even her cookbooks – are too dangerous to handle. Her lab books are kept in lead-lined boxes and those who wish to see them must don protective clothing.

Thanks to the devoted and unwittingly high-risk work of the first atomic scientists, by the early years of the twentieth century it was becoming clear that the Earth was unquestionably venerable, though another half-century of science would have to be done before anyone could confidently say quite how venerable. Science, meanwhile, was about to get a new age of its own – the atomic one.

III

A NEW AGE DAWNS

8

EINSTEIN'S UNIVERSE

As the nineteenth century drew to a close, scientists could reflect with satisfaction that they had pinned down most of the mysteries of the physical world: electricity, magnetism, gases, optics, acoustics, kinetics and statistical mechanics, to name just a few, had all fallen into order before them. They had discovered the X-ray, the cathode ray, the electron and radioactivity, invented the ohm, the watt, the kelvin, the joule, the amp and the little erg.

If a thing could be oscillated, accelerated, perturbed, distilled, combined, weighed or made gaseous they had done it, and in the process produced a body of universal laws so weighty and majestic that we still tend to write them out in capitals: the Electromagnetic Field Theory of Light, Richter's Law of Reciprocal Proportions, Charles's Law of Gases, the Law of Combining Volumes, the Zeroth Law, the Valence Concept, the Laws of Mass Actions, and others beyond counting. The whole world clanged and chuffed with the machinery and instruments that their ingenuity had produced. Many wise people believed that there was nothing much left for science to do.

In 1875, when a young German in Kiel named Max Planck was deciding whether to devote his life to mathematics or to physics, he was urged most heartily not to choose physics because the breakthroughs had all been made there. The coming century, he was assured, would be one of consolidation and refinement, not revolution. Planck didn't listen. He studied theoretical physics and threw himself body and soul into work on entropy, a process at the heart of thermodynamics, which seemed to hold much promise for an ambitious young man.* In 1891 he produced his results and learned to his dismay that the important work on entropy *had* in fact been done already, in this instance by a retiring scholar at Yale University named J. Willard Gibbs.

Gibbs is perhaps the most brilliant person most people have never heard of. Modest to the point of near-invisibility, he passed virtually the whole of his life, apart from three years spent studying in Europe, within a three-block area bounded by his house and the Yale campus in New Haven, Connecticut. For his first ten years at Yale he didn't even bother to draw a salary. (He had independent means.) From 1871, when he joined the university as a professor, to his death in 1903, his courses attracted an average of slightly over one student a semester. His written work was difficult to follow and employed a private form of notation that many found incomprehensible. But buried among his arcane formulations were insights of the loftiest brilliance.

In 1875–8, Gibbs produced a series of papers, collectively titled *On the Equilibrium of Heterogeneous Substances*, which

---

* Specifically, it is a measure of randomness or disorder in a system. Darrell Ebbing, in the textbook *General Chemistry*, very usefully suggests thinking of a deck of cards. A new pack fresh out of the box, arranged by suit and in sequence from ace to king, can be said to be in its ordered state. Shuffle the cards and you put them in a disordered state. Entropy is a way of measuring just how disordered that state is and of determining the likelihood of particular outcomes with further shuffles. To grasp entropy fully, it is also necessary to understand concepts such as thermal non-uniformities, lattice distances and stoichiometric relationships, but that's the general idea.

dazzlingly elucidated the thermodynamic principles of, well, nearly everything – 'gases, mixtures, surfaces, solids, phase changes ... chemical reactions, electrochemical cells, sedimentation, and osmosis', to quote William H. Cropper. In essence, what Gibbs did was show that thermodynamics didn't apply simply to heat and energy at the sort of large and noisy scale of the steam engine, but was also present and influential at the atomic level of chemical reactions. Gibbs's *Equilibrium* has been called 'the *Principia* of thermodynamics', but for reasons that defy speculation Gibbs chose to publish these landmark observations in the *Transactions of the Connecticut Academy of Arts and Sciences*, a journal that managed to be obscure even in Connecticut, which is why Planck did not hear of him until too late.

Undaunted – well, perhaps mildly daunted – Planck turned to other matters.* We shall turn to these ourselves in a moment, but first we must make a slight (but relevant!) detour to Cleveland, Ohio, and an institution then known as the Case School of Applied Science. There, in the 1880s, a physicist of early middle years named Albert Michelson, assisted by his friend the chemist Edward Morley, embarked on a series of experiments that produced curious and disturbing results that would have great ramifications for much of what followed.

What Michelson and Morley did, without actually intending to, was undermine a long-standing belief in something called the luminiferous ether, a stable, invisible, weightless, frictionless and unfortunately wholly imaginary medium that

---

* Planck was often unlucky in life. His beloved first wife died early, in 1909, and the younger of his two sons was killed in the First World War. He also had twin daughters whom he adored. One died giving birth. The surviving twin went to look after the baby and fell in love with her sister's husband. They married and two years later *she* died in childbirth. In 1944, when Planck was eighty-five, an Allied bomb fell on his house and he lost everything – papers, diaries, a lifetime of accumulations. The following year his surviving son was caught in a conspiracy to assassinate Hitler and executed.

was thought to permeate the universe. Conceived by Descartes, embraced by Newton, and venerated by nearly everyone ever since, the ether held a position of absolute centrality in nineteenth-century physics as a way of explaining how light travelled across the emptiness of space. It was especially needed in the 1800s because light and electromagnetism were now seen as waves, which is to say types of vibrations. Vibrations must occur *in* something; hence the need for, and lasting devotion to, an ether. As late as 1909, the great British physicist J. J. Thomson was insisting: 'The ether is not a fantastic creation of the speculative philosopher; it is as essential to us as the air we breathe' – this more than four years after it was pretty incontestably established that it didn't exist. People, in short, were *really* attached to the ether.

If you needed to illustrate the idea of nineteenth-century America as a land of opportunity, you could hardly improve on the life of Albert Michelson. Born in 1852 on the German–Polish border to a family of poor Jewish merchants, he came to the United States with his family as an infant and grew up in a mining camp in California's gold rush country where his father ran a dry goods business. Too poor to pay for college, he travelled to Washington, DC, and took to loitering by the front door of the White House so that he could fall in beside Ulysses S. Grant when the President emerged for his daily constitutional. (It was clearly a more innocent age.) In the course of these walks, Michelson so ingratiated himself with the President that Grant agreed to secure for him a free place at the US Naval Academy. It was there that Michelson learned his physics.

Ten years later, by now a professor at the Case School in Cleveland, Michelson became interested in trying to measure something called the ether drift – a kind of headwind produced by moving objects as they ploughed through space. One of the predictions of Newtonian physics was that the speed of light as it pushed through the ether should vary with respect to an observer depending on whether the observer was moving towards the source of light or away from it, but no one had figured out a way to measure this. It occurred to Michelson that for half the year the Earth is travelling in one direction against the background of stars in the cosmos and for half the year it is moving in the opposite direction, like a runner on an elliptical track. He reasoned that if you took careful enough measurements at opposite seasons, and compared light's travel time between the two, you would have your answer.

Michelson talked Alexander Graham Bell, newly enriched inventor of the telephone, into providing the funds to build an ingenious and sensitive instrument of Michelson's own devising called an interferometer, which could measure the velocity of light with great precision. Then, assisted by the genial but shadowy Morley, Michelson embarked on years of fastidious measurements. The work was delicate and exhausting, and had to be suspended for a time to permit Michelson a brief but comprehensive nervous breakdown, but by 1887 they had their results. They were not at all what the two scientists had expected to find.

As Caltech astrophysicist Kip S. Thorne has written: 'The speed of light turned out to be the same in *all* directions and at *all* seasons.' It was the first hint in 200 years – in exactly 200 years, in fact – that Newton's laws might not apply all the time everywhere. The Michelson–Morley outcome became, in the words of William H. Cropper, 'probably the most famous negative result in the history of physics'. Michelson was awarded a Nobel Prize in Physics for the work – the first American so honoured – but not for twenty years. Meanwhile, the Michelson–Morley experiments would hover unpleasantly, like a musty odour, in the background of scientific thought.

Remarkably, and despite his findings, when the twentieth century dawned Michelson counted himself among those who believed that the work of science was nearly at an end, with 'only a few turrets and pinnacles to be added, a few roof bosses to be carved', in the words of a writer in *Nature*. In fact, of course, the world was about to enter a century of science where

A NEW AGE DAWNS

many people wouldn't understand much of it and none would understand all of it. Scientists would soon find themselves adrift in a bewildering realm of particles and antiparticles, where things pop in and out of existence in spans of time that make nanoseconds look plodding and uneventful, where everything is strange. Science was moving from a world of macrophysics, where objects could be seen and held and measured, to one of microphysics, where events transpire with inconceivable swiftness on scales of magnitude far below the limits of imagining. We were about to enter the quantum age, and the first person to push on the door was the so-far unfortunate Max Planck.

In 1900, now a theoretical physicist at the University of Berlin, and at the somewhat advanced age of forty-two, Planck unveiled a new 'quantum theory', which posited that energy is not a continuous thing like flowing water but comes in individualized packets, which he called quanta. This *was* a novel concept, and a good one. In the short term it would help to provide a solution to the puzzle of the Michelson–Morley experiments in that it demonstrated that light needn't be a wave after all. In the longer term it would lay the foundation for the whole of modern physics. It was, at all events, the first clue that the world was about to change.

But the landmark event – the dawn of a new age – came in 1905 when there appeared in the German physics journal *Annalen der Physik* a series of papers by a young Swiss bureaucrat who had no university affiliation, no access to a laboratory and the regular use of no library greater than that of the national patent office in Bern, where he was employed as a technical examiner third class. (An application to be promoted to technical examiner second class had recently been rejected.)

His name was Albert Einstein, and in that one eventful year he submitted to *Annalen der Physik* five papers, of which three, according to C. P. Snow, 'were among the greatest in the history of physics' – one examining the photoelectric effect by means of Planck's new quantum theory, one on the behaviour of small particles in suspension (what is known as Brownian motion), and one outlining a Special Theory of Relativity. The first won its author a Nobel Prize and explained the nature of light (and also helped to make television possible, among other things).* The second provided proof that atoms do indeed exist – a fact that had, surprisingly, been in some dispute. The third merely changed the world.

Einstein was born in Ulm, in southern Germany, in 1879, but grew up in Munich. Little in his early life suggested the greatness to come. Famously, he didn't learn to speak until he was three. In the 1890s, his father's electrical business failing, the family moved to Milan, but Albert, by now a teenager, went to Switzerland to continue his education – though he failed his college entrance exams on the first try. In 1896 he gave up his German citizenship to avoid military conscription and entered the Zurich Polytechnic Institute on a four-year course designed to churn out high school science teachers. He was a bright but not outstanding student.

In 1900 he graduated and within a few months was beginning to contribute papers to *Annalen der Physik*. His very first paper, on the physics of fluids in drinking straws (of all things), appeared in the same issue as Planck's quantum theory. From 1902 to 1904 he produced a series of papers on statistical mechanics, only to discover that the quietly productive J. Willard Gibbs in Connecticut had done that work as well, in his *Elementary Principles of Statistical Mechanics* of 1901.

---

* Einstein was honoured, somewhat vaguely, 'for services to theoretical physics'. He had to wait sixteen years, until 1921, to receive the award – quite a long time, all things considered, but nothing at all compared with Frederick Reines, who detected the neutrino in 1957 but wasn't given a Nobel until 1995, thirty-eight years later, or the German Ernst Ruska, who invented the electron microscope in 1932 and received his Nobel Prize in 1986, more than half a century after the fact. Since Nobel Prizes are never awarded posthumously, longevity can be as important as ingenuity in securing one.

Albert had fallen in love with a fellow student, a Hungarian named Mileva Marić. In 1901 they had a child out of wedlock, a daughter, who was discreetly put up for adoption. Einstein never saw his child. Two years later, he and Marić were married. In between these events, in 1902, Einstein took a job with the Swiss patent office, where he stayed for the next seven years. He enjoyed the work: it was challenging enough to engage his mind, but not so challenging as to distract him from his physics. This was the background against which he produced the Special Theory of Relativity in 1905.

'On the Electrodynamics of Moving Bodies' is one of the most extraordinary scientific papers ever published, as much for how it was presented as for what it said. It had no footnotes or citations, contained almost no mathematics, made no mention of any work that had influenced or preceded it, and acknowledged the help of just one individual, a colleague at the patent office named Michele Besso. It was, wrote C. P. Snow, as if Einstein 'had reached the conclusions by pure thought, unaided, without listening to the opinions of others. To a surprisingly large extent, that is precisely what he had done.'

His famous equation, $E = mc^2$, did not appear with the paper but came in a brief supplement that followed a few months later. As you will recall from schooldays, E in the equation stands for energy, m for mass and c^2 for the speed of light squared. In simplest terms, what the equation says is that mass and energy have an equivalence. They are two forms of the same thing: energy is liberated matter; matter is energy waiting to happen. Since c^2 (the speed of light times itself) is a truly enormous number, what the equation is saying is that there is a huge amount – a really huge amount – of energy bound up in every material thing.*

---

* How c came to be the symbol for the speed of light is something of a mystery, but David Bodanis suggests it probably came from the Latin *celeritas*, meaning swiftness.

You may not feel outstandingly robust, but if you are an average-sized adult you will contain within your modest frame no less than 7×10^{18} joules of potential energy – enough to explode with the force of thirty very large hydrogen bombs, assuming you knew how to liberate it and really wished to make a point. Everything has this kind of energy trapped within it. We're just not very good at getting it out. Even a uranium bomb – the most energetic thing we have produced yet – releases less than 1 per cent of the energy it could release if only we were more cunning.

Among much else, Einstein's theory explained how radiation worked: how a lump of uranium could throw out constant streams of high-level energy without melting away like an ice cube. (It could do it by converting mass to energy extremely efficiently à la $E = mc^2$.) It explained how stars could burn for billions of years without racing through their fuel. (Ditto.) At a stroke, in a simple formula, Einstein endowed geologists and astronomers with the luxury of billions of years. Above all, the special theory showed that the speed of light was constant and supreme. Nothing could overtake it. It brought light (no pun intended exactly) to the very heart of our understanding of the nature of the universe. Not incidentally, it also solved the problem of the luminiferous ether by making it clear that it didn't exist. Einstein gave us a universe that didn't need it.

Physicists as a rule are not over-attentive to the pronouncements of Swiss patent office clerks and so, despite the abundance of useful tidings they offered, Einstein's papers attracted little notice. Having just solved several of the deepest mysteries of the universe, Einstein applied for a job as a university lecturer and was rejected, and then for one as a high school teacher and was rejected there as well. So he went back to his job as an examiner third class – but of course he kept thinking. He hadn't even come close to finishing yet.

When the poet Paul Valéry once asked Einstein if he kept a notebook to record his ideas, Einstein looked at him with mild but genuine surprise. 'Oh, that's not necessary,' he replied. 'It's so seldom I have one.' I need hardly point out that when he did get one it tended to be good. Einstein's next idea was one of the greatest that anyone has ever had – indeed, the very greatest, according to Boorse, Motz and Weaver in their thoughtful history of atomic science. 'As the creation of a single mind,' they write, 'it is undoubtedly the highest intellectual achievement of humanity,' which is of course as good as a compliment can get.

In 1907, or so it has sometimes been written, Albert Einstein saw a workman fall off a roof and began to think about gravity. Alas, like many good stories this one appears to be apocryphal. According to Einstein himself, he was simply sitting in a chair when the problem of gravity occurred to him.

Actually, what occurred to Einstein was something more like the beginning of a solution to the problem of gravity, since it had been evident to him from the outset that one thing missing from the special theory was gravity. What was 'special' about the special theory was that it dealt with things moving in an essentially unimpeded state. But what happened when a thing in motion – light, above all – encountered an obstacle such as gravity? It was a question that would occupy his thoughts for most of the next decade and lead to the publication in early 1917 of a paper entitled 'Cosmological Considerations on the General Theory of Relativity'.

The Special Theory of Relativity of 1905 was a profound and important piece of work, of course; but, as C. P. Snow once observed, if Einstein hadn't thought of it when he did someone else would have, probably within five years; it was an idea waiting to happen. But the general theory was something else altogether. 'Without it,' wrote Snow in 1979, 'it is likely that we should still be waiting for the theory today.'

With his pipe, genially self-effacing manner and electrified hair, Einstein was too splendid a figure to remain permanently obscure and in 1919, the war over, the world suddenly discovered him. Almost at once his theories of relativity developed a reputation for being impossible for an ordinary person to grasp. Matters were not helped, as David Bodanis points out in his superb book $E = mc^2$, when the *New York Times* decided to do a story, and – for reasons that can never fail to excite wonder – sent the paper's golfing correspondent, one Henry Crouch, to conduct the interview.

Crouch was hopelessly out of his depth and got nearly everything wrong. Among the more lasting errors in his report was the assertion that Einstein had found a publisher daring enough to publish a book that only twelve men 'in all the world could comprehend'. There was no such book, no such publisher, no such circle of learned men, but the notion stuck anyway. Soon the number of people who could grasp relativity had been reduced even further in the popular imagination – and the scientific establishment, it must be said, did little to disturb the myth.

When a journalist asked the British astronomer Sir Arthur Eddington if it was true that he was one of only three people in the world who could understand Einstein's relativity theories, Eddington considered deeply for a moment and replied: 'I am trying to think who the third person is.' In fact, the problem with relativity wasn't that it involved a lot of differential equations, Lorentz transformations and other complicated mathematics (though it did – even Einstein needed help with some of it), but that it was just so thoroughly non-intuitive.

In essence what relativity says is that space and time are not absolute, but relative both to the observer and to the thing being observed, and the faster one moves the more pronounced these effects become. We can never accelerate ourselves to the speed of light, and the harder we try (and the faster we go) the more distorted we will become, relative to an outside observer.

Almost at once, popularizers of science tried to come up with ways to make these concepts accessible to a general audience. One of the more successful attempts – commercially at least – was *The ABC of Relativity* by the mathematician and philosopher Bertrand Russell. In it Russell employed an image that has been used many times since. He asked the reader to envision a train 100 yards long moving at 60 per cent of the speed of light. To someone standing on a platform watching it pass, the train would appear to be only 80 yards long and everything on it would be similarly compressed. If we could hear the passengers on the train speak, their voices would sound slurred and sluggish, like a record played at too slow a speed, and their movements would appear similarly ponderous. Even the clocks on the train would seem to be running at only four-fifths of their normal speed. However – and here's the thing – people on the train would have no sense of these distortions. To them, everything on the train would seem quite normal. It would be us on the platform who looked weirdly compressed and slowed down. It is all to do, you see, with your position relative to the moving object.

This effect actually happens every time you move. Fly across the United States and you will step from the plane a quinzillionth of a second, or something, younger than those you left behind. Even in walking across the room you will very slightly alter your own experience of time and space. It has been calculated that a baseball thrown at 100 miles an hour will pick up 0.000000000002 grams of mass on its way to home plate. So the effects of relativity are real and have been measured. The problem is that such changes are much too small to make the tiniest detectable difference to us in our daily lives. But for other things in the universe – light, gravity, the universe itself – these are matters of consequence.

So if the ideas of relativity seem weird, it is only because we don't experience these sorts of interactions in normal life. However, to turn to Bodanis again, we all commonly encounter other kinds of relativity – for instance, with regard to sound. If you are in a park and someone is playing annoying music, you know that if you move to a more distant spot the music will seem quieter. That's not because the music *is* quieter, of course, but simply that your position relative to it has changed. To something too small or sluggish to duplicate this experience – a snail, say – the idea that a radio speaker could seem to two observers to produce two different volumes of music simultaneously might seem incredible.

The most challenging and non-intuitive of all the concepts in the General Theory of Relativity is the idea that time is part of space. Our instinct is to regard time as eternal, absolute, immutable; to believe that nothing can disturb its steady tick. In fact, according to Einstein, time is variable and ever-changing. It even has shape. It is bound up – 'inextricably interconnected', in Stephen Hawking's expression – with the three dimensions of space in a curious dimension known as spacetime.

Spacetime is usually explained by asking you to imagine something flat but pliant – a mattress, say, or a sheet of stretched rubber – on which is resting a heavy round object, such as an iron ball. The weight of the iron ball causes the material on which it is sitting to stretch and sag slightly. This is roughly analogous to the effect that a massive object such as the Sun (the iron ball) has on spacetime (the material): it stretches and curves and warps it. Now, if you roll a smaller ball across the sheet, it tries to go in a straight line as required by Newton's laws of motion, but as it nears the massive object and the slope of the sagging fabric it rolls downwards, ineluctably drawn to the more massive object. This is gravity – a product of the bending of spacetime. Every object that has mass creates a little depression in the fabric of the cosmos. Thus the universe, as Dennis Overbye has put it, is 'the ultimate sagging mattress'. Gravity on this view is no longer so much a thing as an outcome – 'not a "force" but a byproduct of the warping of spacetime', in the words of the physicist Michio Kaku, who goes on: 'In some sense, gravity does not exist; what moves the planets and stars is the distortion of space and time.'

Of course, the sagging mattress analogy can take us only so

far, because it doesn't incorporate the effect of time. But then, our brains can take us only so far, because it is so nearly impossible to envision a dimension comprising three parts space to one part time, all interwoven like the threads in a plaid fabric. At all events, I think we can agree that this was an awfully big thought for a young man idly staring out a window.

Among much else, Einstein's General Theory of Relativity suggested that the universe must be either expanding or contracting. But Einstein was not a cosmologist and he accepted the prevailing wisdom that the universe was fixed and eternal. More or less reflexively, he dropped into his equations something called the cosmological constant, which arbitrarily counterbalanced the effects of gravity, serving as a kind of mathematical pause button. Books on the history of science always forgive Einstein this lapse, but it was actually a fairly appalling piece of science and he knew it. He called it 'the biggest blunder of my life'.

One other phenomenon predicted by Einstein's theory was the existence of gravitational waves – ripples of gravity spreading outward across the universe at the speed of light, like ripples on a pond, distorting space-time with their passing. Almost exactly a century passed before gravitational waves were finally detected in 2015 by a pair of massive, exceedingly sensitive antennae known jointly as the Laser Interferometer Gravitational-Wave Observatory (or LIGO). The discovery is considered by many to be one of the great moments in modern science. 'This opens up a whole new area for astrophysics,' Matthew Evans of MIT told a reporter for *MIT News*. 'We always look to the sky with telescopes for electromagnetic radiation like light, radio waves, or X-rays. Now gravitational waves are a completely new way in which we can get to know the universe.'

Coincidentally, at about the time that Einstein was affixing a cosmological constant to his theory, at the Lowell Observatory in Arizona an astronomer with the cheerily intergalactic name of Vesto Slipher (who was in fact from Indiana) was taking spectrographic readings of distant stars and discovering that they appeared to be moving away from us. The universe wasn't static. The stars Slipher looked at showed unmistakable signs of a Doppler shift – the same mechanism behind that distinctive stretched out *yee–yummm* sound cars make as they flash past on a racetrack.* The phenomenon also applies to light, and in the case of receding galaxies it is known as a red shift (because light moving away from us shifts towards the red end of the spectrum; approaching light shifts to blue).

Slipher was the first to notice this effect with light and to realize its potential importance for understanding the motions of the cosmos. Unfortunately, no one much noticed him. The Lowell Observatory, as you will recall, was a bit of an oddity thanks to Percival Lowell's obsession with Martian canals, which in the 1910s made it, in every sense, an outpost of astronomical endeavour. Slipher was unaware of Einstein's theory of relativity and the world was equally unaware of Slipher. So his finding had no impact.

Glory instead would pass to a large mass of ego named Edwin Hubble. Hubble was born in 1889, ten years after Einstein, in a small Missouri town on the edge of the Ozarks, and grew up there and in Wheaton, Illinois, a suburb of Chicago. His father was a successful insurance executive, so life was always comfortable, and Edwin enjoyed a wealth of physical endowments,

---

* Named for Johann Christian Doppler, an Austrian physicist, who first noticed the effect in 1842. Briefly, what happens is that as a moving object approaches a stationary one its sound waves become bunched up as they cram up against whatever device is receiving them (your ears, say), just as you would expect of anything that is being pushed from behind towards an immobile object. This bunching is perceived by the listener as a kind of pinched and elevated sound (the *yee*). As the sound source passes, the sound waves spread out and lengthen, causing the pitch to drop abruptly (the *yummm*).

too. He was a strong and gifted athlete, charming, smart and immensely good-looking – 'handsome almost to a fault', in the description of William H. Cropper, 'an Adonis' in the words of another admirer. According to his own accounts, he also managed to fit into his life more or less constant acts of valour – rescuing drowning swimmers, leading frightened men to safety across the battlefields of France, embarrassing world-champion boxers with knockdown punches in exhibition bouts. It all seemed too good to be true. It was. For all his gifts, Hubble was also an inveterate liar.

This was more than a little odd, for Hubble's life was filled from an early age with a level of genuine distinction that was at times almost ludicrously golden. At a single high school track meeting in 1906, he won the pole vault, shotput, discus, hammer throw, standing high jump and running high jump, and was on the winning mile relay team – that is, seven first places in one meeting – and came third in the long jump. In the same year, he set a state record for the high jump in Illinois.

As a scholar he was equally proficient, and had no trouble gaining admission to study physics and astronomy at the University of Chicago (where, coincidentally, the head of the department was now Albert Michelson). There he was selected to be one of the first Rhodes Scholars at Oxford. Three years of English life evidently turned his head, for he returned to Wheaton in 1913 wearing an Inverness cape, smoking a pipe and talking with a peculiarly orotund accent – not quite British but not quite not – that would remain with him for life. Though he later claimed to have passed most of the second decade of the century practising law in Kentucky, in fact he worked as a high school teacher and basketball coach in New Albany, Indiana, before belatedly attaining his doctorate and passing briefly through the Army. (He arrived in France one month before the armistice and almost certainly never heard a shot fired in anger.)

In 1919, now aged thirty, he moved to California and took up a position at the Mount Wilson Observatory near Los Angeles. Swiftly, and more than a little unexpectedly, he became the most outstanding astronomer of the twentieth century.

It is worth pausing for a moment to consider just how little was known of the cosmos at this time. Astronomers today believe there are something in the region of 150 billion to 200 billion galaxies in the visible universe, though some put the number as high as 2 trillion based on extrapolations from surveys by the Hubble Space Telescope. Even at the lower end, that's a much bigger number than merely saying it would lead you to suppose. If galaxies were frozen peas, 150 billion of them would be enough to fill a large auditorium – the old Boston Garden, say, or the Royal Albert Hall. (An astrophysicist named Bruce Gregory has actually computed this.) In 1919, when Edwin Hubble first put his head to the eyepiece, the number of these galaxies that were known to us was exactly one: the Milky Way. Everything else was thought to be either part of the Milky Way itself or one of many distant, peripheral puffs of gas. Hubble quickly demonstrated how wrong that belief was.

Over the next decade, Hubble tackled two of the most fundamental questions of the universe: how old is it, and how big? To answer both it is necessary to know two things – how far away certain galaxies are and how fast they are flying away from us (what is known as their recessional velocity). The red shift gives the speed at which galaxies are retiring, but doesn't tell us how far away they are to begin with. For that you need what are known as 'standard candles' – stars whose brightness can be reliably calculated and used as benchmarks to measure the brightness (and hence relative distance) of other stars.

Hubble's luck was to come along soon after an ingenious woman named Henrietta Swan Leavitt had figured out a way to find these stars. Leavitt worked at the Harvard College Observatory as a computer, as they were known. Computers spent their lives studying photographic plates of stars and making computations – hence the name. It was little more than drudgery by another name, but it was as close as women could get to real astronomy at Harvard – or, indeed, pretty much anywhere – in those days. The system, however unfair, did have certain unexpected benefits: it meant that half the finest minds available were directed to work that would otherwise have attracted little reflective attention and it ensured that women ended up with an appreciation of the fine structure of the cosmos that often eluded their male counterparts.

One Harvard computer, Annie Jump Cannon, used her repetitive acquaintance with the stars to devise a system of stellar classifications so practical that it is still in use today. Leavitt's contribution was even more profound. She noticed that a type of star known as a Cepheid variable (after the constellation Cepheus, where the first was identified) pulsated with a regular rhythm – a kind of stellar heartbeat. Cepheids are quite rare, but at least one of them is well known to most of us. Polaris, the Pole Star, is a Cepheid.

We now know that Cepheids throb as they do because they are elderly stars that have moved past their 'main sequence phase', in the parlance of astronomers, and become red giants. The chemistry of red giants is a little weighty for our purposes here (it requires an appreciation for the properties of singly ionized helium atoms, among quite a lot else), but put simply it means that they burn their remaining fuel in a way that produces a very rhythmic, very reliable brightening and dimming. Leavitt's genius was to realize that by comparing the relative magnitudes of Cepheids at different points in the sky you could work out where they were in relation to each other. They could be used as standard candles – a term she coined and still in universal use. The method provided only relative distances, not absolute distances, but even so it was the first time that anyone had come up with a usable way to measure the large-scale universe.

(Just to put these insights into perspective, it is perhaps worth noting that at the time Leavitt and Cannon were inferring fundamental properties of the cosmos from dim smudges of distant stars on photographic plates, the Harvard astronomer William H. Pickering, who could of course peer into a first-class telescope as often as he wanted, was developing *his* seminal theory that dark patches on the Moon were caused by swarms of seasonally migrating insects.)

Combining Leavitt's cosmic yardstick with Vesto Slipher's handy red shifts, Hubble began to measure selected points in space with a fresh eye. In 1923 he showed that a puff of distant gossamer in the Andromeda constellation known as M31 wasn't a gas cloud at all but a blaze of stars, a galaxy in its own right, 100,000 light years across and at least 900,000 light years away. The universe was vaster – vastly vaster – than anyone had ever supposed. In 1924 Hubble produced a landmark paper, 'Cepheids in Spiral Nebulae' (nebulae, from the Latin for 'clouds', was his word for galaxies), showing that the universe consisted not just of the Milky Way but of lots of independent galaxies – 'island universes' – many of them bigger than the Milky Way and much more distant.

This finding alone would have ensured Hubble's reputation, but he now turned to the question of working out just how large the universe was, and made an even more striking discovery. Hubble began to measure the spectra of distant galaxies – the business that Slipher had begun in Arizona. Using Mount Wilson's new 100-inch Hooker telescope and some clever inferences, by the early 1930s he had worked out that all the galaxies in the sky (except for our own local cluster) are moving away from us. Moreover, their speed and distance were neatly proportional: the further away the galaxy, the faster it was moving.

This was truly startling. The universe was expanding, swiftly and evenly in all directions. It didn't take a huge amount of imagination to read backwards from this and realize that it must therefore have started from some central point. Far from being the stable, fixed, eternal void that everyone had always

A NEW AGE DAWNS

assumed, this was a universe that had a beginning. It might therefore also have an end.

The wonder, as Stephen Hawking has noted, is that no one had hit on the idea of the expanding universe before. A static universe, as should have been obvious to Newton and every thinking astronomer since, would collapse in upon itself. There was also the problem that if stars had been burning indefinitely in a static universe, they'd have made the whole intolerably hot – certainly much too hot for the likes of us. An expanding universe resolved much of this at a stroke.

Hubble was a much better observer than a thinker and didn't immediately appreciate the full implications of what he had found. Partly this was because he was woefully ignorant of Einstein's General Theory of Relativity. This was quite remarkable because, for one thing, Einstein and his theory were world-famous by now. Moreover, in 1929 Albert Michelson – now in his twilight years but still one of the world's most alert and esteemed scientists – accepted a position at Mount Wilson to measure the velocity of light with his trusty interferometer, and must surely have at least mentioned to Hubble the applicability of Einstein's theory to his own findings.

At all events, Hubble failed to make theoretical hay when the chance was there. Instead, it was left to a Belgian priest-scholar (with a PhD from MIT) named Georges Lemaître to bring together the two strands in his own 'fireworks theory', which suggested that the universe began as a geometrical point, a 'primeval atom', which burst into glory and had been moving apart ever since. It was an idea that very neatly anticipated the modern conception of the Big Bang, but was so far ahead of its time that Lemaître seldom gets more than the sentence or two that we have given him here. The world would need additional decades, and the inadvertent discovery of cosmic background radiation by Penzias and Wilson at their hissing antenna in New Jersey, before the Big Bang would begin to move from interesting idea to established theory.

Neither Hubble nor Einstein would be much of a part of that big story. Though no one would have guessed it at the time, both men had done about as much as they were ever going to do. In 1936 Hubble produced a popular book called *The Realm of the Nebulae*, which explained in flattering style his own considerable achievements. Here at last he showed that he had acquainted himself with Einstein's theory – up to a point, anyway: he gave it four pages out of about 200.

Hubble died of a stroke in 1953 at the age of sixty-three. He was never awarded a Nobel Prize because at that time astronomy was considered a different discipline from physics. (That has since changed.) One last small oddity awaited him. For reasons cloaked in mystery, his wife declined to have a funeral and never revealed what she did with his body. More than seventy years later, the whereabouts of the twentieth century's greatest astronomer remain unknown. For a memorial you must look to the sky and the Hubble Space Telescope, launched in 1990 and named in his honour, and still working away productively 550 or so kilometres above us.

9

THE MIGHTY ATOM

While Einstein and Hubble were productively unravelling the large-scale structure of the cosmos, others were struggling to understand something closer to hand but in its way just as remote: the tiny and ever-mysterious atom.

The great Caltech physicist Richard Feynman once observed that if you had to reduce scientific history to one important statement it would be: 'All things are made of atoms.' They are everywhere and they constitute everything. Look around you. It is all atoms. Not just the solid things like walls and tables and sofas, but the air in between. And they are there in numbers that you really cannot conceive.

The basic working arrangement of atoms is the molecule (from the Latin for 'little mass'). A molecule is simply two or more atoms working together in a more or less stable arrangement: add two atoms of hydrogen to one of oxygen and you have a molecule of water. Chemists tend to think in terms of molecules rather than elements in much the way that writers tend to think in terms of words and not letters, so it is molecules they count, and these are numerous to say the least. At sea level, at a temperature of 0 degrees centigrade, one cubic centimetre of air (that is, a space about the size of a sugar cube) will contain 45 billion billion molecules. And they are in every single cubic centimetre you see around you. Think how many cubic centimetres there are in the world outside your window – how many sugar cubes it would take to fill that view. Then think how many it would take to build a universe. Atoms, in short, are very abundant.

They are also fantastically durable. Because they are so long-lived, atoms really get around. Every atom you possess has almost certainly passed through several stars and been part of millions of organisms on its way to becoming you. We are each so atomically numerous and so vigorously recycled at death that a significant number of our atoms – up to a billion for each of us, it has been suggested – probably once belonged to Shakespeare. A billion more each came from Buddha and Genghis Khan and Beethoven, and any other historical figure you care to name. (The personages have to be historical, apparently, as it takes the atoms some decades to become thoroughly redistributed; however much you may wish it, you are not yet one with Elvis Presley.) So we are all reincarnations – though short-lived ones.

When we die, our atoms will disassemble and move off to find new uses elsewhere – as part of a leaf or other human being or drop of dew. Atoms themselves, however, go on practically for ever. Nobody actually knows how long an atom can survive, but according to Martin Rees it is probably about 10^{35} years – a number so big that even I am happy to express it in mathematical notation.

Above all, atoms are tiny – very tiny indeed. Half a million of them lined up shoulder to shoulder could hide behind a human hair. On such a scale an individual atom is essentially impossible to imagine, but we can of course try.

Start with a millimetre, which is a line this long: -. Now imagine that line divided into 1,000 equal widths. Each of those widths is a micron. This is the scale of microorganisms. A typical paramecium, for instance – a tiny, single-celled, freshwater creature – is about 2 microns wide, 0.002 millimetres,

A NEW AGE DAWNS

which is really very small. According to Richard Feynman, if you wanted to see with your naked eye a paramecium swimming in a drop of water, you would have to enlarge the drop until it was some 12 metres across. However, if you wanted to see the atoms in the same drop, you would have to make the drop 24 *kilometres* across.

Atoms, in other words, exist on a scale of minuteness of another order altogether. To get down to the scale of atoms, you would need to take each one of those micron slices and shave it into 10,000 finer widths. *That*'s the scale of an atom: one ten-millionth of a millimetre. It is a degree of slenderness way beyond the capacity of our imaginations, but you can get some idea of the proportions if you bear in mind that one atom is to that millimetre line above as the thickness of a sheet of paper is to the height of the Empire State Building.

It is, of course, the abundance and extreme durability of atoms that make them so useful, and the tininess that makes them so hard to detect and understand. The realization that atoms are these three things – small, numerous, practically indestructible – and that all things are made from them first occurred not to Antoine-Laurent Lavoisier, as you might expect, or even to Henry Cavendish or Humphry Davy, but rather to a spare and lightly educated English Quaker named John Dalton, whom we first encountered in Chapter 7.

Dalton was born in 1766 on the edge of the Lake District, near Cockermouth, to a family of poor and devout Quaker weavers. (Four years later the poet William Wordsworth would also join the world at Cockermouth.) He was an exceptionally bright student – so very bright, indeed, that at the improbably youthful age of twelve he was put in charge of the local Quaker school. This perhaps says as much about the school as about Dalton's precocity, but perhaps not: we know from his diaries that at about this time he was reading Newton's *Principia* – in the original Latin – and other works of a similarly challenging nature. At fifteen, still schoolmastering, he took a job in the nearby town of Kendal, and a decade after that he moved to Manchester, whence he scarcely stirred for the remaining fifty years of his life. In Manchester he became something of an intellectual whirlwind, producing books and papers on subjects ranging from meteorology to grammar. Colour blindness, a condition from which he suffered, was for a long time called Daltonism because of his studies. But it was a plump book called *A New System of Chemical Philosophy*, published in 1808, that established his reputation.

There, in a short chapter of just five pages (out of the book's more than 900), people of learning first encountered atoms in something approaching their modern conception. Dalton's simple insight was that at the root of all matter are exceedingly tiny, irreducible particles. 'We might as well attempt to introduce a new planet into the solar system or annihilate one already in existence, as to create or destroy a particle of hydrogen,' he wrote.

Neither the idea of atoms nor the term itself was exactly new. Both had been developed by the ancient Greeks. Dalton's contribution was to consider the relative sizes and characters of these atoms and how they fit together. He knew, for instance, that hydrogen was the lightest element, so he gave it an atomic weight of 1. He believed also that water consisted of seven parts of oxygen to one of hydrogen, and so he gave oxygen an atomic weight of 7. By such means was he able to arrive at the relative weights of the known elements. He wasn't always terribly accurate – oxygen's atomic weight is actually 16, not 7 – but the principle was sound and formed the basis for all of modern chemistry and much of the rest of modern science.

The work made Dalton famous – albeit in a low-key, English Quaker sort of way. In 1826, the French chemist P. J. Pelletier travelled to Manchester to meet the atomic hero. Pelletier expected to find him attached to some grand institution, so he was astounded to discover him teaching elementary arithmetic

A NEW AGE DAWNS

to boys in a small school on a back street. According to the scientific historian E. J. Holmyard, a confused Pelletier, upon beholding the great man, stammered:

> 'Est-ce que j'ai l'honneur de m'addresser à Monsieur Dalton?' for he could hardly believe his eyes that this was the chemist of European fame, teaching a boy his first four rules. 'Yes,' said the matter-of-fact Quaker. 'Wilt thou sit down whilst I put this lad right about his arithmetic?'

Although Dalton tried to avoid all honours, he was elected to the Royal Society against his wishes, showered with medals and given a handsome government pension. When he died in 1844, 40,000 people viewed the coffin and the funeral cortège stretched for 2 miles. His entry in the *Dictionary of National Biography* is one of the longest, rivalled in length among nineteenth-century men of science only by those of Darwin and Lyell.

For a century after Dalton made his proposal, it remained entirely hypothetical, and a few eminent scientists – notably the Viennese physicist Ernst Mach, for whom is named the speed of sound – doubted the existence of atoms at all. 'Atoms cannot be perceived by the senses . . . they are things of thought,' he wrote. Such was the scepticism with which the existence of atoms was viewed in the German-speaking world in particular that it was said to have played a part in the suicide of the great theoretical physicist and atomic enthusiast Ludwig Boltzmann in 1906.

It was Einstein who provided the first incontrovertible evidence of atoms' existence with his paper on Brownian motion in 1905, but this attracted little attention and in any case Einstein was soon to become consumed with his work on general relativity. So the first real hero of the atomic age, if not the first personage on the scene, was Ernest Rutherford.

Rutherford was born in 1871 in the 'back blocks' of New

Zealand to parents who had emigrated from Scotland to raise a little flax and a lot of children (to paraphrase Steven Weinberg). Growing up in a remote part of a remote country, he was about as far from the mainstream of science as it was possible to be, but in 1895 he won a scholarship that took him to the Cavendish Laboratory at Cambridge University, which was about to become the hottest place in the world to do physics.

Physicists are notoriously scornful of scientists from other fields. When the great Austrian physicist Wolfgang Pauli's wife left him for a chemist, he was staggered with disbelief. 'Had she taken a bullfighter I would have understood,' he remarked in wonder to a friend. 'But a *chemist* . . .'.

It was a feeling Rutherford would have understood. 'All science is either physics or stamp collecting,' he once said, in a line that has been used many times since. There is a certain engaging irony, therefore, that his award of the Nobel Prize in 1908 was in Chemistry, not Physics.

Rutherford was a lucky man – lucky to be a genius, but even luckier to live at a time when physics and chemistry were so exciting and so compatible (his own sentiments notwithstanding). Never again would they quite so comfortably overlap.

For all his success, Rutherford was not an especially brilliant man and was actually pretty terrible at mathematics. Often during lectures he would get so lost in his own equations that he would give up halfway through and tell the students to work it out for themselves. According to his longtime colleague James Chadwick, discoverer of the neutron, he wasn't even particularly clever at experimentation. He was simply tenacious and open-minded.

For brilliance he substituted shrewdness and a kind of daring. His mind, in the words of one biographer, was 'always operating out towards the frontiers, as far as he could see, and that was a great deal further than most other men'. Confronted with an intractable problem, he was prepared to work at it harder and longer than most people and to be more receptive to unorthodox explanations. His greatest breakthrough came because he was prepared to spend long, tedious hours sitting at a screen counting alpha particle scintillations, as they were known – the sort of work that would normally have been farmed out. He was one of the first – possibly the very first – to see that the power inherent in the atom could, if harnessed, make bombs powerful enough to 'make this old world vanish in smoke'.

Physically he was big and booming, with a voice that made the timid shrink. Once, when told that Rutherford was about to make a radio broadcast across the Atlantic, a colleague drily asked: 'Why use radio?' He also had a huge amount of good-natured confidence. When someone remarked to him that he seemed always to be at the crest of a wave, he responded, 'Well, after all, I made the wave, didn't I?' C. P. Snow recalled how, in a Cambridge tailor's, he overheard Rutherford remark cheerily: 'Every day I grow in girth. And in mentality.'

But both girth and fame were far ahead of him in 1895 when he fetched up at the Cavendish.* It was a singularly eventful period in science. In the year of Rutherford's arrival in Cambridge, Wilhelm Roentgen discovered X-rays at the University of Würzburg in Germany; the next year, Henri Becquerel discovered radioactivity. And the Cavendish itself was about to embark on a long period of greatness. In 1897 J. J. Thomson and colleagues would discover the electron there, in 1911 C. T. R. Wilson would produce the first particle detector there (as we shall see), and in 1932 James Chadwick would discover the neutron there. Further still in the future, in 1953 James Watson and Francis Crick would discover the structure of DNA at the Cavendish.

In the beginning Rutherford worked on radio waves, and with some distinction – he managed to transmit a crisp signal

---

* The name comes from the same Cavendishes who produced Henry. This one was William Cavendish, 7th Duke of Devonshire, who was a gifted mathematician and steel baron in Victorian England. In 1870 he gave the university £6,300 to build an experimental laboratory.

more than a mile, a very reasonable achievement for the time – but he gave it up when he was persuaded by a senior colleague that radio had little future. On the whole, however, Rutherford didn't thrive at the Cavendish, and after three years there, feeling he was going nowhere, he took a post at McGill University in Montreal, where he began his long and steady rise to greatness. By the time he received his Nobel Prize (for 'investigations into the disintegration of the elements, and the chemistry of radioactive substances', according to the official citation) he had moved on to Manchester University, and it was there, in fact, that he would do his most important work in determining the structure and nature of the atom.

By the early twentieth century it was known that atoms were made of parts – Thomson's discovery of the electron had established that – but it wasn't known how many parts there were or how they fitted together or what shape they took. Some physicists thought that atoms might be cube-shaped, because cubes can be packed together so neatly without any wasted space. The more general view, however, was that an atom was more like a currant bun or a plum pudding: a dense, solid object that carried a positive charge but that was studded with negatively charged electrons, like the currants in a currant bun.

In 1910, Rutherford (assisted by his student Hans Geiger, who would later invent the radiation detector that bears his name) fired ionized helium atoms, or alpha particles, at a sheet of gold foil. To Rutherford's astonishment, some of the particles bounced back. It was as if, he said, he had fired a 15-inch shell at a sheet of paper and it rebounded into his lap. This was just not supposed to happen. After considerable reflection he realized there could be only one possible explanation: the particles that bounced back were striking something small and dense at the heart of the atom, while the other particles sailed through unimpeded. An atom, Rutherford realized, was mostly empty space, with a very dense nucleus at the centre. This was a most gratifying discovery, but it presented one immediate problem.

By all the laws of conventional physics, atoms shouldn't therefore exist.

Let us pause for a moment and consider the basic structure of the atom as we know it now. Every atom is made from three kinds of elementary particles: protons, which have a positive electrical charge; electrons, which have a negative electrical charge; and neutrons, which have no charge. Protons and neutrons are packed into the nucleus, while electrons spin around outside. The number of protons is what gives an atom its chemical identity. An atom with one proton is an atom of hydrogen, one with two protons is helium, with three protons lithium, and so on up the scale. Each time you add a proton you get a new element. (Because the number of protons in an atom is always balanced by an equal number of electrons, you will sometimes see it written that it is the number of electrons that defines an element; it comes to the same thing. The way it was explained to me is that protons give an atom its identity, electrons its personality.)

Neutrons don't influence an atom's identity, but they do add to its mass. The number of neutrons is generally about the same as the number of protons, but they can vary up and down slightly. Add or subtract a neutron or two and you get an isotope. The terms you hear in reference to dating techniques in archaeology refer to isotopes – carbon-14, for instance, which is an atom of carbon with six protons and eight neutrons (the fourteen being the sum of the two).

Neutrons and protons occupy the atom's nucleus. The nucleus of an atom is tiny – only one-millionth of a billionth of the full volume of the atom – but fantastically dense, since it contains virtually all the atom's mass. As Cropper has put it, if an atom were expanded to the size of a cathedral, the nucleus would be only about the size of a fly – but a fly many thousands of times heavier than the cathedral. It was this spaciousness – this resounding, unexpected roominess – that had Rutherford scratching his head in 1910.

THE MIGHTY ATOM

It is still a fairly astounding notion to consider that atoms are mostly empty space, and that the solidity we experience all around us is an illusion. When two objects come together in the real world – billiard balls are most often used for illustration – they don't actually strike each other. 'Rather,' as Timothy Ferris explains, 'the negatively charged fields of the two balls repel each other . . . [W]ere it not for their electrical charges they could, like galaxies, pass right through each other unscathed.' When you sit in a chair, you are not actually sitting there, but levitating above it at a height of one angstrom (a hundred-millionth of a centimetre), your electrons and its electrons implacably opposed to any closer intimacy.

The picture of an atom that nearly everybody has in mind is of an electron or two flying around a nucleus, like planets orbiting a sun. This image was created in 1904, based on little more than clever guesswork, by a Japanese physicist named Hantaro Nagaoka. It is completely wrong, but durable just the same. As Isaac Asimov liked to note, it inspired generations of science fiction writers to create stories of worlds-within-worlds, in which atoms become tiny inhabited solar systems or our solar system turns out to be merely a mote in some much larger scheme. For many years, even CERN, the European Organization for Nuclear Research, used Nagaoka's image as a logo. In fact, as physicists were soon to realize, electrons are not like orbiting planets at all but more like the blades of a spinning fan, managing to fill every bit of space in their orbits simultaneously (but with the crucial difference that the blades of a fan only *seem* to be everywhere at once; electrons *are*).

Needless to say, very little of this was understood in 1910 or for many years afterwards. Rutherford's finding presented some large and immediate problems, not least that no electron should be able to orbit a nucleus without crashing. Conventional electrodynamic theory demanded that a flying electron should run out of energy very quickly – in only an instant or so – and spiral into the nucleus, with disastrous consequences for both. There was also the problem of how protons, with their positive charges, could bundle together inside the nucleus without blowing themselves and the rest of the atom apart. Clearly, whatever was going on down there in the world of the very small was not governed by the laws that applied in the macro world where our expectations reside.

As physicists began to delve into this subatomic realm, they realized that it wasn't merely different from anything we knew, but different from anything ever imagined. 'Because atomic behaviour is so unlike ordinary experience,' Richard Feynman once observed, 'it is very difficult to get used to and it appears peculiar and mysterious to everyone, both to the novice and to the experienced physicist.' When Feynman made that comment, physicists had had half a century to adjust to the strangeness of atomic behaviour. So think how it must have felt to Rutherford and his colleagues in the early 1910s when it was all brand new.

One of the people working with Rutherford was a mild and affable young Dane named Niels Bohr. In 1913, while puzzling over the structure of the atom, Bohr had an idea so exciting that he postponed his honeymoon to write what became a landmark paper.

Because physicists couldn't see anything so small as an atom, they had to try to work out its structure from how it behaved when they did things to it, as Rutherford had done by firing alpha particles at foil. Sometimes, not surprisingly, the results of these experiments were puzzling. One puzzle that had been around for a long time was to do with spectrum readings of the wavelengths of hydrogen. These produced patterns showing that hydrogen atoms emitted energy at certain wavelengths but not others. It was rather as if someone under surveillance kept turning up at particular locations but was never observed travelling between them. No one could understand why this should be.

It was while puzzling over this problem that Bohr was struck by a solution and dashed off his famous paper. Called 'On the Constitutions of Atoms and Molecules', the paper explained how electrons could keep from falling into the nucleus by suggesting that they could occupy only certain well-defined orbits. According to the new theory, an electron moving between orbits would disappear from one and reappear instantaneously in another *without visiting the space between*. This idea – the famous 'quantum leap' – is of course utterly strange, but it was too good not to be true. It not only kept electrons from spiralling catastrophically into the nucleus, it also explained hydrogen's bewildering wavelengths. The electrons only appeared in certain orbits because they only existed in certain orbits. It was a dazzling insight and it won Bohr the 1922 Nobel Prize in Physics, the year after Einstein received his.

Meanwhile the tireless Rutherford, now back at Cambridge having succeeded J. J. Thomson as head of the Cavendish Laboratory, came up with a model that explained why the nuclei didn't blow up. He saw that the positive charge of the protons must be offset by some type of neutralizing particles, which he called neutrons. The idea was simple and appealing, but not easy to prove. Rutherford's associate, James Chadwick, devoted eleven intensive years to hunting for neutrons before finally succeeding in 1932. He, too, was awarded a Nobel Prize in Physics, in 1935. As Boorse and his colleagues point out in their history of the subject, the delay in discovery was probably a very good thing, as mastery of the neutron was essential to the development of the atomic bomb. (Because neutrons have no charge, they aren't repelled by the electrical fields at the heart of an atom and thus could be fired like tiny torpedoes into an atomic nucleus, setting off the destructive process known as fission.) Had the neutron been isolated in the 1920s, they note, it is 'very likely the atomic bomb would have been developed first in Europe, undoubtedly by the Germans'.

As it was, the Europeans had their hands full trying to understand the strange behaviour of the electron. The principal problem they faced was that the electron sometimes behaved like a particle and sometimes like a wave. This impossible duality drove physicists nearly mad. For the next decade all across Europe they furiously thought and scribbled and offered competing hypotheses. In France, Prince Louis-Victor de Broglie, the scion of a ducal family, found that certain anomalies in the behaviour of electrons disappeared when one regarded them as waves. The observation excited the attention of the Austrian Erwin Schrödinger, who made some deft refinements and devised a handy system called wave mechanics. At almost the same time, the German physicist Werner Heisenberg came up with a competing theory called matrix mechanics. This was so mathematically complex that hardly anyone really understood it, including Heisenberg himself ('I do not even know what a matrix *is*,' Heisenberg despaired to a friend at one point), but it did seem to solve certain problems that Schrödinger's waves failed to explain.

The upshot is that physics had two theories, based on conflicting premises, that produced the same results. It was an impossible situation.

Finally, in 1926, Heisenberg came up with a celebrated compromise, producing a new discipline that came to be known as quantum mechanics. At the heart of it was Heisenberg's Uncertainty Principle, which states that the electron is a particle but a particle that can be described in terms of waves. The uncertainty around which the theory is built is that we can know the path an electron takes as it moves through a space or we can know where it is at a given instant, but we cannot know both.* Any attempt to measure one will unavoidably disturb

---

* There is a little uncertainty about the use of the word uncertainty in regard to Heisenberg's principle. Michael Frayn, in an afterword to his play *Copenhagen*, notes that several words in German – *Unsicherheit*, *Unschärfe*, *Ungenauigkeit* and *Unbestimmtheit* – have been used by various translators, but that none quite equates to the English *uncertainty*. Frayn suggests that *indeterminacy* would be a better word for the principle and *indeterminability* would be better still. Heisenberg himself generally used *Unbestimmtheit*.

the other. This isn't a matter of simply needing more precise instruments; it is an immutable property of the universe.

What this means in practice is that you can never predict where an electron will be at any given moment. You can only list its probability of being there. In a sense, as Dennis Overbye has put it, an electron doesn't exist until it is observed. Or, put slightly differently, until it is observed an electron must be regarded as being 'at once everywhere and nowhere'.

If this seems confusing, you may take some comfort in knowing that it was confusing to physicists, too. Overbye notes: 'Bohr once commented that a person who wasn't outraged on first hearing about quantum theory didn't understand what had been said.' Heisenberg, when asked how one could envision an atom, replied: 'Don't try.'

So the atom turned out to be quite unlike the image that most people had created. The electron doesn't fly around the nucleus like a planet around its sun, but instead takes on the more amorphous aspect of a cloud. The 'shell' of an atom isn't some hard, shiny casing, as illustrations sometimes encourage us to suppose, but simply the outermost of these fuzzy electron clouds. The cloud itself is essentially just a zone of statistical probability marking the area beyond which the electron only very seldom strays. Thus an atom, if you could see it, would look more like a very fuzzy tennis ball than a hard-edged metallic sphere (but not much like either or, indeed, like anything you've ever seen; we are, after all, dealing here with a world very different from the one we see around us).

It seemed as if there was no end of strangeness. For the first time, as James Trefil has put it, scientists had encountered 'an area of the universe that our brains just aren't wired to understand'. Or, as Feynman expressed it, 'things on a small scale behave *nothing* like things on a large scale.' As physicists delved deeper, they realized they had found a world not only where electrons could jump from one orbit to another without travelling across any intervening space, but where matter could pop into existence from nothing at all – 'provided', in the words of Alan Lightman of MIT, 'it disappears again with sufficient haste'.

Perhaps the most arresting of quantum improbabilities is the idea, arising from Wolfgang Pauli's Exclusion Principle of 1925, that certain pairs of subatomic particles, even when separated by the most considerable distances, can each instantly 'know' what the other is doing. Particles have a quality known as spin and, according to quantum theory, the moment you determine the spin of one particle, its sister particle, no matter how distant away, will immediately begin spinning in the opposite direction and at the same rate.

It is as if, in the words of the science writer Lawrence Joseph, you had two identical billiard balls, one in Ohio and the other in Fiji, and that the instant you sent one spinning the other would immediately spin in a contrary direction at precisely the same speed. Remarkably, the phenomenon was proved in 1997 when physicists at the University of Geneva sent photons 7 miles in opposite directions and demonstrated that interfering with one provoked an instantaneous response in the other.

Things reached such a pitch that at one conference Bohr remarked of a new theory that the question was not whether it was crazy, but whether it was crazy enough. To illustrate the non-intuitive nature of the quantum world, Schrödinger offered a famous thought experiment in which a hypothetical cat was placed in a box with one atom of a radioactive substance attached to a vial of hydrocyanic acid. If the particle degraded within an hour, it would trigger a mechanism that would break the vial and poison the cat. If not, the cat would live. But we could not know which was the case, so there was no choice, scientifically, but to regard the cat as 100 per cent alive and 100 per cent dead at the same time. This means, as Stephen Hawking has observed with a touch of understandable excitement, that one cannot 'predict future events exactly if one cannot even measure the present state of the universe precisely!'

THE MIGHTY ATOM

Because of its oddities, many physicists disliked quantum theory, or at least certain aspects of it, and none more so than Einstein. This was more than a little ironic since it was he, in his *annus mirabilis* of 1905, who had so persuasively explained how photons of light could sometimes behave like particles and sometimes like waves – the notion at the very heart of the new physics. 'Quantum theory is very worthy of regard,' he observed politely, but he really didn't like it. 'God doesn't play dice,' he said.*

Einstein couldn't bear the notion that God could create a universe in which some things were for ever unknowable. Moreover, the idea of action at a distance – that one particle could instantaneously influence another trillions of miles away – was a stark violation of the special theory of relativity. Nothing could outrace the speed of light and yet here were physicists insisting that, somehow, at the subatomic level, information could. (No one, incidentally, has ever explained how the particles achieve this feat. Scientists have dealt with this problem, according to the physicist Yakir Aharanov, 'by not thinking about it'.)

Above all, there was the issue that quantum physics introduced a level of untidiness that hadn't previously existed. Suddenly you needed two sets of laws to explain the behaviour of the universe – quantum theory for the world of the very small and relativity for the larger universe beyond. The gravity of relativity theory was brilliant at explaining why planets orbited suns or why galaxies tended to cluster, but turned out to have no influence at all at the particle level. To explain what kept atoms together, other forces were needed, and in the 1930s two were discovered: the strong nuclear force and the weak nuclear force. The strong force binds atoms together; it's what

---

* Or at least, that is how it is nearly always rendered. The actual quote was: 'It seems hard to sneak a look at God's cards. But that He plays dice and uses "telepathic" methods . . . is something that I cannot believe for a single moment.'

allows protons to bed down together in the nucleus. The weak force engages in more miscellaneous tasks, mostly to do with controlling the rates of certain sorts of radioactive decay.

The weak nuclear force, despite its name, is many billions of times stronger than gravity, and the strong nuclear force is more powerful still – vastly so, in fact – but their influence extends to only the tiniest distances. The grip of the strong force reaches out only to about one-hundred-thousandth of the diameter of an atom. That's why the nuclei of atoms are so compacted and dense, and why elements with big, crowded nuclei tend to be so unstable: the strong force just can't hold on to all the protons.

The upshot of all this is that physics ended up with two bodies of laws – one for the world of the very small, one for the universe at large – leading quite separate lives. Einstein disliked that too. He devoted the rest of his life to searching for a way to tie up these loose ends by finding a Grand Unified Theory, and always failed. From time to time he thought he had it, but it always unravelled on him in the end. As time passed he became increasingly marginalized and even a little pitied. Almost without exception, wrote Snow, 'his colleagues thought, and still think, that he wasted the second half of his life.'

Elsewhere, however, real progress was being made. By the mid-1940s scientists had reached a point where they understood the atom at an extremely profound level – as they all too effectively demonstrated in August 1945 by exploding a pair of atomic bombs over Japan.

By this point physicists could be excused for thinking that they had just about conquered the atom. In fact, everything in particle physics was about to get a whole lot more complicated. But before we take up that slightly exhausting story, we must bring another strand of our history up to date by considering an important and salutary tale of avarice, deceit, bad science, several needless deaths and the final determination of the age of the Earth.

10

GETTING THE LEAD OUT

In the late 1940s, a graduate student at the University of Chicago named Clair Patterson (who was, first name notwithstanding, an Iowa farm boy by origin) was using a new method of lead isotope measurement to try to get a definitive age for the Earth at last. Unfortunately, all his rock samples became contaminated – usually wildly so. Most contained something like 200 times the levels of lead that would normally be expected to occur. Many years would pass before Patterson realized that the reason for this lay with a regrettable Ohio inventor named Thomas Midgley, Junior.

Midgley was an engineer by training and the world would no doubt have been a safer place if he had stayed so. Instead, he developed an interest in the industrial applications of chemistry. In 1921, while working for the General Motors Research Corporation in Dayton, Ohio, he investigated a compound called tetraethyl lead (also known, confusingly, as lead tetraethyl), and discovered that it significantly reduced the juddering condition known as engine knock.

Even though lead was widely known to be dangerous, by the early years of the twentieth century it could be found in all manner of consumer products. Food came in cans sealed with lead solder. Water was often stored in lead-lined tanks. Lead arsenate was sprayed on to fruit as a pesticide. Lead even came as part of the composition of toothpaste tubes. Hardly a product existed that didn't bring a little lead into consumers' lives.* However, nothing gave it a greater and more lasting intimacy than its addition to motor fuel.

Lead is a neurotoxin. Get too much of it and you can irreparably damage the brain and central nervous system. Among the many symptoms associated with overexposure are blindness, insomnia, kidney failure, hearing loss, cancer, palsies and convulsions. In its most acute form it produces abrupt and terrifying hallucinations, disturbing to victims and onlookers alike, which generally then give way to coma and death. You really don't want to get too much lead into your system.

On the other hand, lead was easy to extract and work, and almost embarrassingly profitable to produce industrially – and tetraethyl lead did indubitably stop engines from knocking. So in 1923 three of America's largest corporations, General Motors, Du Pont and Standard Oil of New Jersey, formed a joint enterprise called the Ethyl Gasoline Corporation (later shortened to simply Ethyl Corporation) with a view to making as much tetraethyl lead as the world was willing to buy, and that proved to be a very great deal. They called their additive 'ethyl' because it sounded friendlier and less toxic than 'lead', and introduced it for public consumption (in more ways than most people realized) on 1 February 1923.

Almost at once production workers began to exhibit the staggered gait and confused faculties that mark the recently poisoned. Also almost at once, the Ethyl Corporation embarked on a policy of calm but unyielding denial that would serve it well for decades. As Sharon Bertsch McGrayne notes in her

---

* Sometimes until quite recently. Although the US Centers for Disease Control long ago declared that there is no such thing as a safe level of lead, the US Food and Drug Administration only got around to proposing limits on lead in baby food as recently as 2023.

absorbing history of industrial chemistry, *Prometheans in the Lab*, when employees at one plant developed irreversible delusions, a spokesman blandly informed reporters: 'These men probably went insane because they worked too hard.' Altogether, at least fifteen workers died in the early days of production of leaded gasoline, and untold numbers of others became ill, often violently so; the exact numbers are unknown because the company nearly always managed to hush up news of embarrassing leakages, spills and poisonings. At times, however, suppressing the news became impossible – most notably in 1924 when, in a matter of days, five production workers died and thirty-five more were turned into permanent staggering wrecks at a single ill-ventilated facility.

As rumours circulated about the dangers of the new product, ethyl's ebullient inventor, Thomas Midgley, decided to hold a demonstration for reporters to allay their concerns. As he chatted away about the company's commitment to safety, he poured tetraethyl lead over his hands, then held a beaker of it to his nose for sixty seconds, claiming all the while that he could repeat the procedure daily without harm. In fact, Midgley knew only too well the perils of lead poisoning: he had himself been made seriously ill from overexposure a few months earlier and now, except when reassuring journalists, never went near the stuff if he could help it.

Buoyed by the success of leaded petrol, Midgley now turned to another technological problem of the age. Refrigerators in the 1920s were often appallingly risky because they used insidious and dangerous gases that sometimes seeped out. One leak from a refrigerator at a hospital in Cleveland, Ohio, in 1929 killed more than 100 people. Midgley set out to create a gas that was stable, non-flammable, non-corrosive and safe to breathe. With an instinct for the regrettable that was almost uncanny, he invented chlorofluorocarbons, or CFCs.

Seldom has an industrial product been more swiftly or unfortunately embraced. CFCs went into production in the

A NEW AGE DAWNS

early 1930s and found a thousand applications in everything from car air-conditioners to deodorant sprays before it was noticed, half a century later, that they were devouring the ozone in the stratosphere. As you will be aware, this was not a good thing. Ozone is a form of oxygen in which each molecule bears three atoms of oxygen instead of the normal two. It is a bit of a chemical oddity in that at ground level it is a pollutant, while way up in the stratosphere it is beneficial since it soaks up dangerous ultraviolet radiation. Beneficial ozone is not terribly abundant, however. If it were distributed evenly throughout the stratosphere, it would form a layer just 2 millimetres or so thick. That is why it is so easily disturbed. Chlorofluorocarbons are also not very abundant – they constitute only about one part per billion of the atmosphere as a whole – but they are extravagantly destructive. A single kilogram of CFCs can capture and annihilate 70,000 kilograms of atmospheric ozone. CFCs also hang around for a long time – about a century on average – wreaking havoc all the while. And they are great heat sponges. A single CFC molecule is about 10,000 times more efficient at exacerbating greenhouse effects than a molecule of carbon dioxide – and carbon dioxide is of course no slouch itself as a greenhouse gas. In short, chlorofluorocarbons may ultimately prove to be just about the worst invention of the twentieth century.

Midgley never knew this because he died long before anyone realized how destructive CFCs were. His death was itself memorably unusual. After becoming crippled with polio, Midgley invented a contraption involving a series of motorized pulleys that automatically raised or turned him in bed. In 1944, he became entangled in the cords as the machine went into action and was strangled.

If you were interested in finding out the ages of things, the University of Chicago in the 1940s was the place to be. Willard Libby was in the process of inventing radiocarbon dating,

GETTING THE LEAD OUT

allowing scientists to get an accurate reading of the age of bones and other organic remains, something they had never been able to do before. Up to this time, the oldest reliable dates went back no further than the First Dynasty in Egypt – about 3000 BC. No one could confidently say, for instance, when the last ice sheets had retreated or at what time in the past the Cro-Magnon people had decorated the caves of Lascaux in France.

Libby's idea was so useful that he would be awarded a Nobel Prize for it in 1960. It was based on the realization that all living things have within them an isotope of carbon called carbon-14, which begins to decay at a measurable rate the instant they die. Carbon-14 has a half-life of about 5,600 years, so by working out how much of a given sample of carbon had decayed, Libby could get a good fix on the age of an object – though only up to a point. After eight half-lives, only 0.39 per cent of the original radioactive carbon remains, which is too little to make a reliable measurement, so radiocarbon dating works only for objects up to 40,000 or so years old.

Curiously, just as the technique was becoming widespread, certain flaws within it became apparent. To begin with, it was discovered that one of the basic components of Libby's formula, known as the decay constant, was out by about 3 per cent. By this time, however, thousands of measurements had been taken throughout the world. Rather than restate every one, scientists decided to keep the inaccurate constant. 'Thus,' Tim Flannery notes, 'every raw radiocarbon date you read today is given as too young by around 3 per cent.' The problems didn't quite stop there. It was also quickly discovered that carbon-14 samples can be easily contaminated with carbon from other sources – a tiny scrap of vegetable matter, for instance, that has been collected with the sample and not noticed. For younger samples – those under 20,000 years or so – slight contamination does not always matter so much, but for older samples it can be a serious problem because so few remaining atoms are being counted. In the first instance, to borrow from Flannery, it is like miscounting by a dollar when counting to 1,000; in the second it is more like miscounting by a dollar when you have only two dollars to count.

Libby's method was also based on the assumption that the amount of carbon-14 in the atmosphere, and the rate at which it has been absorbed by living things, has been consistent throughout history. In fact it hasn't been. We now know that the volume of atmospheric carbon-14 varies depending on how well or not the Earth's magnetism is deflecting cosmic rays, and that that can vary significantly over time. This means that some carbon-14 dates are more dubious than others. Among the more dubious are dates just around the time that people first came to the Americas, which is one of the reasons the matter is so perennially in dispute.

Finally, and perhaps a little unexpectedly, readings can be thrown out by seemingly unrelated external factors – such as the diets of those whose bones are being tested. One case involved the long-running debate over whether syphilis originated in the New World or the Old. Archaeologists in Hull found that monks in a monastery graveyard had suffered from syphilis, but the initial conclusion that the monks had done so before Columbus's voyage was cast into doubt by the realization that they had eaten a lot of fish, which could make their bones appear to be older than in fact they were. The monks may well have had syphilis, but how it got to them, and when, remains tantalizingly unresolved.

Because of the accumulated shortcomings of carbon-14, scientists devised other methods of dating ancient materials, among them thermoluminescence, which measures electrons trapped in clays, and electron spin resonance, which involves bombarding a sample with electromagnetic waves and measuring the vibrations of the electrons. But even the best of these could not date anything older than about 200,000 years, and they couldn't date inorganic materials like rocks at all, which

GETTING THE LEAD OUT

is of course what you need to do if you wish to determine the age of your planet.

The problems of dating rocks were such that at one point almost everyone in the world had given up on them. Had it not been for a determined English professor named Arthur Holmes, the quest might well have fallen into abeyance altogether.

Holmes was heroic as much for the obstacles he overcame as for the results he achieved. By the 1920s, when he was in the prime of his career, geology had slipped out of fashion – physics was the new excitement of the age – and had become severely underfunded, particularly in Britain, its spiritual birthplace. At Durham University, Holmes was for many years the entire geology department. Often he had to borrow or patch together equipment in order to pursue his radiometric dating of rocks. At one point, his calculations were effectively held up for a year while he waited for the university to provide him with a simple adding machine. Occasionally, he had to drop out of academic life altogether to earn enough to support his family – for a time he ran a curio shop in Newcastle upon Tyne – and sometimes he could not even afford the £5 annual membership fee for the Geological Society.

The technique Holmes used in his work was theoretically straightforward and arose directly from the process first observed by Ernest Rutherford in 1904 by which some atoms decay from one element into another at a rate predictable enough that you can use them as clocks. If you know how long it takes for potassium-40 to become argon-40, and you measure the amounts of each in a sample, you can work out how old a material is. Holmes's contribution was to measure the decay rate of uranium into lead to calculate the age of rocks, and thus – he hoped – of the Earth.

But there were many technical difficulties to overcome. Holmes also needed – or at least would very much have appreciated – sophisticated gadgetry of a sort that could make very fine measurements from tiny samples, and, as we have seen, it was all he could do to get a simple adding machine. So it was quite an achievement when in 1946 he was able to announce with some confidence that the Earth was at least 3 billion years old and possibly rather more. Unfortunately, he now met yet another formidable impediment to acceptance: the conservativeness of his fellow scientists. Although happy to praise his methodology, many maintained that he had found not the age of the Earth but merely the age of the materials from which the Earth had been formed.

It was just at this time that Harrison Brown of the University of Chicago developed a new method for counting lead isotopes in igneous rocks (which is to say, those that were created through heating, as opposed to the laying down of sediments). Realizing that the work would be exceedingly tedious, he assigned it to young Clair Patterson as his dissertation project. Famously, he promised Patterson that determining the age of the Earth with his new method would be 'duck soup'. In fact, it would take years.

Patterson began work on the project in 1948. Compared with Thomas Midgley's colourful contributions to the march of progress, Patterson's discovery of the age of the Earth feels more than a touch anticlimactic. For seven years, first at the University of Chicago and then at the California Institute of Technology (where he moved in 1952), he worked in a sterile lab, making very precise measurements of the lead/uranium ratios in carefully selected samples of old rock.

The problem with measuring the age of the Earth was that you needed rocks that were extremely ancient, containing lead- and uranium-bearing crystals that were about as old as the planet itself – anything much younger would obviously give you misleadingly youthful dates – but really ancient rocks are only rarely found on Earth. In the late 1940s no one altogether understood why this should be. Indeed, and rather extraordinarily, we would be well into the space age before anyone could

GETTING THE LEAD OUT

plausibly account for where all the Earth's old rocks went. (The answer was plate tectonics, which we shall of course get to.) Patterson, meanwhile, was left to try to make sense of things with very limited materials. Eventually, and ingeniously, it occurred to him that he could circumvent the rock shortage by using rocks from beyond Earth. He turned to meteorites.

The assumption he made – rather a large one, but correct as it turned out – was that many meteorites are essentially leftover building materials from the early days of the solar system, and thus have managed to preserve a more or less pristine interior chemistry. Measure the age of these wandering rocks and you would have the age also (near enough) of the Earth. As always, however, nothing was quite as straightforward as such a breezy description makes it sound. Meteorites are not abundant and meteoritic samples not especially easy to get hold of. Moreover, Brown's measurement technique proved finicky in the extreme and needed much refinement. Above all, there was the problem that Patterson's samples were continuously and unaccountably contaminated with large doses of atmospheric lead whenever they were exposed to air. It was this that eventually led him to create a sterile laboratory – the world's first, according to at least one account.

It took Patterson seven years of patient work just to find and measure suitable samples for final testing. In the spring of 1953 he took his specimens to the Argonne National Laboratory in Illinois, where he was granted time on a late-model mass spectrograph, a machine capable of detecting and measuring the minute quantities of uranium and lead locked up in ancient crystals. When at last he had his results, Patterson was so excited that he drove straight to his boyhood home in Iowa and had his mother check him into a hospital because he thought he was having a heart attack.

Soon afterwards, at a meeting in Wisconsin, Patterson announced a definitive age for the Earth of 4,550 million years (plus or minus 70 million years) – 'a figure that stands unchanged

50 years later', as McGrayne admiringly noted. After 200 years of attempts, the Earth finally had an age.

Almost at once, Patterson turned his attention to the question of all that lead in the atmosphere. He was astounded to find that what little was known about the effects of lead on humans was almost invariably wrong or misleading — and not surprisingly, since for forty years every study of lead's effects had been funded exclusively by manufacturers of lead additives.

In one such study, a doctor who had no specialized training in chemical pathology undertook a five-year programme in which volunteers were asked to breathe in or swallow lead in elevated quantities. Then their urine and faeces were tested. Unfortunately, as the doctor appears not to have known, lead is not excreted as a waste product. Rather, it accumulates in the bones and blood — that's what makes it so dangerous — and neither bone nor blood was tested. In consequence, lead was given a clean bill of health. Patterson quickly established that we had a lot of lead in the atmosphere — still do, in fact, since lead never goes away — and that about 90 per cent of it appeared to come from car exhaust pipes; but he couldn't prove it. What he needed was a way to compare lead levels in the atmosphere now with the levels that existed before 1923, when tetraethyl lead began to be commercially produced. It occurred to him that ice cores could provide the answer.

It was known that snowfall in places like Greenland accumulates into discrete annual layers (because seasonal temperature differences produce slight changes in coloration from winter to summer). By counting back through these layers and measuring the amount of lead in each, he could work out global atmospheric lead concentrations at any time for hundreds, or even thousands, of years. The notion became the foundation of ice core studies, on which much modern climatological work is based. What Patterson found was that before 1923 there was almost no lead in the atmosphere, and that since that time lead levels had climbed steadily and dangerously. He now made it his life's quest to get lead taken out of petrol. To that end, he became a constant and often vocal critic of the lead industry and its interests.

It would prove to be a hellish campaign. Ethyl was a powerful global corporation with many friends in high places. (Among its directors have been Supreme Court Justice Lewis Powell and Gilbert Grosvenor of the National Geographic Society.) Patterson suddenly found research funding withdrawn or difficult to acquire. The American Petroleum Institute cancelled a research contract with him, as did the United States Public Health Service, a supposedly neutral government body.

As Patterson increasingly became a liability to his institution, the Caltech trustees were repeatedly pressed by lead industry officials to shut him up or let him go. According to Jamie Lincoln Kitman, writing in *The Nation*, Ethyl executives allegedly offered to endow a chair at Caltech 'if Patterson was sent packing'. Absurdly, he was excluded from a 1971 National Research Council panel appointed to investigate the dangers of atmospheric lead poisoning, even though he was by then unquestionably America's leading expert on atmospheric lead.

To his great credit, Patterson never wavered. Eventually his efforts led to the introduction of the Clean Air Act of 1970 and finally to the removal from sale of all leaded petrol in the United States in 1986. Almost immediately lead levels in the blood of Americans fell by 80 per cent. But because lead is for ever, Americans alive today have several hundred times more lead in their blood than people did a century ago. The United States also banned lead in indoor paint, '44 years after most of Europe', as McGrayne notes. Remarkably, considering its startling toxicity, lead solder was not removed from American food containers until 1993.

As for the other scourge left to us by Thomas Midgley, chlorofluorocarbons, they were banned in 1974 in the United States and from 2010 globally, but they are tenacious little devils and any that were loosed into the atmosphere before then (in deodorants or hairsprays, for instance) will almost certainly be around and devouring ozone long after you and I have shuffled off. Worse, the numbers are still growing through illicit production. No one knows quite where the illegal factories are located or who is responsible, but the impact on global warming of illicit production is estimated to be equivalent to the annual carbon emissions of Switzerland, so hardly trifling.

Clair Patterson died in 1995. He didn't win a Nobel Prize for his work. Geologists never do. Nor, more puzzlingly, did he gain any fame or even much attention from half a century of consistent and increasingly selfless achievement. A good case could be made that he was the most influential geologist of the twentieth century. Yet who has ever heard of Clair Patterson? Most geology textbooks don't mention him. Two popular books on the history of the dating of the Earth actually manage to misspell his name. A reviewer in the journal *Nature* made the additional, rather astounding error of thinking Patterson was a woman.

At all events, thanks to the work of Clair Patterson, by 1953 the Earth at last had an age everyone could agree on. The only problem now was that it was older than the universe that contained it.

11

MUSTER MARK'S QUARKS

In 1911, a British scientist named C. T. R. Wilson was studying cloud formations by tramping regularly to the summit of Ben Nevis, a famously damp Scottish mountain, when it occurred to him that there must be an easier way. Back in the Cavendish Lab in Cambridge he built an artificial cloud chamber – a simple device in which he could cool and moisten the air, creating a reasonable model of a cloud in laboratory conditions.

The device worked very well, but had an additional, unexpected benefit. When he accelerated an alpha particle through the chamber to seed his make-believe clouds, it left a visible trail – like the contrails of a passing airliner. He had just invented the particle detector. It provided convincing evidence that subatomic particles did indeed exist.

Eventually two other Cavendish scientists invented a more powerful proton-beam device, while in California Ernest Lawrence at Berkeley produced his famous and impressive cyclotron, or atom-smasher as such devices were long excitingly known. All of these contraptions worked – and indeed still work – on more or less the same principle, the idea being to accelerate a proton or other charged particle to an extremely high speed along a track (sometimes circular, sometimes linear), then bang it into another particle and see what flies off. That's why they were called atom-smashers. It wasn't science at its subtlest, but it was generally effective.

As physicists built bigger and more ambitious machines, they began to find or postulate particles or particle families seemingly without number: muons, pions, hyperons, mesons, K-mesons, Higgs bosons, intermediate vector bosons, baryons, tachyons. Even physicists began to grow a little uncomfortable. 'Young man,' the Italian physicist Enrico Fermi replied when a student asked him the name of a particular particle, 'if I could remember the names of these particles, I would have been a botanist.'

Today accelerators have names that sound like something Flash Gordon would use in battle: the Super Proton Synchrotron, the Large Electron-Positron Collider, the Relativistic Heavy Ion Collider, and of course the celebrated Large Hadron Collider at CERN in Switzerland. Using huge amounts of energy, they can whip particles into such a state of liveliness that a single proton can do 11,000 laps around a 27-kilometre tunnel in under a second. Fears have been raised that in their enthusiasm scientists might inadvertently create a black hole or even something called 'strangelets', which could, theoretically, interact with other subatomic particles and propagate uncontrollably. If you are reading this, that hasn't happened.

Finding particles takes a certain amount of concentration. They are not just tiny and swift but often also tantalizingly evanescent. Particles can come into being and be gone again in as little as 0.000000000000000000000001 of a second (10^{-24} seconds). Even the most sluggish of unstable particles hang around for no more than 0.0000001 of a second (10^{-7} seconds).

Some particles are almost ludicrously slippery. Every second the Earth is visited by 10,000 trillion trillion tiny, all-but-massless neutrinos (mostly shot out by the nuclear broilings of the Sun) and virtually all of them pass right through the planet and everything that is on it, including you and me, as if it weren't there. To trap just a few of them, scientists need tanks holding up to 57,000 cubic metres of heavy water (that is, water with a relative abundance of deuterium in it) in underground chambers (old mines, usually) where they can't be interfered with by other types of radiation.

Very occasionally, a passing neutrino will bang into one of the atomic nuclei in the water and produce a little puff of energy. Scientists count the puffs and by such means take us very slightly closer to understanding the fundamental properties of the universe.

What it really takes to find particles these days is money, and lots of it. There is a curious inverse relationship in modern physics between the tininess of the thing being sought and the scale of the facilities required to do the searching. CERN, the European Organization for Nuclear Research, is more like a little city than a research park. Founded in 1954, and straddling the border of France and Switzerland, it is a disorderly collection of utilitarian buildings that look like leftovers from a 1960s Festival of Bad Design, but out of sight beneath it is a subterranean tunnel 27 kilometres around that constitutes one of the technological wonders of the modern age.

Breaking up atoms, as James Trefil once noted, is easy; you do it each time you switch on a fluorescent light. Breaking up atomic nuclei, however, requires quite a lot of money and a generous supply of electricity. Getting down to the level of quarks – the particles that make up particles – requires still more: trillions of volts of electricity and the budget of a small Central American state. CERN's Large Hadron Collider took 10,000 scientists and engineers more than ten years and £5 billion to build.

But these numbers are as nothing compared with what could have been achieved by, and spent upon, the vast and now unfortunately never-to-be Superconducting Super-collider, which began construction near Waxahachie, Texas, in the 1980s, before experiencing a supercollision of its own with the United States Congress. The intention of the collider was to let scientists probe 'the ultimate nature of matter', as it is always put, by recreating as nearly as possible the conditions in the universe during its first ten-thousand-billionths of a second. The plan was to fling particles through a tunnel 84 kilometres long, achieving a truly staggering 99 trillion volts of energy. It was a grand scheme, but would have cost $8 billion to build (a figure that eventually rose to $10 billion) and hundreds of millions of dollars a year to run. In perhaps the finest example in history of pouring money into a hole in the ground, Congress spent $2 billion on the project, then cancelled it in 1993 after 22 kilometres of tunnel had been dug.

Particle physics, in short, is a hugely expensive enterprise – but it is a productive one. Today the particle count is over 200, but unfortunately, in the words of the late Richard Feynman, 'it is very difficult to understand the relationships of all these particles, and what nature wants them for, or what the connections are from one to another.'

Inevitably, each time we manage to unlock a box we find that there is another locked box inside. Some people think there are particles called tachyons, which can travel faster than the speed of light. Others long to find gravitons – the seat of gravity. At what point we reach the irreducible bottom is not easy to say. Carl Sagan in *Cosmos* raised the possibility that if you travelled downwards into an electron, you might find that it contained a universe of its own, recalling all those science fiction stories of the 1950s. 'Within it, organized into the local equivalent of galaxies and smaller structures, are an immense number of other, much tinier elementary particles, which are themselves universes at the next level and so on for ever – an infinite downward regression, universes within universes, endlessly. And upward as well.'

For most of us it is a world that surpasses understanding. To read even an elementary guide to particle physics nowadays you must find your way through lexical thickets such as this: 'The charged pion and antipion decay respectively into a muon plus antineutrino and an antimuon plus neutrino with an average lifetime of 2.603×10^{-8} seconds, the neutral pion decays into two photons with an average lifetime of about 0.8×10^{-16} seconds, and the muon and antimuon decay respectively into . . .'. And so it runs on – and this from a book for the general reader by one of the (normally) most lucid of interpreters, Steven Weinberg.

In the 1960s, in an attempt to bring just a little simplicity to matters, the Caltech physicist Murray Gell-Mann invented a new class of particles, essentially, in the words of Weinberg, 'to restore some economy to the multitude of hadrons' – a collective term used by physicists for protons, neutrons and other particles governed by the strong nuclear force. Gell-Mann's theory was that all hadrons were made up of still smaller, even more fundamental particles. His colleague Richard Feynman wanted to call these new basic particles *partons*, as in Dolly, but was overruled. Instead they became known as *quarks*.*

Gell-Mann took the name from a line in *Finnegans Wake*: 'Three quarks for Muster Mark!' (Discriminating physicists rhyme the word with *storks*, as Gell-Mann preferred, and not with *larks*, even though the latter is almost certainly the pronunciation Joyce had in mind.) The fundamental simplicity of quarks was not long-lived. As they became better understood it was necessary to introduce subdivisions. Although quarks are much too small to have colour or taste or any other physical characteristics we would recognize, they became clumped into six categories – up, down, strange, charm, top and bottom – which physicists oddly refer to as their 'flavours', and these are further divided into the colours red, green and blue. (One

---

* I am now told, and am delighted to learn, that physicists do use the term parton for the constituent parts of hadrons. By happy coincidence, the great physicist James Clerk Maxwell (1831-1879) grew up near, and is buried in, the little town of Parton in Dumfries and Galloway, Scotland.

suspects that it was not altogether coincidental that these terms were first applied in California during the age of psychedelia.)

Eventually out of all this emerged what is called the Standard Model, which is a sort of parts kit for the subatomic world. The Standard Model consists of seventeen types of elementary particles, and these fall into two basic groups: fermions (named for Enrico Fermi) and bosons (named for the Indian physicist Satyendra Nath Bose). Fermions are the bricks that build matter, and they are split into two further groups: quarks and leptons. These are held together by particles called gluons; and together quarks and gluons form protons and neutrons, the stuff of the atom's nucleus. Leptons, meanwhile, come in six types (electron, electron neutrino, muon, muon neutrino, tau and tau neutrino), and quarks come in from six to thirty-six types, depending on how you count them. Finally, bosons are particles that produce and carry forces, and include photons and gluons.

It is all, as you can see, just a little unwieldy, but it is the simplest model that can explain all that happens in the world of the very, very small. Most particle physicists feel, as Leon Lederman remarked in a 1985 television documentary, that the Standard Model lacks elegance and simplicity. 'It is too complicated. It has too many arbitrary parameters,' Lederman said. 'We don't really see the creator twiddling twenty knobs to set twenty parameters to create the universe as we know it.' Physics is really nothing more than a search for ultimate simplicity, but so far all we have is a kind of elegant messiness – or as Lederman put it: 'There is a deep feeling that the picture is not beautiful.'

The one piece missing from this subatomic jigsaw was the long-theorized Higgs particle (or boson), which it is safe to say nearly everyone has heard of, but few outside the world of physics can claim to understand. It was theorized in the 1960s by several physicists – eight by one count – though it was Peter Higgs of Edinburgh University whose name became indelibly attached to it (though he was always at pains to share the credit with others). Initially its importance wasn't entirely appreciated, including by Higgs himself, but without the Higgs field and its bosons there would be no stars, planets, galaxies or us, among much else. 'Without the Higgs we would all be puffs of radiation,' CERN scientist Tejinder Virdee cheerfully remarked to me on a visit to CERN in 2009.

Higgs particles make up something called the Higgs field, which permeates the universe and gives other fundamental particles their mass. Higgs particles can be likened to photons, the particles that give us light, but with the difference that you can generate millions and millions of photons by turning a light switch, but to make a single Higgs boson you must build a massive accelerator and produce 125 billion electron volts of energy focused on a single spot – and even then it only happens occasionally and only for an instant.

The Large Hadron Collider was designed to recreate the conditions that existed in the universe during the first millionth of a millionth of a second of its existence – but with the difference that now some of the smartest people on Earth can draw up chairs and watch. Particle physicists divine the secrets of the universe in a startlingly straightforward way: by flinging particles together with extreme violence and seeing what flies off. The process has been likened to firing two Swiss watches into each other and deducing how they work by examining their debris.

At first sight, the LHC itself isn't much to behold. It is essentially just a shiny pipe running through a long, circular concrete tunnel, rather like a pipe you would find under any city street carrying water or sewage. But functionally it is a marvel. To accelerate particles satisfactorily, the medium through which they fly must be as airless as the surface of the Moon and as cold as deep space. Maintaining that over 27 kilometres of tubing – actually 54, because there are two tubes shooting particle beams in opposing directions – is an act of breathtaking precision.

A NEW AGE DAWNS

When fully activated the LHC beams can complete 11,245 circuits around their 27-kilometre-long track every second, moving at 99.9999991 per cent the speed of light. That is so fast that if a light beam and a CERN beam were sent simultaneously to the star Alpha Centauri, 4.3 light years away, the CERN beam would arrive just two seconds after the light beam. Keeping the particles shooting around the racetracks without banging into the sides or flying off into space is a fantastically delicate operation. Getting them to collide on command is even more so.

And all this to track down a particle that was, as far as anyone knew, entirely notional.

So we may imagine the delight when, in 2012, Higgs particles were at last successfully created by the LHC – though these scattered moments of glory were strikingly short-lived. Each new Higgs came into and went out of existence in one-tenth of a billion-trillionth of a second (10^{-22}). In fact, the bosons themselves were never seen. Their existence was deduced from what flew out of the collision.

The following year Peter Higgs and François Englert of the Free University of Brussels were jointly awarded the Nobel Prize in Physics. Peter Higgs died in 2024 aged ninety-four. The Higgs particle, he once said, was the only really good idea he had ever had.

The discovery of the Higgs boson filled in the last missing piece of the Standard Model, but it didn't answer all our questions. For one thing, the Standard Model has nothing at all to say about gravity. Search through it though you will, you won't find anything to explain why, when you place a hat on a table, it doesn't float up to the ceiling. In an attempt to draw everything together, physicists came up with string theory, subsequently refined and expanded into superstring theory. This postulates that all those fundamental entities like quarks and electrons are not point particles, as traditionally envisioned, but 'strings' – vibrating strands of energy that oscillate in multiple dimensions (ten in one version, eleven in another). The manner in which they vibrate determines whether they are an electron, photon or whatever. The great advantage of string theories is that they allow for (indeed, mathematically insist upon) the inclusion of gravitons, the putative particles responsible for gravity, thus tying up the most nagging loose end in the Standard Model.

One puzzle with superstring theory was that it yielded five separate versions, all mathematically sound but in one way or another not quite complete, and no one could say which was the most accurate version. That led the American physicist Edward Witten, of Princeton, to propose that they were not five different theories, but just different aspects of the same theory – rather as in the famous parable of the blind men who try to describe an elephant by each touching a different part. Witten's new idea brought all the versions together into something called M theory. It incorporates surfaces known as membranes – or simply branes to the hipper souls of the world of physics.

All these conceptual advances provided physicists with a comparatively tidy package – a sort of proto-Theory of Everything – but an exceedingly dense and technical one. One consequence is that almost anything scientists say about these theories begins to sound worryingly like the sort of thoughts that would make you edge away if conveyed to you by a stranger on a park bench. Here, for example, is the physicist Michio Kaku explaining the structure of the universe from a superstring perspective:

> The ekpyrotic process begins far in the indefinite past with a pair of flat empty branes sitting parallel to each other in a warped five-dimensional space ... The two branes, which form the walls of the fifth dimension, could have popped out of nothingness as a quantum fluctuation in the even more distant past and then drifted apart.

Matters in physics have now reached such a pitch that, as Paul Davies noted in *Nature*, it is 'almost impossible for the non-scientist to discriminate between the legitimately weird and the outright crackpot'. Karl Popper, whom Steven Weinberg called 'the dean of modern philosophers of science', once suggested that there may not in fact be an ultimate theory for physics – that, rather, every explanation may require a further explanation, producing 'an infinite chain of more and more fundamental principles'. A rival possibility is that such knowledge may simply be beyond us. 'So far, fortunately,' wrote Weinberg in *Dreams of a Final Theory*, 'we do not seem to be coming to the end of our intellectual resources.' Almost certainly this is an area that will see further developments of thought, and almost certainly again these thoughts will be beyond most of us.

While physicists in the middle decades of the twentieth century were looking perplexedly into the world of the very small, astronomers were finding no less arresting an incompleteness of understanding in the universe at large.

When we last met Edwin Hubble, he had determined that nearly all the galaxies in our field of view are flying away from us, and that the speed and distance of this retreat are neatly proportional: the further away the galaxy, the faster it is moving. Hubble realized that this could be expressed with a simple equation, $H_0 = v/d$ (where H_0 is the constant, v is the recessional velocity of a flying galaxy and d its distance away from us). H_0 has been known ever since as the Hubble constant and the whole as Hubble's Law. Using his formula, Hubble calculated that the universe was about 2 billion years old, which was a little awkward because even by the late 1920s it was increasingly evident that many things within the universe – including, probably, the Earth itself – were older than that. Refining this figure has been an ongoing preoccupation of cosmology.

Almost the only thing constant about the Hubble constant

has been the amount of disagreement over what value to give it. In 1956, astronomers discovered that Cepheid variables were more variable than they had thought; they came in two varieties, not one. This allowed them to rework their calculations and come up with a new age for the universe of between 7 billion and 20 billion years – not terribly precise, but at least old enough, at last, to embrace the formation of the Earth.

In the years that followed there erupted a dispute that would run and run, between Allan Sandage, heir to Hubble at Mount Wilson, and Gérard de Vaucouleurs, a French-born astronomer based at the University of Texas. Sandage, after years of careful calculations, arrived at a value for the Hubble constant of 50, giving the universe an age of 20 billion years. De Vaucouleurs was equally certain that the Hubble constant was 100.* This would mean that the universe was only half the size and age that Sandage believed – 10 billion years. Matters took a further lurch into uncertainty when in 1994 a team from the Carnegie Observatories in California, using measures from the Hubble Space Telescope, suggested that the universe could be as little as 8 billion years old – an age even they conceded was younger than some of the stars within the universe. In February 2003, a team from NASA and the Goddard Space Flight Center in Maryland, using a new, far-reaching type of satellite called the

---

* You are, of course, entitled to wonder what is meant exactly by 'a constant of 50' or 'a constant of 100'. The answer lies in astronomical units of measure. Except conversationally, astronomers don't use light years. They use a distance called the *parsec* (a contraction of *parallax* and *second*), based on a universal measure called the stellar parallax and equivalent to 3.26 light years. Really big measures, like the size of a universe, are measured in megaparsecs: 1 megaparsec = 1 million parsecs. The constant is expressed in terms of kilometres per second per megaparsec. Thus when astronomers refer to a Hubble constant of 50, what they really mean is '50 kilometres per second per megaparsec'. For most of us that is of course an utterly meaningless measure; but then, with astronomical measures most distances are so huge as to be meaningless. As of 2024, the value of the Hubble constant ranges from 67.8 to 73 km/s/Mpc, as astronomers express it.

Wilkinson Microwave Anisotropy Probe, announced with confidence that the age of the universe is 13.77 billion years, give or take 50 million years or so, and that is now widely accepted as the best existing answer.

The difficulty in making absolute final determinations is that there are often acres of room for interpretation. Imagine standing in a field at night and trying to decide how far away two distant electric lights are. Using fairly straightforward tools of astronomy, you can easily enough determine that the bulbs are of equal brightness and that one is, say, 50 per cent more distant than the other. But what you can't be certain of is whether the nearer light is, let us say, a 58-watt bulb that is 37 metres away or a 61-watt light that is 36.5 metres away. On top of that you must make allowances for distortions caused by variations in the Earth's atmosphere, by intergalactic dust, by contaminating light from foreground stars and many other factors. The upshot is that your computations are necessarily based on a series of nested assumptions, any of which could be a source of contention. There is also the problem that access to telescopes is always at a premium and, historically, measuring red shifts has been notably costly in telescope time. It could take all night to get a single exposure. In consequence, astronomers have sometimes been compelled (or willing) to base conclusions on notably scanty evidence. In cosmology, as the journalist Geoffrey Carr has suggested, we have 'a mountain of theory built on a molehill of evidence'. Or as Martin Rees has put it more diplomatically: 'Our present satisfaction [with our state of understanding] may reflect the paucity of the data rather than the excellence of the theory.'

The fact is, there is a great deal, even at quite a fundamental level, that we don't know – not least what the universe is made of. When scientists calculate the amount of matter needed to hold things together, they always come up desperately short. The precise proportions tend to vary depending on whose calculations you follow, but roughly 25 per cent of the universe consists of dark matter, 70 per cent of the similarly elusive dark energy, and just 5 per cent of the stuff we are familiar with – what is known formally as baryonic matter.

Dark matter could hardly be more invisible. It is wholly divorced from light: it doesn't emit it, absorb it or reflect it. Harvard professor Lisa Randall has suggested that it would be better called 'transparent matter' because light just passes through it, as it would a pane of glass. Its existence has been a cosmological certainty since the 1960s, when Vera Rubin (1928–2016), an astronomer at the Carnegie Institution in Washington, DC, realized that the galaxies she was studying were spinning so quickly that their outermost stars should have been flung into deep space, like a thrown discus. Yet something – some unseen mass – was stopping that from happening. It was the first formal confirmation of Zwicky's theorized dark matter, and it set in motion a search that continues yet.

Few people have worked longer or more devotedly on the question of what dark matter is than Carlos Frenk of Durham University in England. Now in his seventies, Frenk is a modest, endlessly good-natured, widely admired and extraordinarily dogged fellow – but then anyone searching for dark matter needs to be extraordinarily dogged. Frenk has been at it for more than forty years. 'The universe does not give up its mysteries cheaply,' he says cheerfully. 'It makes you work for them.'

Frenk grew up in Mexico, the son of a Jewish refugee from Nazi Germany and a Catholic Spanish-Mexican mother, in a large, prosperous family of high achievers. After studying physics in Mexico and then at Cambridge, he spent much of his early career working at the University of California at Berkeley before settling at Durham in the mid-1980s. Today he is Ogden Professor of Fundamental Physics and head of the Institute for Computational Cosmology at Durham University. He has often been mentioned as a Nobel Prize candidate for his work on galaxy formation.

'When I first started, dark matter wasn't even accepted by

the majority of astronomers, and cosmology barely existed as a field,' he told me when we met in the summer of 2024 at King's College, Cambridge, where he is an honorary Fellow. 'Dark matter then was mostly referred to as "missing matter".'

At Berkeley, Frenk was part of a team – popularly known as the Gang of Four – which came together to work out what the missing matter consisted of. Candidate particles fell into two broad groups, hot dark matter and cold dark matter. Those designations don't actually have anything to do with how the particles would feel if you touched them, but rather denote the speeds at which they were presumed to move.

Their plan was to replicate the first moments of the universe through computer modelling – something that had never been done before. The idea was to feed all the theorized ingredients and parameters into a big computer (including their best guess at dark matter) in the hope that what emerged would match the universe that we find ourselves in today. It was a wildly ambitious idea, but as Carlos recalls, 'We were young and ambitious ourselves and it was an entirely new field. There were no authorities we had to please, no eminent scientists we had to look up to.'

Simulating the dawn of creation was, to put it mildly, an exacting challenge, and involved much writing and rewriting of code, wrestling with equations, numberless trial runs, and endless tweaking of assumptions. This was in the early 1980s when computers were vastly more basic than now. 'You can hardly believe how much harder it was back then,' Carlos says. 'When we started, one of my colleagues, Marc Davis, had just spent five years making a map of the known galaxies. Today a computer can do the same work in ten minutes.'

For dark matter, the Berkeley team chose neutrinos, which were known to pervade the universe and whose characteristics put them into the category of hot dark matter.

The work took two years of intensive effort, and when at last the simulation was run it was a complete failure. 'Our modelling created a universe that looked nothing like the real one,' Frenk recalls ruefully now. 'It was very disappointing. The rational side of my brain said, "Well, at least we have advanced science." But my heart said' – and here he flashed a broad grin – 'Oh, shit.'

It turned out that although the universe is absolutely packed with neutrinos, they still are not abundant enough or massive enough to account for all the dark matter you need to make a universe.

'So as a next step,' Carlos went on, 'we decided to create a simulated universe using cold dark matter, purely with the expectation of ruling it out. In fact, to our astonishment, what came out had an uncanny resemblance to the real universe.' A new, somewhat ironic term was born for the kind of particles that dark matter must be made of: WIMPS (for Weakly Interacting Massive Particles).

And that is about where matters still stand. Despite four decades of intensive searching at facilities like CERN, we've still no clear idea what dark matter is. That human ingenuity could work out in such exacting detail how we went from there being nothing at all to there being a universe in all its complex glory, full of blazing stars and swirling galaxies, gossamer nebulae, black holes and all the rest, is one of the supreme intellectual achievements of all time. The composition of dark matter is, in a sense, just detail.

But it is, for obvious reasons, a detail that nags. As time has dragged on without any positive answers, many other possibilities have been proposed for what might constitute dark matter – axions,* neutralinos, heavy sterile neutrinos,

---

* The name axion, rather amazingly, is taken from a laundry product called Axion, and given to the notional particle by the American Nobel laureate Frank Wilczek, who was taken with the name on childhood trips to the supermarket with his mother and thought then, 'If I ever get a chance to name a particle I'll call it axion.'

self-interacting dark matter, light dark matter, ultralight dark matter, particles hiding in another dimension, primordial black holes, ultralight bosons, perhaps even an entire 'dark sector' composed of many different types of dark matter particles.

Nearly all these alternative proposals, Carlos believes, are simply born of impatience, and are often dreamed up by 'very smart guys with not much to do, so they come up with ideas that are based on nothing much or nothing at all. We live in a world of instant gratification, and that doesn't work in physics. People need to remember the example of the Higgs boson. It took fifty years to find it – much longer than we have been looking for dark matter. So we need to be patient.'

I asked him when he thought we would have an answer. 'I think we will discover dark matter within five years,' he said. 'Mind you, I have been saying that for the last twenty. The serious answer to that question is that it is impossible to predict.'

Dark matter is commonly bracketed with dark energy, though there is no certainty that the two are in any way connected. It is a much more recent concept. The term was coined in 1998 by an astrophysicist at the University of Chicago named Michael Turner after it was realized that the expansion of the universe is in fact speeding up – something counter to all expectations. The universe is a big place, needless to say, so it takes a lot of energy to cause the whole thing to expand, which is why dark energy is calculated to make up about 70 per cent (and perhaps even more) of everything there is, everywhere.

Ironically, despite forty years of trying to pin down dark matter, it is with dark energy that Carlos and his colleagues have been finding breakthroughs – indeed, a 'paradigm shift', in his words. Durham University is part of an international collaboration behind the Dark Energy Spectroscopic Instrument, or DESI, which has been studying the cosmos in unprecedented detail via a special modified telescope at the Kitt Peak National Observatory in Arizona.

'It is allowing us to measure the expansion rate of the universe with great precision, and what we have found already is that the expansion is slowing,' Carlos told me. 'It's still expanding, but more slowly. This means that one day in the very distant future – we haven't calculated exactly when yet – it will collapse, and then probably expand again in another big bang. It may have done this many times already. What is significant is that it makes the cosmological constant unnecessary – because universal expansion is not in fact constant. It changes over time. Why it changes is another mystery to solve, but that will be for a future generation.'

The upshot of all this is that we live in a universe filled with matter and energy we can't see or identify, operating in conformance with physical laws whose properties we don't truly understand. In short, we find ourselves in the paradoxical situation that the more we learn about the universe the less well we seem to understand it.

And on that rather unsettling note, let's return to Planet Earth and consider something that we *do* understand – though by now you perhaps won't be surprised to hear that we don't understand it completely, and what we do understand we haven't understood for long.

12

THE EARTH MOVES

In one of his last professional acts before his death in 1955, Albert Einstein wrote a short but glowing foreword to a book by a geologist named Charles Hapgood entitled *Earth's Shifting Crust: A Key to Some Basic Problems of Earth Science*. Hapgood's book was a steady demolition of the idea that continents were in motion. In a tone that all but invited the reader to join him in a tolerant chuckle, Hapgood observed that a few gullible souls had noticed 'an apparent correspondence in shape between certain continents'. It would appear, he went on, 'that South America might be fitted together with Africa, and so on . . . It is even claimed that rock formations on opposite sides of the Atlantic match.'

Mr Hapgood briskly dismissed any such notions, noting that the geologists K. E. Caster and J. C. Mendes had done extensive fieldwork on both sides of the Atlantic and had established beyond question that no such similarities existed. Goodness knows what outcrops Messrs Caster and Mendes had looked at, because in fact many of the rock formations on both sides of the Atlantic *are* the same – not just very similar but the same.

This was not an idea that flew with Mr Hapgood, or many other geologists of his day. The theory Hapgood alluded to was one first propounded in 1908 by an amateur American geologist named Frank Bursley Taylor. Taylor came from a wealthy family and had both the means and the freedom from academic constraints to pursue unconventional lines of enquiry. He was one of those struck by the similarity in shape between the facing coastlines of Africa and South America, and from this observation he developed the idea that the continents had once slid around. He suggested – presciently, as it turned out – that the crunching together of continents could have thrust up the world's mountain chains. He failed, however, to produce much in the way of evidence, and the theory was considered too crackpot to merit serious attention.

In Germany, however, Taylor's idea was picked up, and effectively appropriated, by a theorist named Alfred Wegener, a meteorologist at the University of Marburg. Wegener investigated the many plant and fossil anomalies that did not fit comfortably into the standard model of Earth history and realized that very little of it made sense if conventionally interpreted. Animal fossils repeatedly turned up on opposite sides of oceans that were clearly too wide to swim. How, he wondered, did marsupials travel from South America to Australia? How did identical snails turn up in Scandinavia and New England?

And how, come to that, did one account for coal seams and other semi-tropical remnants in frigid spots like Spitsbergen, over 600 kilometres north of Norway, if they had not somehow migrated there from warmer climes?

Wegener developed the theory that the world's continents had once existed as a single land mass he called Pangaea, where flora and fauna had been able to mingle before splitting apart and floating off to their present positions. He set the idea out in a book called *Die Entstehung der Kontinente und Ozeane*, or *The Origin of Continents and Oceans*, which was published in German in 1912 and – despite the outbreak of the First World War in the meantime – in English three years later.

Because of the war, Wegener's theory didn't attract much notice at first, but by 1920, when he produced a revised and expanded edition, it quickly became a subject of discussion. Everyone agreed that continents moved – but up and down, not sideways. The process of vertical movement, known as isostasy, was a foundation of geological belief for generations, though no one had any really good theories as to how or why it happened. One idea, which remained in textbooks well into my own schooldays, was the 'baked apple' theory propounded by the Austrian Eduard Suess just before the turn of the century. This suggested that as the molten Earth had cooled, it had become wrinkled in the manner of a baked apple, creating ocean basins and mountain ranges. Never mind that James Hutton had shown long before that any such static arrangement would eventually result in a featureless spheroid as erosion levelled the bumps and filled in the divots.

There was also the problem, demonstrated by Rutherford and Soddy early in the century, that earthly elements hold huge reserves of heat – much too much to allow for the sort of cooling and shrinking Suess suggested. And anyway, if Suess's theory was correct, then mountains should be evenly distributed across the face of the Earth, which patently they were not, and of more or less the same ages; yet by the early 1900s it was already evident that some ranges, like the Urals and Appalachians, were hundreds of millions of years older than others, like the Alps and Rockies. Clearly the time was ripe for a new theory. Unfortunately, Alfred Wegener was not the man geologists wished to provide it.

For a start, his radical notions questioned the foundations of their discipline, seldom an effective way to generate warmth in an audience. Such a challenge would have been painful enough coming from a geologist, but Wegener had no background in geology. He was a meteorologist, for goodness' sake. A weatherman – a German weatherman. These were not remediable deficiencies.

And so geologists took every pain they could to dismiss his evidence and belittle his suggestions. To get around the problems of fossil distributions, they posited ancient 'land bridges' wherever they were needed. When an ancient horse named *Hipparion* was found to have lived in France and Florida at the same time, a land bridge was drawn across the Atlantic. When it was realized that ancient tapirs had existed simultaneously in South America and Southeast Asia, a land bridge was drawn there, too. Soon maps of prehistoric seas were almost solid with hypothesized land bridges – from North America to Europe, from Brazil to Africa, from Southeast Asia to Australia, from Australia to Antarctica. These connective tendrils had not only conveniently appeared whenever it was necessary to move a living organism from one land mass to another, but then had obligingly vanished without leaving a trace of their former existence. None of this, of course, was supported by so much as a grain of evidence – nothing so wrong could be – yet it was geological orthodoxy for the next half-century.

Even land bridges couldn't explain some things. One species of trilobite that was well known in Europe was also found to have lived on Newfoundland – but only on one side. No one could persuasively explain how it had managed to cross 3,000 kilometres of hostile ocean but then failed to find its way around the corner of an island 300 kilometres wide. Even more awkwardly anomalous was another species of trilobite found in Europe and the Pacific northwest of America but nowhere in between, which would have required not so much a land bridge as a flyover. Yet as late as 1964, when the *Encyclopaedia Britannica* discussed the rival theories it was Wegener's that was held to be full of 'numerous grave theoretical difficulties'.

To be sure, Wegener made mistakes. He asserted that Greenland is drifting west by about 1.6 kilometres a year, a clear nonsense. (It's more like a centimetre.) Above all, he could offer no convincing explanation for how the land masses moved about. To believe in his theory you had to accept that massive continents somehow pushed through solid crust, like a farm plough through soil, without leaving any furrow in their wake.

Nothing then known could plausibly explain what motored these colossal movements.

It was Arthur Holmes, the English geologist who did so much to determine the age of the Earth, who came up with a suggestion. Holmes was the first scientist to understand that radioactive warming could produce convection currents within the Earth. In theory, these could be powerful enough to slide continents around on the surface. In his popular and influential textbook *Principles of Physical Geology*, first published in 1944, Holmes laid out a continental drift theory that was, in its fundamentals, the theory that prevails today. It was still a radical proposition for the time and widely criticized, particularly in the United States, where resistance to drift lasted longer than elsewhere. One reviewer there fretted, without any sense of irony, that Holmes presented his arguments so clearly and compellingly that students might actually come to believe them.

Elsewhere, however, the new theory drew steady if cautious support. In 1950, a vote at the annual meeting of the British Association for the Advancement of Science showed that about half of those present now embraced the idea of continental drift. (Hapgood soon after cited this figure as proof of how tragically misled British geologists had become.) Curiously, Holmes himself sometimes wavered in his conviction. In 1953 he confessed: 'I have never succeeded in freeing myself from a nagging prejudice against continental drift; in my geological bones, so to speak, I feel the hypothesis is a fantastic one.' Continental drift was not entirely without support in the United States. Reginald Daly of Harvard spoke for it, but he, you may recall, was the man who suggested that the Moon had been formed by a cosmic impact and his ideas tended to be considered interesting, even worthy, but a touch too exuberant for serious consideration. And so most American academics stuck to the belief that the continents had occupied their present positions for ever and that their surface features could be attributed to something other than lateral motions. Interestingly, oil company geologists had known for years that if you wanted to find oil you had to allow for precisely the sort of surface movements that were implied by plate tectonics. But oil geologists didn't write academic papers; they just found oil.

There was one other major problem with Earth theories that no one had resolved, or even come close to resolving. That was the question of where all the sediments went. Every year the Earth's rivers carried massive volumes of eroded material – 500 million tonnes of calcium, for instance – to the seas. If you multiplied the rate of deposition by the number of years it had been going on, you arrived at a disturbing figure: there should be about 20 kilometres of sediments on the ocean bottoms – or, put another way, the ocean bottoms should by now be well above the ocean tops. Scientists dealt with this paradox in the handiest possible way. They ignored it. But eventually there came a point when they could ignore it no longer.

In the Second World War, a Princeton University mineralogist named Harry Hess was put in charge of an attack transport ship, the USS *Cape Johnson*. Aboard this vessel was a fancy new depth sounder called a fathometer, which was designed to facilitate inshore manoeuvres during beach landings, but Hess realized that it could equally well be used for scientific purposes and never switched it off, even when far out at sea, even in the heat of battle. What he found was entirely unexpected. If the ocean floors were ancient, as everyone assumed, they should be thickly blanketed with sediments, like the mud on the bottom of a river or lake. But Hess's readings showed that the ocean floor offered anything but the gooey smoothness of ancient silts. It was scored everywhere with canyons, trenches and crevasses and dotted with volcanic seamounts that he called guyots after an earlier Princeton geologist named Arnold Guyot. All this was a puzzle, but Hess had a war to take part in, and put such thoughts to the back of his mind.

After the war, Hess returned to Princeton and the preoccupations of teaching, but the mysteries of the sea floor continued to occupy a space in his thoughts. Meanwhile, throughout the 1950s oceanographers were undertaking more and more sophisticated surveys of the ocean floors. In so doing, they found an even bigger surprise: the mightiest and most extensive mountain range on Earth was – mostly – under water. It traced a continuous path along the world's seabeds, rather like the pattern on a tennis ball. If you began at Iceland and travelled south, you could follow it down the centre of the Atlantic Ocean, around the bottom of Africa, and across the Indian and Southern Oceans and into the Pacific just below Australia; there it angled across the Pacific as if making for Baja California before shooting up the west coast of the United States to Alaska. Occasionally its higher peaks poked above the water as an island or archipelago – the Azores and Canaries in the Atlantic, Hawaii in the Pacific, for instance – but mostly it was buried under thousands of fathoms of salty sea, unknown and unsuspected. When all its branches were added together, the network extended to 75,000 kilometres.

A very little of this had been known for some time. People laying ocean-floor cables in the nineteenth century had realized that there was some kind of mountainous intrusion in the mid-Atlantic from the way the cables ran, but the continuous nature and overall scale of the chain was a stunning surprise. Moreover, it contained physical anomalies that couldn't be explained. Down the middle of the mid-Atlantic ridge was a canyon – a rift – up to 20 kilometres wide for its entire 19,000-kilometre length. This seemed to suggest that the Earth was splitting apart at the seams, like a nut bursting out of its shell. It was an absurd and unnerving notion, but the evidence couldn't be denied.

Then in 1960 core samples showed that the ocean floor was quite young at the mid-Atlantic ridge but grew progressively older as you moved away from it to east or west. Harry Hess considered the matter and realized that this could mean only one thing: new ocean crust was being formed on either side of the central rift, then being pushed away from it as more new crust came along behind. The Atlantic floor was effectively two large conveyor belts, one carrying crust towards North America, the other carrying crust towards Europe. The process became known as sea-floor spreading.

When the crust reached the end of its journey at the boundary with continents, it plunged back into the Earth in a process known as subduction. That explained where all the sediment went. It was being returned to the bowels of the Earth. It also explained why ocean floors everywhere were so comparatively youthful. None had ever been found to be older than about 175 million years, which was a puzzle because continental rocks were often billions of years old. Now Hess could see why. Ocean rocks lasted only as long as it took them to travel to shore. It was a beautiful theory that explained a great deal. Hess elaborated his arguments in an important paper, which was almost universally ignored. Sometimes the world just isn't ready for a good idea.

Meanwhile, two researchers, working independently, were making some startling findings by drawing on a curious fact of Earth history that had been discovered several decades earlier. In 1906, a French physicist named Bernard Brunhes had found that the planet's magnetic field reverses itself from time to time, and that the record of these reversals is permanently fixed in certain rocks at the time of their birth. Specifically, tiny grains of iron ore within the rocks point to wherever the magnetic poles happen to be when they are formed, then stay pointing in that direction as the rocks cool and harden. In effect, they 'remember' where the magnetic poles were at the moment of their creation. For years this was little more than a curiosity, but in the 1950s Patrick Blackett of the University of London and S. K. Runcorn of the University of Newcastle studied the ancient magnetic patterns frozen in British rocks and were startled, to say the very least, to find them indicating that at some time in the distant past Britain had spun on its axis and travelled some distance to the north, as if it had somehow come loose from its moorings. Moreover, they also discovered that if you placed a map of Europe's magnetic patterns alongside an American one from the same period, they fit together as neatly as two halves of a torn letter. It was uncanny. Their findings were ignored too.

It finally fell to two men from Cambridge University, a geophysicist named Drummond Matthews and a graduate student of his named Fred Vine, to draw all the strands together. In 1963, using magnetic studies of the Atlantic Ocean floor, they demonstrated conclusively that the sea floors were spreading in precisely the manner Hess had suggested and that the continents were in motion too. An unlucky Canadian geologist named Lawrence Morley came up with the same conclusion at the same time, but couldn't find anyone to publish his paper. In what has become a famous snub, the editor of the *Journal of Geophysical Research* told him: 'Such speculations make interesting talk at cocktail parties, but it is not the sort of thing that ought to be published under serious scientific aegis.' One geologist later described it as 'probably the most significant paper in the earth sciences ever to be denied publication'.

At all events, mobile crust was an idea whose time had finally come. A symposium of many of the most important figures in the field was convened in London under the auspices of the Royal Society in 1964, and suddenly, it seemed, everyone was a convert. The Earth, the meeting agreed, was a mosaic of interconnected segments whose various stately jostlings accounted for much of the planet's surface behaviour.

The name 'continental drift' was fairly swiftly discarded when it was realized that the whole crust was in motion and not just the continents, but it took a while to settle on a name for the individual segments. At first people called them 'crustal blocks' or sometimes 'paving stones'. Not until late 1968, with the publication of an article by three American seismologists in the *Journal of Geophysical Research*, did the segments receive the name by which they have since been known: plates. The same article called the new science plate tectonics.

Old ideas die hard, and not everyone rushed to embrace the exciting new theory. Well into the 1970s, one of the most popular and influential geological textbooks, *The Earth* by the venerable Harold Jeffreys, strenuously insisted that plate tectonics was a physical impossibility, just as it had in the first edition way back in 1924. It was equally dismissive of convection and sea-floor spreading. And in *Basin and Range*, published in 1980, John McPhee noted that even then one American geologist in eight still didn't believe in plate tectonics.

Today we know that the Earth's surface is made up of eight to twelve big plates (depending on how you define big) and twenty or so smaller ones, and that they all move in different directions and at different speeds. Some plates are large and comparatively inactive, others small but energetic. They bear only an incidental relationship to the land masses that sit upon them. The North American plate, for instance, is much larger than the continent with which it is associated. It roughly traces the outline of the continent's western coast (which is why that area is so seismically active, because of the bump and crush of the plate boundary), but ignores the eastern seaboard altogether and instead extends halfway across the Atlantic to the mid-ocean ridge. Iceland is split down the middle, which makes it tectonically half American and half European. New Zealand, meanwhile, is part of the immense Indian Ocean plate even though it is nowhere near the Indian Ocean. And so it goes for most plates.

The connections between modern land masses and those of the past were found to be infinitely more complex than anyone had imagined. Kazakhstan, it turns out, was once attached to Norway and New England. One corner of Staten Island, but only a corner, is European. So is part of Newfoundland. Pick up a pebble from a Massachusetts beach and its nearest kin will now be in Africa. The Scottish Highlands and much of

Scandinavia are substantially American. Some of the Shackleton Range of Antarctica, it is thought, may once have belonged to the Appalachians of the eastern USA. Rocks, in short, get around.

The constant turmoil keeps the plates from fusing into a single immobile plate. Assuming things continue much as at present, the Atlantic Ocean will expand until eventually it is much bigger than the Pacific. Much of California will float off and become a kind of Madagascar of the Pacific. Africa will push northward into Europe, squeezing the Mediterranean out of existence and thrusting up a chain of mountains of Himalayan majesty running from Paris to Calcutta. Australia will colonize the islands to its north and connect by some isthmian umbilicus to Asia. These are future outcomes, but not future events. The events are happening now. As we sit here, continents are adrift, like leaves on a pond. Thanks to Global Positioning Systems we can see that Europe and North America are parting at about the speed a fingernail grows – roughly 2 metres in a human lifetime. If you were prepared to wait long enough, you could ride from Los Angeles all the way up to San Francisco. It is only the brevity of lifetimes that keeps us from appreciating the changes. Look at a globe and what you are seeing really is a snapshot of the continents as they have been for just one-tenth of 1 per cent of the Earth's history.

Earth is alone among the rocky planets in having plate tectonics, and why this should be is a bit of a mystery. It may be that we have just the right materials in just the right measures to keep the Earth bubbling away. It is thought – though it is really nothing more than a thought – that tectonics is an important part of the planet's organic well-being. As the physicist and writer James Trefil has put it: 'It would be hard to believe that the continuous movement of tectonic plates has no effect on the development of life on Earth.' He has suggested that the challenges induced by tectonics – changes in climate, for instance – were an important spur to the development of intelligence. Others believe the driftings of the continents may have produced at least some of the Earth's various extinction events. In November 2002 Tony Dickson of Cambridge University produced a report, published in the journal *Science*, proposing that there may well be a relationship between the history of rocks and the history of life. What Dickson established was that the chemical composition of the world's oceans has altered abruptly and dramatically at times throughout the past half-billion years and that these changes often correlate with important events in biological history – the huge outburst of tiny organisms that created the chalk cliffs of England's south coast, the sudden fashion for shells among marine organisms during the Cambrian period, and so on. No one can say what causes the oceans' chemistry to change so dramatically from time to time, but the opening and shutting of ocean ridges would be an obvious possible culprit.

At all events, plate tectonics explained not only the surface dynamics of the Earth – how an ancient *Hipparion* got from France to Florida, for example – but also many of its internal actions. Earthquakes, the formation of island chains, the carbon cycle, the locations of mountains, the coming of ice ages, the origins of life itself – there was hardly a matter that wasn't directly influenced by this remarkable new theory. Geologists, as McPhee has noted, found themselves in the giddying position where 'the whole earth suddenly made sense.'

But only up to a point. The distribution of continents in former times is much less neatly resolved than most people outside geophysics think. Although textbooks give confident-looking representations of ancient land masses with names like Laurasia, Gondwana, Rodinia and Pangaea, these are sometimes based on conclusions that don't altogether hold up. As George Gaylord Simpson observes in *Fossils and the History of Life*, species of plants and animals from the ancient world have a habit of appearing inconveniently where they shouldn't and failing to be where they ought.

The outline of Gondwana, a once-mighty continent connecting Australia, Africa, Antarctica and South America, was based in large part on the distributions of a genus of ancient tongue fern called *Glossopteris*, which was found in all the right places. However, much later *Glossopteris* was also discovered in parts of the world that had no known connection to Gondwana. Similarly, a Triassic reptile called lystrosaurus has been found from Antarctica all the way to Asia, supporting the idea of a former connection between those continents, but it has never turned up in South America or Australia, which are believed to have been part of the same continent at the same time.

There are also many surface features that tectonics can't explain. Take Denver. It is, as everyone knows, a mile high, but that rise is comparatively recent. When dinosaurs roamed the Earth, Denver was part of an ocean bottom many thousands of metres lower. Yet the rocks on which Denver sits are not fractured or deformed in the way they would be if Denver had been pushed up by colliding plates, and anyway Denver was too far from the plate edges to be susceptible to their actions. It would be as if you pushed against the edge of a rug hoping to raise a ruck at the opposite end. Mysteriously and over millions of years, it appears that Denver has been rising, like baking bread. So, too, has much of southern Africa; a portion of it 1,600 kilometres across has risen about one and a half kilometres in 100 million years without any known associated tectonic activity. Australia, meanwhile, has been tilting and sinking. Over the past 100 million years, as it has drifted north towards Asia, its leading edge has sunk by nearly 200 metres. It appears that Indonesia is very slowly drowning, and dragging Australia down with it. Nothing in the theories of tectonics can explain any of this.

Alfred Wegener never lived to see his ideas vindicated. On an expedition to Greenland in 1930, he set out alone, on his fiftieth birthday, to check out a supply drop. He never returned. He was found a few days later, frozen to death on the ice. He was buried on the spot and lies there yet, but about a metre closer to North America than on the day he died.

Einstein also failed to live long enough to see that he had backed the wrong horse. In fact, he died at Princeton, New Jersey, in 1955, before Charles Hapgood's rubbishing of continental drift theories was even published.

The other principal player in the emergence of tectonics theory, Harry Hess, was also at Princeton at the time, and would spend the rest of his career there. One of his students was a bright young fellow named Walter Alvarez, who would eventually change the world of science in a quite different way. As for geology itself, its cataclysms had only just begun, and it was young Alvarez who helped to start the process.

IV

DANGEROUS PLANET

13

BANG!

People knew for a long time that there was something odd about the earth beneath Manson, Iowa. In 1905, a man drilling a well for the town water supply reported bringing up a lot of strangely deformed rock – 'crystalline clast breccia with a melt matrix' and 'overturned ejecta flap', as it was later described in an official report. The water was odd, too. It was almost as soft as rainwater. Naturally occurring soft water had never been found in Iowa before.

Though Manson's strange rocks and silken waters were matters of curiosity, forty-one years would pass before a team from the University of Iowa got around to making a trip to the community, then as now a town of about 2,000 people in the northwest part of the state. In 1953, after sinking a series of experimental bores, university geologists agreed that the site was indeed anomalous and attributed the deformed rocks to some ancient, unspecified volcanic action. This was in keeping with the wisdom of the day, but it was also about as wrong as a geological conclusion can get.

The trauma to Manson's geology had come not from within the Earth, but from at least 100 million miles beyond. Some time in the very ancient past, when Manson stood on the edge of a shallow sea, a rock about a mile and a half across, weighing 12 trillion tons and travelling at perhaps 200 times the speed of sound, ripped through the atmosphere and punched into the Earth with a violence and suddenness that we can scarcely imagine. Where Manson now stands became in an instant a hole 3 miles deep and more than 20 miles across. The limestone that elsewhere gives Iowa its hard, mineralized water was obliterated and replaced by the shocked basement rocks that so puzzled the water driller in 1905.

The Manson impact was the biggest thing that has ever occurred on the mainland United States – of any type, ever.*

Unfortunately for lovers of spectacle, much of the ejected material fell straight back in, like a splash in a puddle, and 2.5 million years of passing ice sheets filled the rest of the crater to the very top with rich glacial till, so that today the landscape at Manson, and for miles around, is as flat as a table top. Which is of course why hardly anyone one has ever heard of the Manson crater.

To most people in Manson the biggest thing ever to happen was a tornado that rolled up Main Street in 1979, tearing apart the business district and much of the town beyond. One of the advantages of all that surrounding flatness is that you can see danger from a long way off. Virtually the whole town turned out and watched for half an hour as the tornado came towards them, hoping it would veer off, then prudently fled when it did not. Three people, alas, didn't move quite fast enough and were killed.

Every June now Manson holds a festival called Crater Days, which was dreamed up as a way of helping people forget that unhappy anniversary. It doesn't really have anything to do

---

* At least as far as can be discerned from the geological record. Bigger impacts occurred early on in Earth's history, but the evidence for them has been wiped out by the endless churn of the Earth's crust.

with the crater. Nobody's figured out a way to capitalize on an impact site that isn't visible.

'Very occasionally we get people coming in and asking where they should go to see the crater and we have to tell them that there is nothing to see,' Ann Schlapkohl, the town's friendly librarian, told me when I visited in 2001. 'Then they go away kind of disappointed.' However, most people, including most Iowans, have never heard of the Manson crater. But for one brief period in the 1980s, Manson was the most geologically exciting place on Earth.

The story begins in the early 1950s when a bright young geologist named Eugene Shoemaker paid a visit to Meteor Crater in Arizona. Today Meteor Crater is the most famous impact site on Earth and a popular tourist attraction. In those days, however, it didn't receive many visitors and was still often referred to as Barringer Crater, after a wealthy mining engineer named Daniel M. Barringer who had staked a claim on it in 1903. Barringer believed that the crater had been formed by a 10-million-ton meteor, heavily freighted with iron and nickel, and it was his confident expectation that he would make a fortune digging it out. Unaware that the meteor and everything in it would have been vaporized on impact, he wasted a fortune, and the next twenty-six years, cutting tunnels that yielded nothing.

By the standards of today, crater research in the early 1900s was a trifle unsophisticated, to say the least. The leading early investigator, G. K. Gilbert of Columbia University, modelled the effects of impacts by flinging marbles into pans of oatmeal. (For reasons I cannot supply, Gilbert conducted these experiments not in a laboratory at Columbia but in a hotel room.) Somehow, from this Gilbert concluded that the Moon's craters were indeed formed by impacts – in itself quite a radical notion for the time – but that the Earth's were not. Most scientists refused to go even that far. To them, the Moon's craters were evidence of ancient volcanoes and nothing more. The few craters that remained evident on the Earth (most had been eroded away) were generally attributed to other causes or treated as fluky rarities.

By the time Shoemaker came along, a common view was that Meteor Crater had been formed by an underground steam explosion. Shoemaker knew nothing about underground steam explosions – he couldn't: they don't exist – but he did know all about blast zones. One of his first jobs out of college had been to study explosion rings at the Yucca Flats nuclear test site in Nevada. He concluded, as Barringer had before him, that there was nothing at Meteor Crater to suggest volcanic activity, but that there were huge distributions of other stuff – anomalous fine silicas and magnetites principally – that suggested an impact from space. Intrigued, he began to study the subject in his spare time.

Working first with his colleague Eleanor Helin and later with his wife, Carolyn, and associate David Levy, Shoemaker began a systematic survey of the inner solar system. They spent one week each month at the Palomar Observatory in California looking for objects, asteroids primarily, whose trajectories carried them across the Earth's orbit.

'At the time we started, only slightly more than a dozen of these things had ever been discovered in the entire course of astronomical observation,' Shoemaker recalled some years later in a television interview. 'Astronomers in the twentieth century essentially abandoned the solar system,' he added. 'Their attention was turned to the stars, the galaxies.'

What Shoemaker and his colleagues found was that there was more risk out there – a great deal more – than anyone had ever imagined.

Asteroids, as most people know, are rocky objects orbiting in loose formation in a belt between Mars and Jupiter. In illustrations they are always shown as existing in a jumble, but in fact the solar system is quite a roomy place and the average asteroid actually will be about 1.5 million kilometres from its nearest neighbour. Nobody knows even approximately how many asteroids there are tumbling through space, but the number is thought to be probably not less than a billion. They are presumed to be a planet that never quite made it, owing to the unsettling gravitational pull of Jupiter, which kept – and keeps – them from coalescing.

When asteroids were first detected in the 1800s – the very first was discovered on the first day of the century by a Sicilian named Giuseppe Piazzi – they were thought to be planets, and the first two were named Ceres and Pallas. It took some inspired deductions by the astronomer William Herschel to work out that they were nowhere near planet-sized but much smaller. He called them asteroids – Greek for 'starlike' – which was slightly unfortunate as they are not like stars at all. Sometimes now they are more accurately called planetoids.

Finding asteroids became a popular activity in the 1800s and by the end of the century about 1,000 were known. The problem was that no one was systematically recording them. By the early 1900s, it had often become impossible to know whether an asteroid that popped into view was new or simply one that had been noted earlier and then lost track of. By this time, too, astrophysics had moved on so much that few astronomers wanted to devote their lives to anything as mundane as rocky planetoids. Only a few, notably Gerard Kuiper, the Dutch-born astronomer for whom is named the Kuiper Belt of comets, took any interest in the solar system at all. Thanks to his work at the McDonald Observatory in Texas, followed later by work done by others at the Minor Planet Center in Cincinnati and the Spacewatch project in Arizona, a long list of lost asteroids was gradually whittled down until by the close of the twentieth century only one known asteroid was unaccounted for – an object called 719 Albert. Last seen in October 1911, it was finally tracked down in 2000 after being missing for eighty-nine years.

So, from the point of view of asteroid research, the twentieth century was essentially just a long exercise in book-keeping. It is really only in the last few years that astronomers have begun to count and keep an eye on the rest of the asteroid community. As of July 2024, the number of recorded 'minor planets' (which is to say asteroids plus a very few odds and ends like Pluto and Ceres) stood at about 1.4 million, but with a billion all told, the count still has a long way to go. Matters are certain to improve in the coming years as some massive new telescopes come online – the Vera C. Rubin Observatory in Chile found 2,100 new asteroids in just two months after it became operational in May 2025 – but even on the most optimistic forecasts it will be many decades before we get near a full census. In a sense it hardly matters. Identifying an asteroid doesn't make it safe. Even if every asteroid in the solar system had a name and known orbit, no one could say what perturbations might send any of them hurtling towards us. We can't forecast rock disturbances on our own surface. Put those rocks adrift in space and what they might do is beyond guessing. Any asteroid out there that has our name on it is very likely to have no other.

Think of the Earth's orbit as a kind of orbital freeway on which we are the only vehicle, but which is crossed regularly by pedestrians who don't know enough to look before stepping off the shoulder. All we know is that at some point, at uncertain intervals, they trundle across the road along which we are cruising at over 100,000 kilometres an hour. As Steven Ostro of the Jet Propulsion Laboratory has put it, 'Suppose that there was a button you could push and you could light up all the Earth-crossing asteroids larger than about ten metres, there would be over a hundred million of these objects in the sky.' In short, you would see not a couple of thousand distant twinkling stars, but millions upon millions upon millions of nearer, randomly moving objects – 'all of which are capable of colliding with the Earth and all of which are moving on slightly different courses through the sky at different rates. It would be deeply unnerving.' Well, be unnerved, because it is there. We just can't see it.

Altogether it is thought – though it is really only a guess, based on extrapolating from cratering rates on the moon – that some 2,000 asteroids big enough to imperil civilized existence (i.e., more than a kilometre across) regularly cross our orbit, and at least 10 per cent of these, according to NASA's Jet Propulsion Laboratory, are yet to be spotted and identified. Altogether, as of 2024, the International Astronomical Union's Minor Planet Center had identified some 19,000 Near Earth Objects at least 30 metres across, but even a small asteroid – the size of a house, say – could destroy a city. The number of these relative tiddlers in Earth-crossing orbits is almost certainly in the hundreds of thousands and possibly in the millions, and they are nearly impossible to track.

The first one wasn't spotted until 1991, and that was after it had already gone by. Named 1991 BA, it was noticed as it sailed past us at a distance of 170,000 kilometres – in cosmic terms the equivalent of a bullet passing through one's sleeve without touching the arm. Three years later, another, somewhat larger asteroid missed us by just 65,000 miles – the closest pass yet recorded. It, too, was not seen until it had passed and would have arrived without warning. Between 2016 and 2021, more than 300 asteroids passed Earth within one lunar distance – that is, the distance between the Earth and the Moon. That's close.

Easily the weirdest of all these unexpected visitors was a mysterious interstellar object called Oumuamua, which flew past the Earth in October 2017. Oumuamua (meaning 'advance traveller' or 'scout' in Hawaiian) was spotted by Robert Weryk, a Canadian astronomer working at an observatory on Maui. Oumuamua didn't move through space like anything ever seen before. It was travelling at 200,000 miles an hour, or some four times the normal speed of an asteroid. It didn't whip around the Sun in an elliptical path, as would be expected, but kept going in a straight line. Indeed, as it passed the Sun it sped up and whooshed away, as if opening a throttle.

What was startling was that it wasn't noticed until *after* it had passed through Earth's orbit. Scientists still don't know whether it was a comet, a lost asteroid, a ball of ice crystals or something else altogether, and they cannot begin to account for its movements. Avi Loeb, a Harvard astrophysicist, argued that it could only be a manufactured object – something created by an advanced civilization and now probably travelling through space as a piece of debris in the way that our own two Voyager spacecraft will soon become ghost objects forever drifting through the cosmos.

While Gene Shoemaker was trying to get people galvanized about the potential dangers of the inner solar system, another development – wholly unrelated on the face of it – was quietly unfolding in Italy with the work of a young geologist from the Lamont Doherty Laboratory at Columbia University. In the early 1970s, Walter Alvarez was doing fieldwork in a comely defile known as the Bottaccione Gorge, near the Umbrian hill town of Gubbio, when he grew curious about a thin band of reddish clay that divided two ancient layers of limestone – one from the Cretaceous period, the other from the Tertiary. This is a point known to geology as the KT boundary* and it marks the time, 65 million years ago, when the dinosaurs and roughly half the world's other species of animals abruptly vanish from the fossil record. Alvarez wondered what it was about a thin lamina of clay, barely 6 millimetres thick, that could account for such a dramatic moment in Earth's history.

At the time, the conventional wisdom about the dinosaur extinction was the same as it had been in Charles Lyell's day a century earlier – namely, that the dinosaurs had died out over millions of years. But the thinness of the clay layer clearly suggested that in Umbria, if nowhere else, something rather more

---

* It is KT rather than CT because *C* had already been appropriated for *Cambrian*. Depending on which source you credit, the *K* comes from either the Greek *kreta* or the German *Kreide*. Both conveniently mean chalk, which is also what *Cretaceous* means.

abrupt had happened. Unfortunately, in the 1970s no tests existed for determining how long such a deposit might have taken to accumulate.

In the normal course of things, Alvarez almost certainly would have had to leave the problem at that; but luckily he had an impeccable connection to someone outside his discipline who could help – his father, Luis. Luis Alvarez was an eminent nuclear physicist; he had won the Nobel Prize in Physics the previous decade. He had always been mildly scornful of his son's attachment to rocks, but this problem intrigued him. It occurred to him that the answer might lie in dust from space.

Every year the Earth accumulates some 30,000 tonnes of 'cosmic spherules' – space dust, in plainer language – which would be quite a lot if you swept it into one pile, but is infinitesimal when spread across the globe. Scattered through this thin dusting are exotic elements not normally much found on Earth. Among these is the element iridium, which is 1,000 times more abundant in space than in the Earth's crust (because, it is thought, most of the iridium on Earth sank to the core when the planet was young).

Luis Alvarez knew that a colleague of his at the Lawrence Berkeley Laboratory in California, Frank Asaro, had developed a technique for measuring very precisely the chemical composition of clays using a process called neutron activation analysis. This involved bombarding samples with neutrons in a small nuclear reactor and carefully counting the gamma rays that were emitted; it was extremely finicky work. Previously Asaro had used the technique to analyse pieces of pottery, but Alvarez reasoned that if they measured the amount of one of the exotic elements in his son's soil samples and compared that with its annual rate of deposition, they would know how long it had taken the samples to form. On an October afternoon in 1977, Luis and Walter Alvarez dropped in on Asaro and asked him if he would run the necessary tests for them.

It was really quite a presumptuous request. They were asking Asaro to devote months to making the most painstaking measurements of geological samples merely to confirm what seemed entirely self-evident to begin with – that the thin layer of clay had been formed as quickly as its thinness suggested. Certainly, no one expected his survey to yield any dramatic breakthroughs.

'Well, they were very charming, very persuasive,' Asaro recalled in an interview. 'And it seemed an interesting challenge, so I agreed to try. Unfortunately, I had a lot of other work on, so it was eight months before I could get to it.' He consulted his notes from the period. 'On June 21, 1978, at 1.45 p.m., we put a sample in the detector. It ran for 224 minutes and we could see we were getting interesting results, so we stopped it and had a look.'

The results were so unexpected, in fact, that the three scientists at first thought they had to be wrong. The amount of iridium in the Alvarez sample was more than 300 times normal levels – far beyond anything they might have predicted. Over the following months Asaro and his colleague Helen Michel worked up to thirty hours at a stretch ('Once you started you couldn't stop,' Asaro explained) analysing samples, always with the same results. Tests on other samples – from Denmark, Spain, France, New Zealand, Antarctica – showed that the iridium deposit was worldwide and greatly elevated everywhere, sometimes by as much as 500 times normal levels. Clearly something big and abrupt, and probably cataclysmic, had produced this arresting spike.

After much thought, the Alvarezes concluded that the most plausible explanation – plausible to them, at any rate – was that the Earth had been struck by an asteroid or comet. The idea that the Earth might be subjected to devastating impacts from time to time was not quite as new as is now sometimes suggested. As far back as 1942, a Northwestern University astrophysicist named Ralph B. Baldwin had suggested such a possibility in an article in *Popular Astronomy* magazine. (He published the article there because no academic publisher was prepared to run it.) And at least two well-known scientists, the astronomer Ernst Öpik and the chemist and Nobel laureate Harold Urey, had also voiced support for the notion at various times. Even among palaeontologists it was not unknown. In 1956 a professor at Oregon State University, M. W. de Laubenfels, writing in the *Journal of Paleontology*, had actually anticipated the Alvarez theory by suggesting that the dinosaurs may have been dealt a death blow by an impact from space, and in 1970 the president of the American Paleontological Society, Dewey J. McLaren, proposed at the group's annual conference the possibility that an extraterrestrial impact may have been the cause of an earlier event known as the Frasnian extinction.

As if to underline just how un-novel the idea had become by this time, in 1979 a Hollywood studio actually produced a movie called *Meteor* ('It's five miles wide . . . It's coming at 30,000 m.p.h. — and there's no place to hide!') starring Henry Fonda, Natalie Wood, Karl Malden and a very large rock.

So when, in the first week of 1980, at a meeting of the American Association for the Advancement of Science, the Alvarezes announced their belief that the dinosaur extinction had not taken place over millions of years as part of some slow inexorable process, but suddenly in a single explosive event, it shouldn't have come as a shock.

But it did. It was received everywhere, but particularly in the palaeontological world, as an outrageous heresy.

'Well, you have to remember,' Asaro recalled, 'that we were amateurs in this field. Walter was a geologist specializing in palaeomagnetism, Luis was a physicist and I was a nuclear chemist. And now here we were telling palaeontologists that we had solved a problem that had eluded them for over a century. It's not terribly surprising that they didn't embrace it immediately.' As Luis Alvarez joked: 'We were caught practising geology without a licence.'

But there was also something much deeper and more fundamentally abhorrent in the impact theory. The belief that terrestrial processes were gradual had been elemental in natural history since the time of Lyell. By the 1980s, catastrophism had been out of fashion for so long that it had become literally unthinkable. For most geologists the idea of a devastating impact was, as Eugene Shoemaker noted, 'against their scientific religion'.

Nor did it help that Luis Alvarez was openly contemptuous of palaeontologists and their contributions to scientific knowledge. 'They're really not very good scientists. They're more like stamp collectors,' he wrote in the *New York Times*, in an article that stings yet.

Opponents of the Alvarez theory produced any number of alternative explanations for the iridium deposits – for instance, that they were generated by prolonged volcanic eruptions in India called the Deccan Traps ('trap' comes from a Swedish word for a type of lava; 'Deccan' is the name of the area today) – and above all insisted that there was no proof that the dinosaurs disappeared abruptly from the fossil record at the iridium boundary. One of the most vigorous opponents was Charles Officer of Dartmouth College. He insisted that the iridium had been deposited by volcanic action even while conceding in a newspaper interview that he had no actual evidence of it. As late as 1988, more than half of all American palaeontologists contacted in a survey continued to believe that the extinction of the dinosaurs was in no way related to an asteroid or cometary impact.

The one thing that would most obviously support the Alvarezes' theory was the one thing they didn't have – an impact site. Enter Eugene Shoemaker. Shoemaker had an Iowa connection – his daughter-in-law taught at the University of Iowa – and he was familiar with the Manson crater from his own studies. Thanks to him, all eyes now turned to Iowa.

Geology is a profession that varies from place to place. In Iowa, a state that is largely flat and stratigraphically uneventful, it tends to be comparatively serene. There are no alpine peaks or grinding glaciers, no great deposits of oil or precious metals, not a hint of a pyroclastic flow. If you are a geologist employed by the state of Iowa, a big part of the work you do is to evaluate Manure Management Plans, which all the state's 'animal confinement operators' – hog farmers, to the rest of us – are required to file periodically. There are 24 million hogs in Iowa (up from 15 million when this book was first written), so a *lot* of manure to manage; indeed, Iowa's farms produce 109 *billion* pounds of animal waste per year, far more than Iowa's people do. Trying to keep Iowa's groundwater healthy is vital work, but with the best will in the world it's not exactly dodging lava bombs on Mount Pinatubo or scrabbling over crevasses on the Greenland ice sheet in search of ancient life-bearing quartzes. So we may well imagine the flutter of excitement that swept through the Iowa Department of Natural Resources when in the mid-1980s the world's geological attention focused on Manson and its crater.

Trowbridge Hall in Iowa City is a turn-of-the-century pile of red brick that houses the University of Iowa's Earth Sciences Department and – way up in a kind of garret – the geologists of the Iowa Department of Natural Resources. No one now can remember quite when, still less why, the state geologists were placed in an academic facility, but you get the impression that the space was conceded grudgingly, for the offices when I first went there were cramped and low-ceilinged and not very accessible. When being shown the way, I half expected to be taken out on to a roof ledge and helped in through a window.

Brian Witzke and Ray Anderson spent many happy and productive years amid disordered heaps of papers, journals, furled charts and hefty specimen stones (geologists are never at a loss for paperweights), though both are now retired.

'Suddenly we were at the centre of things,' Anderson told me, gleaming at the memory of it, when I first met him and Witzke in their offices on a dismal, rainy morning in June 2001. 'It was a wonderful time.'

I asked them about Gene Shoemaker, a man who seems to have been universally revered. 'He was just a great guy,' Witzke replied without hesitation. 'If it hadn't been for him, the whole thing would never have gotten off the ground. Even with his support, it took two years to get it up and running. Drilling's an expensive business – about thirty-five dollars a foot back then, more now, and we needed to go down three thousand feet.'

'Sometimes more than that,' Anderson added. 'Sometimes more than that,' Witzke agreed. 'And at several locations. So you're talking a lot of money. Certainly more than our budget would allow.'

So a collaboration was formed between the Iowa Geological Survey and the US Geological Survey. 'At least, we *thought* it was a collaboration,' said Anderson, producing a small pained smile.

'It was a real learning curve for us,' Witzke went on. 'There was actually quite a lot of bad science going on throughout the period – people rushing in with results that didn't always stand up to scrutiny.' One of those moments came at the annual meeting of the American Geophysical Union in 1985, when Glenn Izett and C. L. Pillmore of the US Geological Survey announced that the Manson crater was of the right age to have been involved with the dinosaurs' extinction. The declaration attracted a good deal of press attention but was unfortunately premature. A more careful examination of the data revealed that Manson was not only too small, but also 9 million years too early.

The first Anderson or Witzke learned of this setback was when they arrived at a conference in South Dakota and found people coming up to them with sympathetic looks and saying: 'We hear you lost your crater.' It was news to them that Izett and the other USGS scientists had just announced refined figures revealing that Manson couldn't after all have been the extinction crater.

'It was pretty stunning,' recalls Anderson. 'I mean, we had this thing that was really important and then suddenly we didn't have it any more. But even worse was the realization that the people we thought we'd been collaborating with hadn't bothered to share with us their new findings.'

'Why not?'

He shrugged. 'Who knows? Anyway, it was a pretty good insight into how unattractive science can get when you're playing at a certain level.'

Though now retired, Brian and Ray still do some teaching at the University of Iowa as adjunct professors, and they remain passionate about the Manson crater. I caught up with them at Witzke's home in Iowa City in the late summer of 2024. They are still a peerless double act, cheerfully talking over or elaborating upon each other's remarks. And they clearly are very fond and respectful of each other, though they have a long-standing disagreement over whether the asteroid landed in water or on dry land.

Not a huge amount has changed with respect to the Manson impact since we first met. The date of the event has been revised backwards by about a million years, from 74.8 million to 75.9 million years ago (plus or minus 100,000 years), but otherwise the state of our knowledge is much as it was. The big excitement for Iowa geology in recent years was the discovery in 2013 of another significant impact crater in the northeast corner of the state directly under the town of Decorah. It dates from about 470 million years ago, and would have been regionally devastating, but not big enough to cause permanent extinctions.

The biggest perennial mystery with Manson is why the first drillers showed such extraordinary perseverance in their efforts to find water beneath Manson.

'The first well, in 1905, went down more than a thousand feet before they finally found water,' Witzke told me. 'Normally drillers would have given up long before that. A few years later they drilled another deep hole and found water again.'

'Since then,' Anderson went on, 'we have drilled holes all over the place and never struck water again. In the 1980s a petroleum company looking for oil deposits drilled even more holes, including one 17,000 feet deep, and they never struck water either. So it was really a miracle that the original drillers found water twice in two tries.'

With the Manson impact ruled out as the cause of dinosaur extinctions, the search moved elsewhere. By chance, in 1990 one of the searchers, Alan Hildebrand of the University of Arizona, met a reporter from the *Houston Chronicle* who happened to know about a large, unexplained ring formation nearly 200 kilometres across under Mexico's Yucatán Peninsula at Chicxulub, near the city of Progreso, about 950 kilometres due south of New Orleans. The formation had been found by Pemex, the Mexican oil company, in 1952 – the year, coincidentally, that Gene Shoemaker first visited Meteor Crater in Arizona – but the company's geologists had concluded that it was volcanic, in line with the thinking of the day. Hildebrand travelled to the site and decided fairly swiftly that they had their crater. By early 1991 it had been established to nearly everyone's satisfaction that Chicxulub was the impact site.

Still, many people didn't quite grasp what an impact could do. As Stephen Jay Gould recalled in one of his essays: 'I remember harboring some strong initial doubts about the efficacy of such an event . . . [W]hy should an object only six miles across wreak such havoc upon a planet with a diameter of eight thousand miles?'

Conveniently, a natural test of the theory arose soon after when the Shoemakers and Levy discovered Comet Shoemaker–Levy 9, which they soon realized was headed for Jupiter. For the first time, humans would be able to witness a cosmic collision – and witness it very well, thanks to the new Hubble Space Telescope. Most astronomers, according to Curtis Peebles, expected little, particularly as the comet was not a coherent sphere but a string of twenty-one fragments. 'My sense', wrote one, 'is that Jupiter will swallow these comets up without so much as a burp.' One week before the impact, *Nature* ran an article, 'The Big Fizzle Is Coming', predicting that the impact would constitute nothing more than a meteor shower.

The impacts began on 16 July 1994, went on for a week and were bigger by far than anyone – with the possible exception of Gene Shoemaker – expected. One fragment, known as Nucleus G, struck with the force of about 6 million megatonnes – seventy-five times all the nuclear weaponry in existence at that time. Nucleus G was only about the size of a small mountain, but it created wounds in the Jovian surface the size of Earth. It was the final blow for critics of the Alvarez theory.

Luis Alvarez never knew of the discovery of the Chicxulub crater or of the Shoemaker–Levy comet, as he died in 1988. Shoemaker also died early. On the third anniversary of the Jupiter collision, he and his wife were in the Australian outback, where they went every year to search for impact sites. On a dirt track in the Tanami Desert – normally one of the emptiest places on Earth – they came over a slight rise just as another vehicle was coming from the other direction and crashed head-on. Shoemaker was killed instantly, his wife injured. Some of his ashes were sent to the Moon aboard the Lunar Prospector spacecraft. The rest were scattered around Meteor Crater.

Anderson and Witzke no longer had the crater that killed the dinosaurs, 'but we still had the largest and most perfectly preserved impact crater in the mainland United States', Anderson said. (A little verbal dexterity is required to keep Manson's superlative status. Other craters are larger – notably Chesapeake Bay, which was recognized as being an impact site in 1994 – but they are either offshore or deformed.) 'Chicxulub is buried under two to three kilometres of limestone and mostly offshore, which makes it difficult to study,' Anderson went on, 'while Manson is really quite accessible. It's because it is buried that it is actually comparatively pristine.'

I asked them how much warning we would receive if a similar hunk of rock were coming towards us today.

'Oh, probably none,' said Anderson breezily. 'It wouldn't be visible to the naked eye until it warmed up and that wouldn't happen until it hit the atmosphere, which would be about one second before it hit the Earth. You're talking about something moving many tens of times faster than the fastest bullet. Unless it had been seen by someone with a telescope, and that's by no means a certainty, it would take us completely by surprise.'

How devastating an impactor is depends on a lot of variables – angle of entry, velocity and trajectory, whether the collision is head-on or from the side, and the mass and density of the impacting object, among much else. But all big impacts have one feature in common, which is that the bulk of the ejecta, no matter the angle of impact, flies straight up into the air and then drops back into the hole – albeit greatly jumbled up and transformed (or 'shocked' in the jargon of geologists) by the extreme heat of the event.

'Those are the telltale signs of an impact,' Ray explains. 'You don't get those effects from anything that happens naturally on Earth.'

By calculating the amount of energy released from the Manson impact, Anderson and Witzke have worked out the probable effects of such an impact happening today, and it is chilling.

An asteroid or comet travelling at cosmic velocities would enter the Earth's atmosphere at such a speed that the air beneath it couldn't get out of the way and would be compressed, as in a bicycle pump. As anyone who has used such a pump knows, compressed air grows swiftly hot, and the temperature below it would rise to some 60,000 kelvin, or ten times the surface temperature of the Sun. In this instant of its arrival in our atmosphere, everything in the meteor's path – people, houses, factories, cars – would crinkle and vanish like cellophane in a flame.

One second after entering the atmosphere, the meteorite would slam into the Earth's surface where the people of Manson had a moment before been going about their business. The meteorite itself would vaporize instantly, but the blast would blow out 1,000 cubic kilometres of rock, earth and superheated gases. Every living thing within 250 kilometres that hadn't been killed by the heat of entry would now be killed by the blast.

For those outside the zone of immediate devastation, the first inkling of catastrophe would be a flash of blinding light – the brightest ever seen by human eyes – followed an instant to a minute or two later by an apocalyptic sight of unimaginable grandeur: a roiling wall of darkness reaching high into the heavens, filling one entire field of view and travelling at some 800 miles an hour. Its approach would be eerily silent since it would be moving far beyond the speed of sound. Anyone in a tall building in Omaha or Des Moines, say, who chanced to look in the right direction would see a bewildering veil of turmoil followed by instantaneous oblivion. Over an area stretching from Denver to Detroit and encompassing what had once been Chicago, St Louis, Kansas City, the Twin Cities – the whole of the Midwest, in short – nearly every standing thing would swiftly be flattened or on fire, and nearly every living thing would be dead. People up to 1,500 kilometres away, from Saskatchewan in the north to Texas and Georgia in the south, would be knocked off their feet and cut or clobbered by a blizzard of flying projectiles.

But that's just the initial shock wave. No one can do more than guess what the associated damage would be, other than that it would be brisk and global. Within hours, a cloud of blackness would cover the Earth and burning rock and other debris would be pelting down everywhere, setting much of the planet ablaze. It has been estimated that at least a billion and a half people would be dead by the end of the first day. The massive disturbances to the ionosphere would knock out communications systems everywhere, so survivors would have no idea what was happening elsewhere or where to turn. It would hardly matter. As one commentator has put it, fleeing would mean 'selecting a slow death over a quick one. The death toll would be very little affected by any plausible relocation effort, since Earth's ability to support life would be universally diminished.'

The amount of soot and floating ash from the impact and following fires would blot out the sun certainly for months, possibly for years, disrupting growing cycles. In 2001 researchers at the California Institute of Technology analysed helium isotopes from sediments left from the later KT impact and concluded that it affected the Earth's climate for about 10,000 years. This was actually used as evidence to support the notion that the extinction of dinosaurs was swift and emphatic – and so it was, in geological terms. We can only guess how well, or whether, humanity would cope with such an event.

And in all likelihood, remember, this would come without warning, out of a clear sky.

But let's suppose we did see the object coming. What would we do? Everyone assumes we would send up a nuclear warhead and blast it to smithereens. There are some problems with that idea, however. First, as John S. Lewis has noted, our missiles are not designed for space work. They haven't the oomph to escape Earth's gravity, and even if they did there are no mechanisms to guide them across tens of millions of kilometres of space.

In 2022, as an experiment, NASA crashed a spacecraft called DART (for Double Asteroid Redirection Test) into an asteroid named Dimorphos, slightly altering its course of travel, giving hope that any such asteroid heading for Earth could be blown apart or nudged aside – assuming it was spotted in enough time to prepare and execute a mission. But, as previous near misses have shown, that is a dangerously big assumption. It has been suggested that even a year's warning would probably be insufficient to take appropriate action. The greater likelihood, however, is that we wouldn't see any object – even a comet – until it was about six months away, which would be much too late. Shoemaker–Levy 9 had been orbiting Jupiter in a fairly conspicuous manner since 1929, but it was over half a century before anyone noticed.

Because these things are so difficult to compute and must incorporate such a significant margin of error, even if we knew an object was heading our way we might not know until very late on – the last couple of weeks, say – whether collision was certain. For most of the time of the object's approach we could well exist in a kind of cone of uncertainty. It would certainly be the most interesting few months in the history of the world. And imagine the party if it passed safely.

'So how often does something like the Manson impact happen?' I asked Anderson and Witzke before leaving.

'Oh, about once every million years on average,' said Witzke.

'And remember,' added Anderson, 'this was a relatively minor event. Do you know how many extinctions were associated with the Manson impact?'

'No idea,' I replied.

'None,' he said, with a strange air of satisfaction. 'Not one.'

Of course, Witzke and Anderson added hastily and more or less in unison, there would have been terrible devastation across much of the Earth, as just described, and complete annihilation for hundreds of miles around ground zero. But life is hardy, and when the smoke cleared there were enough lucky survivors from every species that none permanently perished.

The good news, it appears, is that it takes an awful lot to extinguish a species. The bad news is that the good news can never be counted on. Worse still, it isn't actually necessary to look to space for petrifying danger. As we are about to see, Earth can provide plenty of danger of its own.

14

THE FIRE BELOW

In the summer of 1971, a young geologist named Mike Voorhies was scouting around on some grassy farmland in eastern Nebraska, not far from the little town of Orchard where he had grown up. Passing through a steep-sided gully, he spotted a curious glint in the brush above and clambered up to have a look. What he had seen was the perfectly preserved skull of a young rhinoceros, which had been washed out by recent heavy rains.

A few yards beyond, it turned out, was one of the most extraordinary fossil beds ever discovered in North America: a dried-up waterhole that had served as a mass grave for scores of animals – rhinoceroses, zebra-like horses, sabre-toothed deer, camels, turtles. All had died from some mysterious cataclysm just under 12 million years ago in the time known to geology as the Miocene. In those days Nebraska stood on a vast, hot plain very like the Serengeti of Africa today. The animals had been found buried under volcanic ash up to 3 metres deep. The puzzle of it was that there were not, and never had been, any volcanoes in Nebraska.

Today, the site of Voorhies' discovery is called Ashfall Fossil Beds State Park. It has a stylish visitors' centre and museum, with thoughtful displays on the geology of Nebraska and the history of the fossil beds. The centre incorporates a lab with a glass wall through which visitors can watch palaeontologists cleaning bones. Working alone in the lab on the morning I passed through was a cheerfully grizzled-looking fellow in a blue work shirt whom I recognized as Mike Voorhies from a BBC *Horizon* documentary in which he had featured. They don't get a huge number of visitors to Ashfall Fossil Beds State Park – it's slightly in the middle of nowhere – and Voorhies seemed pleased to show me around. He took me to the spot atop a 6-metre-high ravine where he had made his find.

'It was a dumb place to look for bones,' he said happily. 'But I wasn't looking for bones. I was thinking of making a geological map of eastern Nebraska at the time, and really just kind of poking around. If I hadn't gone up this ravine or the rains hadn't just washed out that skull, I'd have walked on by and this would never have been found.' He indicated a roofed enclosure nearby, which had become the main excavation site. There, some 200 animals had been found lying together in a jumble.

I asked him in what way it was a dumb place to hunt for bones. 'Well, if you're looking for bones, you really need exposed rock. That's why most palaeontology is done in hot, dry places. It's not that there are more bones there. It's just that you have some chance of spotting them. In a setting like this' – he made a sweeping gesture across the vast and unvarying prairie – 'you wouldn't know where to begin. There could be really magnificent stuff out there, but there's no surface clues to show you where to start looking.'

At first they thought the animals were buried alive, and Voorhies stated as much in a *National Geographic* article in 1981. 'The article called the site a "Pompeii of prehistoric animals",' he told me, 'which was unfortunate because just afterwards we realized that the animals hadn't died suddenly at all. They were all suffering from something called hypertrophic pulmonary osteodystrophy, which is what you would get if you were breathing a lot of abrasive ash – and they must have been breathing a lot of it because the ash was feet thick for hundreds of miles.'

He picked up a chunk of greyish, clay-like dirt and crumbled it into my hand. It was powdery but slightly gritty. 'Nasty stuff to have to breathe,' he went on, 'because it's very fine but also quite sharp. So anyway they came here to this watering hole, presumably seeking relief, and died in some misery. The ash would have ruined everything. It would have buried all the grass and coated every leaf and turned the water into an undrinkable grey sludge. It couldn't have been very agreeable at all.'

The *Horizon* documentary had suggested that the existence of so much ash in Nebraska was a surprise. In fact, Nebraska's huge ash deposits had been known about for a long time. For almost a century they had been mined to make household cleaning powders like Comet and Ajax. But, curiously, no one had ever thought to wonder where all the ash came from.

'I'm a little embarrassed to tell you,' Voorhies said, smiling briefly, 'that the first I thought about it was when an editor at the *National Geographic* asked me the source of all the ash and I had to confess that I didn't know. Nobody knew.'

Voorhies sent samples to colleagues all over the western United States asking if there was anything about it that they recognized. Several months later a geologist named Bill Bonnichsen from the Idaho Geological Survey got in touch and told him that the ash matched a volcanic deposit from a place called Bruneau-Jarbidge in southwest Idaho. The event that killed the plains animals of Nebraska was a volcanic explosion on a scale previously unimagined – but big enough to leave an ash layer 3 metres deep some 1,600 kilometres away in eastern Nebraska. It turned out that under the western United States there was a huge cauldron of magma, a colossal volcanic hot spot, which erupted cataclysmically every so often. The hot spot is still there. These days we call it Yellowstone National Park.

We know amazingly little about what happens beneath our

feet. It is fairly remarkable to think that Ford has been building cars and Nobel committees awarding prizes for longer than we have known that the Earth has a core. And of course the idea that the continents move about on the surface like lily pads has been common wisdom for only a few decades. 'Strange as it may seem,' wrote Richard Feynman, 'we understand the distribution of matter in the interior of the Sun far better than we understand the interior of the Earth.'

The distance from the surface of Earth to the middle is 6,370 kilometres, which isn't so very far. It has been calculated that if you sank a well to the centre and dropped a brick down it, it would take only forty-five minutes for it to hit the bottom (though at that point it would be weightless since all the Earth's gravity would be all around it rather than only beneath it). Our own attempts to penetrate towards the middle have been modest indeed. One or two South African gold mines reach to a depth of over 3 kilometres, but most mines on Earth go no more than about 400 metres beneath the surface. If the planet were an apple, we wouldn't yet have broken through the skin. Indeed, we haven't even come close.

Until only a little over a century ago, what the best-informed scientific minds knew about Earth's interior was not much more than what a coal miner knew – namely, that you could dig down through soil for a distance and then you'd hit rock, and that was about it. Then, in 1906, an Irish geologist named R. D. Oldham, while examining some seismograph readings from an earthquake in Guatemala, noticed that certain shock waves had penetrated to a point deep within the Earth and then bounced off at an angle, as if they had encountered some kind of barrier. From this he deduced that the Earth has a core. Three years later, a Croatian seismologist named Andrija Mohorovičić was studying graphs from an earthquake in Zagreb when he noticed a similar odd deflection, but at a shallower level. He had discovered the boundary between the crust and the layer immediately below, the mantle; this zone has been known ever since as the Mohorovičić discontinuity, or Moho for short.

We were beginning to get a vague idea of the Earth's layered interior – though it really was only vague. Not until 1936 did a Danish scientist named Inge Lehmann, studying seismographs of earthquakes in New Zealand, discover that there were two cores – an inner one, which we now believe to be solid, and an outer one (the one that Oldham had detected), which is thought to be liquid and the seat of magnetism.

At just about the time that Lehmann was refining our basic understanding of the Earth's interior by studying the seismic waves of earthquakes, two geologists at Caltech in California were devising a way to make comparisons between one earthquake and the next. They were Charles Richter and Beno Gutenberg, though for reasons that have nothing to do with fairness the scale became known almost at once as Richter's alone. (They were nothing to do with Richter, either. A modest fellow, he never referred to the scale by his own name but always called it 'the Magnitude Scale'.)

The Richter scale has always been widely misunderstood by non-scientists, though it is perhaps a little less so now than in its early days when visitors to Richter's office often asked to see his celebrated scale, thinking it was some kind of machine. The scale is, of course, more an idea than a thing, an arbitrary measure of the Earth's tremblings based on surface measurements. It rises exponentially, so that a 7.3 quake is ten times more powerful than a 6.3 earthquake and 100 times more powerful than a 5.3 earthquake. Theoretically, at least, there is no upper limit for an earthquake – nor, come to that, a lower limit. The scale is a simple measure of force, but says nothing about damage. A magnitude-7 quake happening deep in the mantle – say, 650 kilometres down – might cause no surface damage at all, while a significantly smaller one happening just 6 or 7 kilometres under the surface could wreak widespread devastation. Much, too, depends on the nature of the subsoil, the quake's duration, the frequency and severity of aftershocks, and the physical setting of the affected area. All this means that the most fearsome quakes are not necessarily the most forceful, though force obviously counts for a lot.

The largest earthquake since the scale's invention was in the Pacific Ocean off the coast of Chile in 1960, which was initially logged at 8.6 magnitude but later revised upwards by some authorities (including the US Geological Survey) to a truly grand-scale 9.5. As you will gather from this, measuring earthquakes is not always an exact science, particularly when it involves interpreting readings from remote locations. The Great Chilean earthquake, as it was known, not only caused widespread damage across coastal South America, but also set off a tsunami that rolled nearly 10,000 kilometres across the Pacific and slapped away much of downtown Hilo, Hawaii, destroying 500 buildings and killing sixty people. Similar wave surges claimed yet more victims as far away as Japan and the Philippines.

Nearly as big, and far more devastating, was the Indian Ocean earthquake of 26 December 2004, which was centred off the coast of Sumatra, Indonesia, but had a fault line nearly 800 miles long. It was measured at 9.2 to 9.3 on the Richter scale, lasted for ten minutes, and created a tsunami up to 30 metres high that slammed into the coastlines of fourteen countries, killing 228,000 people, making it by far the deadliest in history. The San Francisco earthquake of 1906, for comparison, measured an estimated 7.8 on the Richter scale and lasted less than thirty seconds.

Earthquakes are fairly common. Every day, on average, somewhere in the world there are two of magnitude 2.0 or greater – that's enough to give anyone nearby a pretty good jolt. Although they tend to cluster in certain places – notably around the rim of the Pacific – they can occur almost anywhere. In the United States, only Florida, eastern Texas and the upper Midwest seem – so far – to be almost entirely immune. New

England has had two quakes of magnitude 6.0 or greater (in 1638 and 1755). In April 2002, the region experienced a 5.1 magnitude shaking in a quake near Lake Champlain on the New York–Vermont border, causing extensive local damage and (I can personally attest) knocking pictures from walls and children from beds as far away as New Hampshire.

The most common types of earthquakes are those where two plates meet, as in California along the San Andreas Fault. As the plates push against each other, pressures build up until one or the other gives way. In general, the longer the interval between quakes, the greater the pent-up pressure and thus the greater the scope for a really big jolt. This is a particular worry for Tokyo, which Bill McGuire, a hazards specialist at University College London, has described as 'the city waiting to die' (not a motto you will find on many tourism leaflets). Tokyo stands on the meeting point of three tectonic plates in a country already well known for its seismic instability. On 9 March 2011, as many will recall, Japan suffered a devastating 9.1-magnitude earthquake along its east coast which, with its attendant tsunami, killed almost 20,000 people, swamped the Fukushima nuclear reactor and caused more than $300 billion in costs to the economy.

The great fear is that an earthquake of similar magnitude will hit Tokyo, with its much greater populace. The Great Kanto quake of 1923 in Tokyo killed at least 142,000 people at a time when the city's population was just 3 million. Today greater Tokyo is more than ten times that size. Since the Kanto quake, Tokyo has been eerily quiet, so the strain beneath the surface has been building for over a century. Japanese seismologists have put the chances of another Kanto-sized earthquake in Tokyo by mid-century at 70 per cent.

Even more unnerving, because they are less well understood and capable of occurring anywhere at any time, are the rarer shakings of the type known as intraplate quakes. These happen away from plate boundaries, which makes them wholly unpredictable. And because they come from a much greater depth, they tend to propagate over much wider areas. The most notorious such quakes ever to hit the United States were a series of three in New Madrid, Missouri, in the winter of 1811–12. The adventure started just after midnight on 16 December when people were awakened first by the noise of panicking farm animals (the restiveness of animals before quakes is not an old wives' tale but is in fact well established, though not at all understood) and then by an almighty rupturing noise from deep within the Earth.

Emerging from their houses, locals found the land rolling in waves up to a metre high and opening up in fissures several metres deep. A strong smell of sulphur filled the air. The shaking lasted for four minutes, with the usual devastating effects to property. Among the witnesses was the artist John James Audubon, who happened to be in the area. The quake radiated outwards with such force that it knocked down chimneys in Cincinnati over 600 kilometres away and, according to at least one account, 'wrecked boats in East Coast harbors and . . . even collapsed scaffolding erected around the Capitol Building in Washington, DC'.

On 23 January and 4 February further quakes of similar magnitude followed. New Madrid has been peaceful ever since – but not surprisingly, since such episodes have never been known to happen in the same place twice. As far as we know, they are as random as lightning. The next one could be under Chicago or Paris or Kinshasa. No one can even begin to guess. And what causes these massive intraplate rupturings? Something deep within the Earth. More than that, we don't know.

By the 1960s scientists had grown sufficiently frustrated by how little they understood of the Earth's interior that they decided to try to do something about it. Specifically, they got the idea to drill through the ocean floor (the continental crust was too thick) to the Moho discontinuity and to extract a piece of the Earth's mantle for examination at leisure. The thinking was that if they could understand the nature of the rocks inside the Earth, they might begin to understand how they interacted, and thus possibly be able to predict earthquakes and other unwelcome events.

The project became known, all but inevitably, as the Mohole, and it was pretty well disastrous. The hope was to lower a drill through over 4,000 metres of Pacific Ocean water off the coast of Mexico and drill some 5,000 metres through relatively thin crustal rock. Drilling from a ship in open waters is, in the words of one oceanographer, 'like trying to drill a hole in the sidewalks of New York from atop the Empire State Building using a strand of spaghetti'. Every attempt ended in failure. The deepest they penetrated was only about 180 metres. The Mohole became known as the No Hole. In 1966, exasperated with ever-rising costs and no results, Congress killed the project.

Four years later, Soviet scientists decided to try their luck on dry land. They chose a spot on Russia's Kola Peninsula, near the Finnish border, and set to work with the hope of drilling to a depth of 15 kilometres. The work proved harder than expected, but the Soviets were commendably persistent. When at last they gave up, nineteen years later, they had drilled to a depth of 12,262 metres. Bearing in mind that the crust of the Earth represents only about 0.3 per cent of the planet's volume and that the Kola hole had not cut even one-third of the way through the crust, we can hardly claim to have conquered the interior.

Even though the hole was modest, nearly everything about what it revealed surprised the researchers. Seismic wave studies had led the scientists to predict, and pretty confidently, that they would encounter sedimentary rock to a depth of 4,700 metres, followed by granite for the next 2,300 metres and basalt from there on down. In the event, the sedimentary layer was 50 per cent deeper than expected and the basaltic layer was never found at all. Moreover, the world down there was far warmer than anyone had expected, with a temperature at 10,000 metres of 180 degrees centigrade, nearly twice the forecast level. Most surprising of all was that the rock at depth was saturated with water – something that had not been thought possible.

Because we can't see into the Earth we have to use other techniques, which mostly involve reading waves as they travel through the interior, to find out what is there. We know a little bit about the mantle from what are known as kimberlite pipes, where diamonds are formed. What happens is that deep in the Earth there is an explosion that fires, in effect, a cannonball of magma to the surface at supersonic speeds. It is a totally random event. A kimberlite pipe could explode in your back garden as you read this. Because they come up from such depths – up to 200 kilometres down – kimberlite pipes bring up all kinds of things not normally found on or near the surface: a rock called peridotite, crystals of olivine and – just occasionally, in about one pipe in 100 – diamonds. Lots of carbon comes up with kimberlite ejecta, but most is vaporized or turns to graphite. Only occasionally does a hunk of it shoot up at just the right speed and cool down with the necessary swiftness to become a diamond. It was such a pipe that made South Africa the most productive diamond-mining country in the world, but there may be others even bigger that we don't know about. Geologists know that somewhere in the vicinity of northeastern Indiana there is evidence of a pipe or group of pipes that may be truly colossal. Diamonds up to 20 carats or more have been found at scattered sites throughout the region. But no one has ever found the source. As John McPhee notes, it may be buried under glacially deposited soil, like the Manson crater in Iowa, or under the Great Lakes.

So how much do we know about what's inside the Earth? Very little. Scientists are generally agreed that the world beneath us is composed of four layers – a rocky outer crust, a mantle of hot, viscous rock, a liquid outer core and a solid inner core.* We know that the surface is dominated by silicates, which are relatively light and not heavy enough to account for the planet's overall density. Therefore there must be heavier stuff inside. We know that to generate our magnetic field somewhere in the interior there must be a concentrated belt of metallic elements in a liquid state. That much is universally accepted. Almost everything beyond that — how the layers interact, what causes them to behave in the way they do, what they will do at any time in the future — is a matter of at least some uncertainty, and generally quite a lot of uncertainty.

Even the one part of it we can see, the crust, is a matter of some fairly strident debate. Nearly all geology texts tell you that continental crust is 5 to 10 kilometres thick under the oceans, about 40 kilometres thick under the continents and 65–95 kilometres thick under big mountain chains, but there are many puzzling variabilities within these generalizations. The crust beneath the Sierra Nevada Mountains, for instance, is only about 30–40 kilometres thick, and no one knows why. By all the laws of geophysics the Sierra Nevadas should be sinking, as if into quicksand. (Some people think they may be.)

How and when the Earth got its crust are questions that divide geologists into two broad camps — those who think it happened abruptly, early in the Earth's history, and those who think it happened gradually and rather later. Strength of feeling runs deep on such matters. Richard Armstrong of Yale proposed an early-burst theory in the 1960s, then spent the rest

---

* For those who crave a more detailed picture of the Earth's interior, here are the dimensions of the various layers, using average figures: From 0 to 40 kilometres is the crust. From 40 to 400 kilometres is the upper mantle. From 400 to 650 kilometres is a transition zone between the upper and lower mantle. From 650 to 2,700 kilometres is the lower mantle. From 2,700 to 2,890 kilometres is the 'D' layer. From 2,890 to 5,150 kilometres is the outer core, and from 5,150 to 6,370 kilometres is the inner core.

of his career fighting those who did not agree with him. He died of cancer in 1991, but shortly before his death he 'lashed out at his critics in a polemic in an Australian earth science journal that charged them with perpetuating myths', according to a report in *Earth* magazine. 'He died a bitter man,' reported a colleague.

The crust and part of the outer mantle together are called the lithosphere (from the Greek *lithos*, meaning stone), which in turn floats on top of a layer of softer rock called the asthenosphere (from Greek words meaning 'without strength'), but such terms are never entirely satisfactory. To say that the lithosphere floats on top of the asthenosphere suggests a degree of easy buoyancy that isn't quite right. Similarly, it is misleading to think of the rocks as flowing in anything like the way we think of materials flowing on the surface. The rocks are viscous, but only in the same way that glass is. It may not look it, but all the glass on Earth is flowing downwards under the relentless drag of gravity. It is often said that if you remove a pane of really old glass from the window of a European cathedral it will be noticeably thicker at the bottom than at the top. In fact, although glass does indeed flow, it does so much too slowly to be visible in anything made by humans. The hour hand on a clock moves about 10,000 times faster than the 'flowing' rocks of the mantle. The movements occur not just laterally, as the Earth's plates move across the surface, but up and down too, as rocks rise and fall under the churning process known as convection.

Convection as a process was first deduced by the eccentric Count von Rumford at the end of the eighteenth century. Sixty years later, an English vicar named Osmond Fisher presciently suggested that the Earth's interior might well be fluid enough for the contents to move about, but that idea took a very long time to gain support.

In about 1970, when geophysicists realized just how much turmoil was going on down there, it came as a considerable shock. As Shawna Vogel put it in the book *Naked Earth: The New Geophysics*: 'It was as if scientists had spent decades figuring out the layers of the Earth's atmosphere – troposphere, stratosphere and so forth – and then had suddenly found out about wind.'

How deep the convection process goes has been a matter of controversy ever since. Some say it begins 650 kilometres down, others more than 3,000 kilometres below us. The problem, as James Trefil has observed, is that 'there are two sets of data, from two different disciplines, that cannot be reconciled.' Geochemists say that certain elements on the planet's surface cannot have come from the upper mantle, but must have come from deeper within the Earth. Therefore, the materials in the upper and lower mantle must at least occasionally mix. Seismologists insist that there is no evidence to support such a thesis.

So all that can be said is that at some slightly indeterminate point as we head towards the centre of the Earth we leave the asthenosphere and plunge into pure mantle. Considering that it accounts for 82 per cent of the Earth's volume and 65 per cent of its mass, the mantle doesn't attract a great deal of attention, largely because the things that interest earth scientists and general readers alike happen either deeper down (as with magnetism) or nearer the surface (as with earthquakes). We know that to a depth of about 150 kilometres the mantle consists predominantly of a type of rock known as peridotite, but what fills the next 2,650 kilometres is uncertain. According to a *Nature* report, it seems not to be peridotite. More than this we do not know.

Beneath the mantle are the two cores, a solid inner core and a liquid outer one. Needless to say, our understanding of the nature of these cores is indirect, but scientists can make some reasonable assumptions. They know that the pressures at the centre of the Earth are sufficiently high – something over 3 million times those found at the surface – to turn any rock there solid. They also know from the Earth's history (among other clues) that the inner core is very good at retaining its heat. Although it is little more than a guess, it is thought that in over 4 billion years the temperature at the core has fallen by no more than 110 degrees centigrade. No one knows exactly how hot the Earth's core is, but estimates range from something over 4,000 degrees to over 7,000 degrees centigrade – about as hot as the surface of the Sun.

The outer core is in many ways even less well understood, though everyone is in agreement that it is fluid and that it is the seat of magnetism. The theory was put forward by E. C. Bullard of Cambridge University in 1949 that this fluid part of the Earth's core revolves in a way that makes it, in effect, an electrical motor, creating the Earth's magnetic field. The assumption is that the convecting fluids in the Earth act somehow like the currents in wires. Exactly what happens isn't known, but it is felt pretty certain that it is connected with the core spinning and with its being liquid. Bodies that don't have a liquid core – the Moon and Mars, for instance – don't have magnetism.

We know that the Earth's magnetic field changes in power from time to time: during the age of the dinosaurs, it was up to three times as strong as it is now. We also know that it reverses itself every 500,000 years or so on average, though that average hides a huge degree of unpredictability. The last reversal was about 750,000 years ago. Sometimes it stays put for millions of years – 37 million years appears to be the longest stretch – and at other times it has reversed after as little as 20,000 years. Altogether, in the last 100 million years it has reversed itself about 200 times, and we don't have any real idea why. This has been called 'the greatest unanswered question in the geological sciences'.

We may be going through a reversal now. The Earth's magnetic field has diminished by perhaps as much as 6 per cent in the last century alone. Any diminution in magnetism is likely to be bad news, because magnetism, apart from holding notes to refrigerators and keeping our compasses pointing the right way, plays a vital role in keeping us alive. Space is full of dangerous cosmic rays which, in the absence of magnetic protection, would tear through our bodies, leaving much of our DNA in useless shreds. When the magnetic field is working, these rays are safely herded away from the Earth's surface and into two zones in near space called the Van Allen belts. They also interact with particles in the upper atmosphere to create the bewitching veils of light known as the auroras.

A big part of the reason for our ignorance is that traditionally there has been little effort to coordinate what's happening on top of the Earth with what's going on inside it. According to Shawna Vogel: 'Geologists and geophysicists rarely go to the same meetings or collaborate on the same problems.'

Perhaps nothing better demonstrates our inadequate grasp of the dynamics of the Earth's interior than how badly we are caught out when it plays up, and it would be hard to come up with a more salutary reminder of the limitations of our understanding than the eruption of Mount St Helens in Washington State in 1980.

At that time, the lower forty-eight states of the Union had not seen a volcanic eruption for over sixty-five years. Therefore, most of the government volcanologists called in to monitor and forecast St Helens' behaviour had seen only Hawaiian volcanoes in action, and they, it turned out, were not the same thing at all.

St Helens started its ominous rumblings on 20 March. Within a week it was erupting magma, albeit in modest amounts, up to 100 times a day, and being constantly shaken with earthquakes. People were evacuated to what was assumed to be a safe distance of 13 kilometres. As the mountain's rumblings grew, St Helens became a tourist attraction for the world. Newspapers gave daily reports on the best places to get a view. Television crews repeatedly flew in helicopters to the summit and people were even seen clambering over the mountain. On one day, more than seventy copters and light aircraft circled the peak. But as the days passed and the rumblings failed to develop into anything dramatic, people grew restless and the view became general that the volcano wasn't going to blow after all.

On 19 April the northern flank of the mountain began to bulge conspicuously. Remarkably, no one in a position of responsibility saw that this strongly signalled a lateral blast. The seismologists resolutely based their conclusions on the behaviour of Hawaiian volcanoes, which don't blow out sideways. Almost the only person who believed that something really bad might happen was Jack Hyde, a geology professor at a community college in Tacoma. He pointed out that St Helens didn't have an open vent, as Hawaiian volcanoes have, so any pressure building up inside was bound to be released dramatically and probably catastrophically. However, Hyde was not part of the official team and his observations attracted little notice.

At 8.32 a.m. on a Sunday morning, 18 May, the north side of the volcano collapsed, sending an enormous avalanche of dirt and rock rushing down the mountain slope at nearly 150 miles an hour. It was the biggest landslide in human history and carried enough material to bury the whole of Manhattan to a depth of 120 metres. A minute later, its flank severely weakened, St Helens exploded with the force of 25,000 Hiroshima-sized atomic bombs, shooting out a murderous hot cloud at up to 650 miles an hour – much too fast, clearly, for anyone nearby to outrace it.* Many people who were thought to be in safe areas, often far out of sight of the volcano, were overtaken. Fifty-seven people were killed. Twenty-three of the bodies were never found. The toll would have been much higher had it not been a Sunday. On any weekday, many lumber workers

---

* Estimates of the forcefulness of the St Helens blast vary greatly, from about 1,500 Hiroshima-bomb equivalents to 27,000. The figure here is the one used by the US government.

would have been working within the death zone. As it was, people were killed 30 kilometres away.

The luckiest person on that day was a graduate student named Harry Glicken. He had been manning an observation post 9 kilometres from the mountain, but he had a college placement interview on 18 May in California, and so had left the site the day before the eruption. His place was taken by David Johnston. Johnston was the first to report the volcano exploding; moments later he was dead. His body was never found. Glicken's luck, alas, was temporary. Eleven years later he was one of forty-three scientists and journalists fatally caught up in a lethal outpouring of superheated ash, gases and molten rock — what is known as a pyroclastic flow — at Mount Unzen in Japan when yet another volcano was catastrophically misread.

Volcanologists may or may not be the worst scientists in the world at making predictions, but they are without question the worst in the world at realizing how bad their predictions are. Less than two years after the Unzen catastrophe, another group of volcano-watchers, led by Stanley Williams of the University of Arizona, descended into the rim of an active volcano called Galeras in Colombia. Despite the deaths of recent years, only two of the sixteen members of Williams's party wore safety helmets or other protective gear. The volcano erupted, killing six of the scientists, along with three tourists who had followed them, and seriously injuring several others, including Williams himself.

But to return to Washington. Mount St Helens lost 400 metres of peak, and 600 square kilometres of forest were devastated. Enough trees to build 150,000 homes (or 300,000 according to some reports) were blown away. The damage was placed at $2.7 billion. A giant column of smoke and ash rose to a height of 18,000 metres in less than ten minutes. An airliner some 48 kilometres away reported being pelted with rocks.

Ninety minutes after the blast, ash began to rain down on

THE FIRE BELOW

Yakima, Washington, a community of 50,000 people about 130 kilometres away. As you would expect, the ash turned day to night and got into everything, clogging motors, generators and electrical switching equipment, choking pedestrians, blocking filtration systems and generally bringing things to a halt. The airport shut down and highways in and out of the city were closed.

All this was happening, you will note, just downwind of a volcano that had been rumbling menacingly for two months. Yet Yakima had no volcano emergency procedures. The city's emergency broadcast system, which was supposed to swing into action during a crisis, did not go on the air because 'the Sunday-morning staff did not know how to operate the equipment'. For three days Yakima was paralysed and cut off from the world, its airport closed, its approach roads impassable. Altogether the city received just over 1.5 centimetres of ash after the eruption of Mount St Helens. Now bear that in mind, please, as we consider what a Yellowstone blast would do.

15

DANGEROUS BEAUTY

In the 1960s, while studying the volcanic history of Yellowstone National Park, Bob Christiansen of the US Geological Survey became puzzled about something that, oddly, had not troubled anyone before: he couldn't find the park's volcano. It had been known for a long time that Yellowstone was volcanic in nature – that's what accounted for all its geysers and other steamy features – and the one thing about volcanoes is that they are generally pretty conspicuous. But Christiansen couldn't find the Yellowstone volcano anywhere. In particular, what he couldn't find was a structure known as a caldera.

Most of us, when we think of volcanoes, think of the classic cone shape of a Fuji or a Kilimanjaro, which is created when erupting magma accumulates in a symmetrical mound. These can form remarkably quickly. In 1943, at Paricutín in Mexico, a farmer was startled to see smoke rising from a patch on his land. In one week he was the bemused owner of a cone 152 metres high. Within two years it had topped out at almost 430 metres and was more than 800 metres across. Altogether there are some 10,000 of these intrusively visible volcanoes on Earth, all but a few hundred of them extinct. But there is a second, less celebrated type of volcano that doesn't involve mountain-building. These are volcanoes so explosive that they burst open in a single mighty rupture, leaving behind a vast subsided pit, the caldera (from a Latin word for cauldron). Yellowstone obviously was of this second type, but Christiansen couldn't find the caldera anywhere.

By coincidence, just at this time NASA decided to test some new high-altitude cameras by taking photographs of Yellowstone, copies of which a thoughtful official passed on to the park authorities on the assumption that they might make a nice display for one of the visitor centres. As soon as Christiansen saw the photos he realized why he had failed to spot the caldera: virtually the whole park – 9,000 square kilometres – was caldera. The explosion had left a crater nearly 65 kilometres across – much too huge to be perceived from anywhere at ground level. At some time in the past Yellowstone must have blown up with a violence far beyond the scale of anything known to humans.

Yellowstone, it turns out, is a supervolcano. It sits on top of an enormous hot spot, a reservoir of molten rock that begins at least 200 kilometres down in the Earth and rises to near the surface, forming what is known as a superplume. The heat from the hot spot is what powers all of Yellowstone's vents, geysers, hot springs and popping mud pots. Beneath the surface is a magma chamber that is about 72 kilometres across – roughly the same dimensions as the park – and about 13 kilometres thick at its thickest point. Imagine a pile of TNT about the size of Rhode Island and reaching 13 kilometres into the sky, and you have some idea of what visitors to Yellowstone are shuffling around on top of. The pressure that such a pool of magma exerts on the crust above has lifted Yellowstone and its surrounding territory about half a kilometre higher than they would otherwise be. If it blew, the cataclysm is pretty well beyond imagining.

Superplumes of the type on which Yellowstone sits are rather like martini glasses – thin on the way up, but spreading out as they near the surface to create vast bowls of unstable magma. Some of these bowls can be up to 1,900 kilometres across. They don't always erupt explosively, but sometimes burst forth in a vast, continuous outpouring – a flood – of molten rock, as happened with the Deccan Traps in India 65 million years ago. These covered an area of over 500,000 square kilometres and probably contributed to the demise of the dinosaurs – they certainly didn't help – with their noxious outpourings of gases.

Such plumes are not all that rare. There are about thirty active ones on the Earth at the moment and they are responsible for many of the world's best-known islands and island chains – Iceland, Hawaii, the Azores, Canaries and Galápagos archipelagos, little Pitcairn in the middle of the South Pacific, and many others – but apart from Yellowstone they are all oceanic. No one knows how or why Yellowstone's ended up beneath a continental plate. Only two things are certain: that the crust at Yellowstone is thin and that the world beneath it is hot. But whether the crust is thin because of the hot spot or whether the hot spot is there because the crust is thin is a matter of heated (as it were) debate. The continental nature of the crust makes a huge difference to its eruptions. Whereas the other supervolcanoes tend to bubble away steadily and in a comparatively benign fashion, Yellowstone is capable of blowing explosively. It doesn't happen often, but when it does you want to stand well back.

Since its first known eruption 16.5 million years ago it has blown up about 100 times, but the most recent three events are the ones that get written about. The last eruption was 1,000 times as big as that of Mount St Helens; the one before that was 280 times as big, and the one before *that* was so big nobody knows exactly how big it was. It was at least 2,500 times as big as St Helens, but perhaps 8,000 times as monstrous.

We have absolutely nothing to compare it to. The biggest blast in recent times was that of Krakatau in Indonesia in August 1883, which made a bang that reverberated around the world for nine days, and made water slosh as far away as the English Channel. But if you imagine the volume of ejected material from Krakatau as being about the size of a golf ball, then that from the biggest of the Yellowstone blasts would be the size of a sphere you could just about hide behind. On this scale, the Mount St Helens eruption would be no more than a pea.

The Yellowstone eruption of 2 million years ago put out enough ash to bury New York State to a depth of 20 metres or California to a depth of 6 metres. This was the ash that made Mike Voorhies' fossil beds in eastern Nebraska. That blast occurred in what is now Idaho, but over millions of years, at a rate of about 2.5 centimetres a year, the Earth's crust has travelled over it, so that today it is directly under northwest Wyoming. (The hot spot itself stays in one place, like an acetylene torch aimed at a ceiling.) In its wake it leaves the sort of rich volcanic plains that are ideal for growing potatoes, as Idaho's farmers long ago discovered. In another 2 million years, geologists like to joke, Yellowstone will be producing French fries for McDonald's and the people of Billings, Montana, will be stepping around geysers.

The ash fall from the last Yellowstone eruption covered all or parts of nineteen western states (plus parts of Canada and Mexico) – nearly the whole of the United States west of the Mississippi. This, bear in mind, is the breadbasket of America, an area that produces roughly half the world's cereals. And ash, it is worth remembering, is not like a big snowfall that will melt in the spring. If you wanted to grow crops again, you would have to find some place to put all the ash. It took thousands of workers eight months to clear 1.8 billion tonnes of debris from the 6.5 hectares of the World Trade Center site in New York. Imagine what it would take to clear Kansas.

And that's not even to consider the climatic consequences. The last supervolcano eruption on Earth was at Toba, in northern Sumatra, 74,000 years ago. No one knows quite how big it was, but it was a whopper. Greenland ice cores show that the Toba blast was followed by at least six years of 'volcanic winter', and goodness knows how many poor growing seasons after that. The event, it is thought, may have carried humans right to the brink of extinction, and kept them in that perilous state for 20,000 years. That is, needless to say, a long time to spend recovering from a single volcanic blast.

All this was hypothetically interesting until 1973, when an odd occurrence made it suddenly momentous: water in Yellowstone Lake, in the heart of the park, began to run over the banks at the lake's southern end, flooding a meadow, while at the opposite end of the lake the water mysteriously flowed away. Geologists did a hasty survey and discovered that a large area of the park had developed an ominous bulge. This was lifting up one end of the lake and causing the water to run out at the other, as would happen if you lifted one side of a child's paddling pool. By 1984, the whole central region of the park – over 100 square kilometres – was more than a metre higher than it had been in 1924, when the park was last formally surveyed. Then, in 1985, the central part of the park subsided by 20 centimetres. The geologists realized that only one thing could cause this – a restless magma chamber. Yellowstone wasn't the site of an ancient supervolcano; it was the site of an active one. It was also at about this time that they were able to work out that Yellowstone averaged one massive eruption every 600,000 years. The last one was 630,000 years ago. Yellowstone, it would seem, is due.

'It may not feel like it, but you're standing on the largest active volcano in the world,' Paul Doss, Yellowstone National Park geologist, told me after climbing off an enormous Harley-Davidson motorcycle outside the park headquarters at Mammoth Hot Springs early on a lovely morning in June 2001. A native of Indiana, Doss was then (and still is) an amiable, soft-spoken, extremely thoughtful man who looked nothing like a National Park Service employee. He had a greying beard and hair tied back in a long ponytail. A small sapphire stud graced one ear. A slight paunch strained against his crisp Park Service uniform. He looked more like a blues musician than a government employee. In fact, he is a blues musician (harmonica). But his passion in life is geology. 'And I've got the best place in the world to do it,' he told me as we set off in a bouncy, battered four-wheel-drive vehicle in the general direction of Old Faithful after he had invited me to accompany him for a day in the field.

Yellowstone, I hardly need point out, is sensationally beautiful, with plump, stately mountains, bison-specked meadows, tumbling streams, a sky-blue lake, wildlife beyond counting. 'It really doesn't get any better than this if you're a geologist,' Doss said. 'You've got rocks up at Beartooth Gap that are nearly three billion years old – three-quarters of the way back to the Earth's beginning – and then you've got mineral springs here' – he pointed at the sulphurous hot springs from which Mammoth takes its title – 'where you can see rocks as they are being born. And in between there's everything you could possibly imagine. I've never been any place where geology is more evident – or prettier.'

As we drove, he pointed out a distant gap in a range of mountains to the west known as the Gallatins. 'That gap is sixty or maybe seventy miles across. For a long time nobody could understand why that gap was there, and then Bob Christiansen realized that it had to be because the mountains were just blown away. When you've got sixty miles of mountains just obliterated, you know you're dealing with something pretty potent. It took Christiansen six years to figure it all out.'

I asked him what caused Yellowstone to blow when it did.

'Nobody knows. Volcanoes are strange things. Vesuvius, in Italy, was active for three hundred years until an eruption in 1944 and then it just stopped. It's been silent ever since. Some volcanologists think that it is recharging in a big way, which is a little worrying because three million people live on or around it. But nobody knows.'

'And how much warning would you get if Yellowstone was going to go?'

He shrugged. 'Nobody was around the last time it blew, so nobody knows what the warning signs are. Probably you would have swarms of earthquakes and some surface uplift and possibly some changes in the patterns of behaviour of the geysers and steam vents.'

The trouble, he explained, is that nearly all the things that would constitute warning signs already exist in some measure at Yellowstone. 'Earthquakes are generally a precursor of volcanic eruptions, but the park already has lots of earthquakes – as many as three thousand a year. Most of them are too small to be felt, but they are earthquakes nonetheless.'

A change in the pattern of geyser eruptions might also be taken as a clue, he said, but these too vary unpredictably. Once the most famous geyser in the park was Excelsior Geyser. It used to erupt regularly and spectacularly to heights of 100 metres, but in 1890 it just stopped. Then in 1985 it erupted again, though only to a height of 25 metres. Steamboat Geyser is the biggest geyser in the world when it blows, shooting water 120 metres into the air, but the intervals between its eruptions have ranged from as little as four days to almost fifty years. 'If it blew today and again next week, that wouldn't tell us anything at all about what it might do the following week or the week after or twenty years from now,' Doss said. 'The whole park is so volatile that it's essentially impossible to draw conclusions from almost anything that happens.'

Today, twenty-three years later, Paul Doss still has an ear stud and a ponytail, still rides a monster Harley and plays blues harmonica, but he does so some 1,500 miles to the east in Evansville, Indiana, where he was just about to retire as a professor of geology at the University of Southern Indiana when I caught up with him. I was surprised to learn that Doss had left the Park Service not long after our first meeting to return to academia. 'It was a tough decision, but my wife and I both felt we needed to be closer to our ageing parents, and I realized I missed teaching.'

But Doss still maintains close links with Yellowstone. He conducts summer seminars in the park and works with a charity that gives troubled military veterans a few days of nature therapy – 'talking and hiking and just enjoying the great outdoors', as he puts it.

Today Doss is a lot more sanguine about the dangers of a catastrophic eruption. 'The technology we have now is vastly improved from what it was twenty years ago,' he told me. 'Today there are instruments deployed all around the park which would easily pick up any unusual movements of magma.'

He doesn't see any reason to fear a catastrophic eruption. 'The 600,000-year interval that always gets talked about is based on only the three most recent eruptions, not the ones that went before. That doesn't make a trend. It's not like it's a ticking clock. In fact, nearly all the variables we measure now point to a magmatic system that is cooling and becoming less likely to erupt in a big way. If anything happened it all, it is much more likely to be not an explosion but basaltic lava flows like they get in Hawaii and Iceland. That would be disruptive, for sure, but it wouldn't be anything like as lethal.'

Yellowstone has had lots of those over the ages, but none at all for about 70,000 years. The Yellowstone authorities are at pains to stress that the risks are tiny – and make it clear that they find the scaremongering overstated. 'Sometimes,' their website notes, 'people spread misinformation even though they think they know the facts.' The USGS estimates that the odds of a Yellowstone mega-eruption in any given year are just 0.00014 per cent. A survey in 2022 using new techniques found that the reservoir of molten magma on which Yellowstone sits is considerably bigger than previously thought. That isn't to say that it is more volatile than previously, just that there is more of it than realized. It is still thought to be well below the critical threshold at which volcanoes erupt. Specifically, between

16 and 20 per cent of the magma is molten; a level of about 35 per cent is thought necessary to trigger an eruption.

In the meantime, there are plenty of other dangers in and around Yellowstone, as was made devastatingly evident on the night of 17 August 1959 at a place called Hebgen Lake just outside the park. At twenty minutes to midnight on that date, Hebgen Lake suffered a catastrophic quake. It was magnitude 7.5, not vast as earthquakes go, but so abrupt and wrenching that it collapsed an entire mountainside. It was the height of the summer season, though fortunately not so many people went to Yellowstone in those days as now. Eighty million tonnes of rock, moving at more than 100 miles an hour, just fell off the mountain, travelling with such force and momentum that the leading edge of the landslide ran 120 metres up a mountain on the other side of the valley. Along its path lay part of the Rock Creek Campground. Twenty-eight campers were killed, nineteen of them buried too deep ever to be found again. The devastation was swift but heartbreakingly fickle. Three brothers, sleeping in one tent, were spared. Their parents, sleeping in another tent beside them, were swept away and never seen again.

Hydrothermal explosions are also a significant risk. They can happen any time, pretty much anywhere and without any predictability. 'You know, by design we funnel visitors into thermal basins,' Doss told me after we had watched Old Faithful blow. 'It's what they come to see. Did you know there are more geysers and hot springs at Yellowstone than in all the rest of the world combined – some 10,000 altogether – and the number is growing all the time?'

His point was dramatically demonstrated in the summer of 2024, when just before 10 a.m. on a calm July morning a massive explosion rent the air and sent a tower of steam, mud and rocks – much wider, higher and darker than a typical Yellowstone geyser – 100 feet or more into the air. The explosion wrecked a stretch of boardwalk, pelted the surrounding area with rocks the size of baseballs and covered several tourists from head to toe in muddy debris, but amazingly no serious injuries were reported.

'It was a miracle that no one was hurt,' Doss reflected, 'but actually these things happen frequently. They just don't often happen right beside a boardwalk. Usually they're in the back country.

'In the very ancient past,' he went on, 'there have been explosions that have made holes a mile across. And nobody can tell you where or when that might happen again. You just have to hope that you're not standing there when it does.'

Big rockfalls are also a danger. There was a big one at Gardiner Canyon in 1999, but again fortunately no one was hurt. Late in the afternoon, Doss and I stopped at a place where there was a rock overhang poised above a busy park road. Cracks were clearly visible. 'It could go at any time,' Doss said thoughtfully.

'You're kidding,' I said. There wasn't a moment when there weren't two cars passing beneath it, all filled with, in the most literal sense, happy campers.

'Oh, it's not likely,' he added. 'I'm just saying it *could*. Equally, it could stay like that for decades. There's just no telling. People have to accept that there is risk in coming here. That's all there is to it.'

As we walked back to his vehicle to head back to Mammoth Hot Springs, Doss added: 'But the thing is, most of the time bad things don't happen. Rocks don't fall. Earthquakes don't occur. New vents don't suddenly open up. For all the instability, it's mostly remarkably and amazingly tranquil.'

'Like the Earth itself,' I remarked.

'Precisely,' he agreed.

The risks at Yellowstone apply to park employees as much as to visitors. Doss had got a horrific sense of that in his first week on the job five years before we first met. Late one night, three young summer employees were engaging in an illicit activity known as 'hot-potting' – swimming or basking in warm pools. Though the park, for obvious reasons, doesn't publicize it, not all the pools at Yellowstone are dangerously hot. Some are extremely agreeable to lie in, and it was the habit of some of the summer employees to have a dip late at night, even though it was against the rules to do so. Foolishly, the threesome had failed to take a torch, which was extremely dangerous because much of the soil around the warm pools is crusty and thin and one can easily fall through into a scalding vent below. In any case, as they made their way back to their dorm they came across a stream that they had had to leap over earlier. They backed up a few paces, linked arms and, on the count of three, took a running jump. In fact, it wasn't the stream at all. It was a very hot pool. In the dark they had lost their bearings. One of the trio died a few hours later. The other two were lucky to survive.

Much more recently, in 2022, a visitor from California fell or jumped into a boiling pool at West Thumb and was almost completely dissolved. His death went unnoticed until part of a foot was found floating on the surface more than two weeks later. Despite explicit warning signs, some people wander off the boardwalk or footpaths to get a closer look, unaware that the crusty soil is thin and can easily give way, leaving them ankle-deep or worse in scalding mud. It's actually something of a miracle that only twenty-two people are known to have been killed by Yellowstone's boiling waters since the park was created in 1872.

I thought about this during a revisit to the park in the autumn of 2024, when I stopped at a place called Emerald Pool in the Upper Geyser Basin, not far from the site of the sudden dramatic eruption of a couple of months earlier. Emerald Pool is not an imposing sight. It is only 20 or 30 feet across and isn't even emerald, just mostly a murky orange, but it has history.

In 1965, a husband and wife team of biologists named Thomas and Louise Brock, while on a summer study trip, had done a crazy thing. They had scooped up some of the yellowy-brown scum that rimmed the pool and examined it for life. To their, and eventually the wider world's, deep surprise, it was full of living microbes. They had found the world's first extremophiles – organisms that could live in water that had previously been assumed to be much too hot or acid or choked with toxic chemicals to bear life. Emerald Pool, remarkably, was all these things, yet at least two types of unicellular organism, *Sulpholobus acidocaldarius* and *Thermophilus aquaticus*, as they became known, found it congenial. It had always been supposed that nothing could survive above temperatures of 50 degrees centigrade, but here were organisms basking in rank, acidic waters nearly twice that hot.*

For almost twenty years, one of the Brocks' two new bacteria, *Thermophilus aquaticus*, remained a laboratory curiosity – until a scientist in California named Kary B. Mullis realized that heat-resistant enzymes within it could be used to create a bit of chemical wizardry known as a polymerase chain reaction, which allows scientists to generate lots of DNA from very small amounts – as little as a single molecule in ideal conditions. It's a kind of genetic photocopying, and it became the basis for all subsequent genetic science, from academic studies to police forensic work – but above all, in the popular consciousness, for the laboratory tests that are the basic universal test for Covid infection. The invention won Mullis the Nobel Prize in Chemistry in 1993. He died in 2019.

Meanwhile, scientists were finding even hardier microbes, now known as hyperthermophiles, which demand temperatures of 80 degrees centigrade or more. The most heat-tolerant organism found so far is *Pyrolobus fumarii*, which dwells in

---

* Since then, many more heat-resistant microbes have been found throughout Yellowstone National Park, but even so it is thought that 99 per cent of the park's extremophiles have yet to be described.

the walls of ocean vents where the temperature can reach 113 degrees centigrade.

Still other microbes have been found living comfortably in environments that would seem impossibly challenging for life – in pools of pure sulphuric acid, in the crushing depths of the Mariana Trench, almost 7 kilometres down in the Earth's crust, where they sustain themselves on hydrogen or methane and are so metabolically sluggish that they are thought to live for thousands of years. The hardiest organism of all appears to be *Deinococcus radiodurans*, which has been found surviving on the inside walls of nuclear reactors and on the outer surfaces of the International Space Station. It would seem to be almost indestructible.

Life, it turns out, is infinitely more resourceful and adaptable than anyone had ever supposed. This is a very good thing for, as we are about to see, we live in a world that doesn't altogether seem to want us here.

V

LIFE ITSELF

16

LONELY PLANET

It isn't easy being an organism. In the whole universe, as far as we yet know, there is only one place, an inconspicuous outpost of the Milky Way called Earth, that will sustain you, and even it can be pretty grudging.

From the bottom of the deepest ocean trench to the top of the highest mountain, the zone that covers nearly the whole of known life is only around 20 kilometres thick – not much when set against the roominess of the cosmos at large.

For humans it is even worse because we happen to belong to the portion of living things that took the rash but venturesome decision 400 million years ago to crawl out of the seas and become land-based and oxygen-breathing. In consequence, no less than 99.5 per cent of the world's habitable space by volume, according to one estimate, is fundamentally – in practical terms completely – off-limits to us.

It isn't simply that we can't breathe in water, but that we couldn't bear the pressures. Because water is about 1,300 times heavier than air, pressures rise swiftly as you descend – by the equivalent of one atmosphere for every 10 metres of depth. On land, if you rise to the top of a 150-metre eminence – Cologne Cathedral or the Washington Monument, say – the change in pressure is so slight as to be indiscernible. At the same depth under water, however, your veins would collapse and your lungs would compress to the approximate dimensions of a Coke can.

Amazingly, people do voluntarily dive to such depths, and beyond, and appear to find it exhilarating. But the risks are formidable. At least eleven people have died in attempts to break deep-dive records. The deepest scuba dive yet recorded was by Ahmed Gabr, an Egyptian, who reached a depth of 332.35 metres in 2014 off the coast of Egypt. For a free dive – i.e., on a single breath – the deepest descent was achieved in 2012 by Herbert Nitsch, of Austria, who reached a stunning depth of 249.5 metres in Greece, but at a terrible personal price. On the ascent, he blacked out from narcosis and suffered a series of paralysing strokes in the days following. Over a long period, he had to re-learn how to walk and talk, and more than ten years afterwards it is reported that he still has difficulty maintaining his balance. Yet he continues to free dive. Nitsch can hold his breath for over nine minutes, which is thought also to be a world record. His 2012 descent didn't count, unfortunately, because he passed out and had to be rescued just 10 metres from the surface.

Nearly everyone assumes that the human body would crumple under the immense pressures of the deep ocean. In fact, this appears not to be the case. Because we are made largely of water ourselves, and water is 'virtually incompressible', in the words of Frances Ashcroft of Oxford University, 'the body remains at the same pressure as the surrounding water, and is not crushed at depth'. It is the gases inside your body, particularly in the lungs, that cause the trouble. These do compress, though at what point the compression becomes fatal is not known. Until quite recently it was thought that anyone diving to 100 metres or so would die painfully as his or her lungs imploded or chest wall collapsed, but the free divers have repeatedly proved otherwise. It appears, according to Ashcroft, that 'humans may be more like whales and dolphins than had been expected'.

Plenty else can go wrong, however. In the days of diving suits – the sort that were connected to the surface by long hoses – divers sometimes experienced a dreaded phenomenon known as 'the squeeze'. This occurred when the surface pumps failed, leading to a catastrophic loss of pressure in the suit. The air would leave the suit with such violence that the hapless diver would be, all too literally, sucked up into the helmet and hosepipe. When hauled to the surface, 'all that is left in the suit are his bones and some rags of flesh', the biologist J. B. S. Haldane wrote in 1947, adding for the benefit of doubters, 'This has happened.'

(Incidentally, the original diving helmet, designed in 1823 by an Englishman named Charles Deane, was intended not for diving but for fire-fighting. It was called a 'smoke helmet', but, being made of metal, it was hot and cumbersome; as Deane soon discovered, fire-fighters had no particular eagerness to enter burning structures in any form of attire, but most especially not in something that heated up like a kettle and made them clumsy into the bargain. In an attempt to save his investment, Deane tried it under water and found it was ideal for salvage work.)

The real terror of the deep, however, is the bends – not so much because they are unpleasant, though of course they are, as because they are so much more likely. The air we breathe is 80 per cent nitrogen. Put the human body under pressure, and that nitrogen is transformed into tiny bubbles that migrate into the blood and tissues. If the pressure is changed too rapidly – as with a too-quick ascent by a diver – the bubbles trapped within the body will begin to fizz in exactly the manner of a freshly opened bottle of champagne, clogging tiny blood vessels, depriving cells of oxygen and causing pain so excruciating that sufferers are prone to bend double in agony – hence 'the bends'.

The bends have been an occupational hazard for sponge and pearl divers since time immemorial, but didn't attract much attention in the Western world until the nineteenth century, and then it was among people who didn't get wet at all (or at least, not very wet and not generally much above the ankles). They were caisson workers. Caissons were enclosed dry chambers built on river beds to facilitate the construction of bridge piers. They were filled with compressed air, and often when the workers emerged after an extended period of working under this artificial pressure they experienced mild symptoms like tingling or itchy skin. But an unpredictable few felt more insistent pain in the joints and occasionally collapsed in agony, sometimes never to get up again.

It was all most puzzling. Sometimes the workers would go to bed feeling fine, but wake up paralysed. Sometimes they wouldn't wake up at all. Ashcroft relates a story concerning the directors of a new tunnel under the Thames who held a celebratory banquet as the tunnel neared completion. To their consternation, their champagne failed to fizz when uncorked in the compressed air of the tunnel. However, when at length they emerged into the fresh air of a London evening, the bubbles sprang instantly to fizziness, memorably enlivening the digestive process.

Apart from avoiding high-pressure environments altogether, only two strategies are reliably successful against the bends. The first is to suffer only a very short exposure to the changes in pressure. That is why the free divers, if they are careful, can descend to depths of 150 metres or more without ill effect. They don't stay down long enough for the nitrogen in their system to dissolve into their tissues. The other solution is to ascend by careful stages. This allows the little bubbles of nitrogen to dissipate harmlessly.

A great deal of what we know about surviving at extremes is owed to the extraordinary father and son team of John Scott and J. B. S. Haldane. Even by the demanding standards of British intellectuals, the Haldanes were outstandingly eccentric. The senior Haldane was born in 1860 to an aristocratic Scottish family (his brother was Viscount Haldane), but spent most of his career in comparative modesty as a professor of physiology at Oxford. He was famously absent-minded. Once, after his wife had sent him upstairs to change for a dinner party, he failed to return and was discovered asleep in bed in his pyjamas. When roused, Haldane explained that he had found himself disrobing and assumed it was bedtime. His idea of a holiday was to travel to Cornwall to study hookworm in miners. Aldous Huxley, the novelist grandson of T. H. Huxley, who lived with the Haldanes for a time, parodied him, a touch mercilessly, as the scientist Edward Tantamount in the novel *Point Counter Point*.

Haldane's gift to diving was to work out the rest intervals necessary to manage an ascent from the depths without getting the bends, but his interests ranged across the whole of physiology, from studying altitude sickness in climbers to the problems of heatstroke in desert regions. He had a particular interest in the effects of toxic gases on the human body. To understand more exactly how carbon monoxide leaks killed miners, he methodically poisoned himself, carefully taking and measuring his own blood samples the while. He quit only when he was on the verge of losing all muscle control and his blood saturation level had reached 56 per cent – a level, as Trevor Norton notes in his entertaining history of diving, *Stars Beneath the Sea*, only fractionally removed from nearly certain lethality.

Haldane's son Jack, known to posterity as J. B. S., was a remarkable prodigy who took an interest in his father's work almost from infancy. At the age of three he was overheard demanding peevishly of his father, 'But is it oxyhaemoglobin or carboxyhaemoglobin?' Throughout his youth, the young Haldane helped his father with experiments. By the time he was a teenager, the two often tested gases and gas masks together, taking it in turns to see how long it took them to pass out.

Though J. B. S. Haldane never took a degree in science (he studied Classics at Oxford), he became a brilliant scientist in his own right, mostly working for the government at Cambridge. The biologist Peter Medawar, who spent his life around mental Olympians, called him 'the cleverest man I ever knew'. Huxley parodied the younger Haldane too, in his novel *Antic Hay*, but also used his ideas on genetic manipulation of humans as the basis for the plot of *Brave New World*. Among many other achievements, Haldane played a central role in marrying Darwinian principles of evolution to the genetic work of Gregor Mendel to produce what is known to geneticists as the Modern Synthesis.

Perhaps uniquely among human beings, the younger Haldane found the First World War 'a very enjoyable experience' and freely admitted that he 'enjoyed the opportunity of killing people'. He was himself wounded twice. After the war he became a successful popularizer of science and wrote twenty-three books (as well as over 400 scientific papers). His books are still thoroughly readable and instructive, though not always easy to find. He also became an enthusiastic Marxist. It has been suggested, not altogether cynically, that this was out of a purely contrarian instinct and that if he had been born in the Soviet Union he would have been a passionate monarchist. At all events, most of his articles first appeared in the Communist *Daily Worker*.

Whereas his father's principal interests concerned miners and poisoning, the younger Haldane became obsessed with saving submariners and divers from the unpleasant consequences of their work. With Admiralty funding, he acquired a decompression chamber that he called the 'pressure pot'. This was a metal cylinder into which three people at a time could be sealed and subjected to tests of various types, all painful and nearly all dangerous. Volunteers might be required to sit in ice water while breathing 'aberrant atmosphere', or subjected to rapid changes of pressurization. In one experiment, Haldane himself simulated a dangerously hasty ascent to see what would happen. What happened was that the dental fillings in his teeth exploded. 'Almost every experiment', Norton writes, 'ended with someone having a seizure, bleeding or vomiting.' The chamber was virtually soundproof, so the only way for occupants to signal unhappiness or distress was to tap insistently on the chamber wall or to hold up notes to a small window.

On another occasion, while poisoning himself with elevated levels of oxygen, Haldane had a fit so severe that he crushed several vertebrae. Collapsed lungs were a routine hazard. Perforated eardrums were quite common, too; but, as Haldane reassuringly noted in one of his essays, 'the drum generally heals up; and if a hole remains in it, although one is somewhat deaf, one can blow tobacco smoke out of the ear in question, which is a social accomplishment'.

What was extraordinary about this was not that Haldane was willing to subject himself to such risk and discomfort in the pursuit of science, but that he had no trouble talking colleagues and loved ones into climbing into the chamber, too. Sent on a simulated descent, his wife once had a fit that lasted thirteen minutes. When at last she stopped bouncing across the floor, she was helped to her feet and sent home to cook dinner. Haldane happily employed whoever happened to be around, including on one memorable occasion a former Prime Minister of Spain, Juan Negrín. Dr Negrín complained afterwards of minor tingling and 'a curious velvety sensation on the lips' but otherwise seems to have escaped unharmed. He may have considered himself very lucky. A similar experiment with oxygen deprivation left Haldane without feeling in his buttocks and lower spine for six years.

Among Haldane's many specific preoccupations was nitrogen intoxication. For reasons that are still poorly understood, at depths beyond about 30 metres nitrogen becomes a powerful intoxicant. Under its influence divers had been known to offer their air hoses to passing fish or to decide to try to have a smoke break. It also produced wild mood swings. In one test, Haldane noted, the subject 'alternated between depression and elation, at one moment begging to be decompressed because he felt "bloody awful" and the next minute laughing and attempting to interfere with his colleague's dexterity test'. In order to measure the rate of deterioration in the subject, a scientist had to go into the chamber with the volunteer to conduct simple mathematical tests. But after a few minutes, as Haldane later recalled, 'the tester was usually as intoxicated as the testee, and often forgot to press the spindle of his stopwatch, or to take proper notes.' The cause of the inebriation is even now a mystery. It is thought that it may be the same thing that causes alcohol intoxication, but as no one knows for certain what causes *that*, we are none the wiser. At all events, without the greatest care, it is easy to get in trouble once you leave the surface world.

Which brings us back (well, nearly) to our earlier observation that the Earth is not the easiest place to be an organism, even if it is the only place. Of the small portion of the planet's surface that is dry enough to stand on, a surprisingly large amount is too hot or cold or steep or lofty or arid to be of much use to us. Partly, it must be conceded, this is our fault. In terms of adaptability, humans are pretty amazingly useless. Like most animals, we don't much like really hot places, but because we sweat so freely and easily succumb to strokes, we are especially vulnerable. In the worst circumstances – on foot without water in a hot desert – most people will grow delirious and keel over, possibly never to rise again, in no more than seven or eight hours. We are no less helpless in the face of cold. Like all mammals, humans are good at generating heat; but – because we are so nearly hairless – we are not good at keeping it. Even in quite mild weather half the calories you burn go to keep your body warm. Of course, we can counter these frailties to a large extent by employing clothing and shelter, but even so the portions of the Earth on which we are prepared or able to live are modest indeed: just 12 per cent of the total land area, and only 4 per cent of the whole surface if you include the seas.

Yet when you consider conditions elsewhere in the known universe, the wonder is not that we use so little of our planet but that we have managed to find a planet of which we can use even a bit. You have only to look at our own solar system – or, come to that, the Earth at certain periods in its own history – to appreciate that most places are much harsher and much less amenable to life than our mild, blue, watery globe.

So far (as of mid-2025), scientists have discovered about 6,000 Earth-like planets outside the solar system, out of the 10 billion trillion or so that are thought to be out there, so humans can hardly claim to speak with authority on the matter; but it appears that if you wish to have a planet suitable for life, you have to be just awfully lucky, and the more advanced the life, the luckier you have to be. Various observers have identified about two dozen particularly fortunate breaks we have had on Earth, but this is a flying survey so we'll distil them down to the principal four.

Excellent location. We are, to an almost uncanny degree, the right distance from the right sort of star, one that is big enough to radiate lots of energy, but not so big as to burn itself out swiftly. It is a curiosity of physics that the larger a star is, the more rapidly it burns. Had our sun been ten times as massive, it would have exhausted itself after 10 million years instead of 10 billion and we wouldn't be here now. We are also fortunate to orbit where we do. Too much nearer, and everything on Earth would have boiled away. Much further away, and everything would have frozen.

In 1978, an astrophysicist named Michael Hart made some calculations and concluded that the Earth would have been uninhabitable had it been just 1 per cent further from or 5 per cent closer to the Sun. That's not much, and in fact it wasn't enough. The figures have since been refined and made a little more generous – 5 per cent nearer and 15 per cent further are thought to be more accurate assessments for our zone of habitability – but that is still a narrow belt.*

To appreciate just how narrow, you have only to look at Venus. Venus is only 25 million miles closer to the Sun than we are – not a great distance even in terms of our own solar system. The Sun's warmth reaches it just two minutes before it touches us. In size and composition, Venus is very like the Earth, but the small difference in orbital distance made all the difference to how it turned out. It appears that during the early years of the solar system Venus was only slightly warmer than the Earth and probably had oceans. But those few degrees of extra warmth meant that Venus could not hold on to its surface water, with disastrous consequences for its climate.

As its water evaporated, the hydrogen atoms escaped into space and the oxygen atoms combined with carbon to form a dense atmosphere of the greenhouse gas carbon dioxide. Venus became stifling. Although people of my age will recall a time when astronomers hoped that Venus might harbour life beneath its padded clouds, possibly even a kind of tropical verdure, we now know that it is much too fierce an environment for any kind of life that we can reasonably conceive of. Its surface temperature is a roasting 470 degrees centigrade, which is hot enough to melt lead, and the atmospheric pressure at the surface is ninety times that of Earth, more than any human body could withstand. We lack the technology to make suits or even spaceships that would allow us to visit. Our knowledge of Venus's surface is based essentially on distant radar and ultraviolet imagery and some startled squawks from an unmanned Soviet probe that was dropped hopefully into the clouds in

---

* The discovery of extremophiles in the boiling mudpots of Yellowstone and of similar organisms elsewhere made scientists realize that actually life of a type could range much further than that – even perhaps beneath the icy skin of Pluto. What we are talking about here are the conditions that would produce reasonably complex surface creatures.

1972 and functioned for barely an hour before permanently shutting down.

So that's what happens when you move two light minutes closer to the Sun. Travel further out and the problem becomes not heat but cold, as Mars frigidly attests. It, too, was once a much more congenial place, but couldn't retain a usable atmosphere and turned into a frozen waste.

But just being the right distance from the Sun cannot be the whole story, for otherwise the Moon would be forested and fair, which patently it is not. For that you need to have:

The right kind of planet. I don't imagine even many geophysicists, when asked to count their blessings, would include living on a planet with a molten interior, but it's a pretty near-certainty that without all that magma swirling around beneath us we wouldn't be here now. Apart from much else, our lively interior created the outpourings of gas that helped to build an atmosphere and provided us with the magnetic field that shields us from cosmic radiation. It also gave us plate tectonics, which continually renews and rumples the surface. If the Earth were perfectly smooth, it would be covered everywhere with water to a depth of 4 kilometres. There might be life in that lonesome ocean, but there certainly wouldn't be football.

In addition to having a beneficial interior, we also have the right elements in the correct proportions. In the most literal way, we are made of the right stuff. This is so crucial to our well-being that we are going to discuss it more fully in a minute, but first we need to consider the two remaining factors, beginning with another one that is often overlooked:

We're a twin planet. Not many of us normally think of the Moon as a companion planet, but that is, in effect, what it is. Most moons are tiny in relation to their master planet. The Martian satellites of Phobos and Deimos, for instance, are only about 10 kilometres in diameter. Our Moon, however, is more than a quarter the diameter of Earth, which makes ours the only planet in the solar system with a sizeable moon in comparison to itself. Without the Moon's steadying influence, the Earth would wobble like a dying top, with goodness knows what consequences for climate and weather. The Moon's steady gravitational influence keeps the Earth spinning at the right speed and angle to provide the sort of stability necessary for the long and successful development of life. This won't go on for ever. The Moon is slipping from our grasp at a rate of about 4 centimetres a year. In another 2 billion years it will have receded so far that it won't keep us steady and we will have to come up with some other solution, but in the meantime you should think of it as much more than just a pleasant feature in the night sky.

For a long time, astronomers assumed that either the Moon and the Earth formed together or that the Earth captured the Moon as it drifted by. We now believe, as you will recall from an earlier chapter, that about 4.4 billion years ago a Mars-sized object slammed into Earth, blowing out enough material to create the Moon from the debris. This was obviously a very good thing for us – but especially so as it happened such a long time ago. If it had happened in 1896 or last Wednesday, clearly we wouldn't be nearly so pleased about it. Which brings us to our fourth and in many ways most crucial consideration:

Timing. The universe is an amazingly fickle and eventful place and our existence within it is a wonder. If a long and unimaginably complex sequence of events stretching back 4.6 billion years or so hadn't played out in a particular manner at particular times – if, to take just one obvious instance, the dinosaurs hadn't been wiped out by a meteor when they were – you might well be a few centimetres long, with whiskers and a tail, and reading this in a burrow.

We don't really know, because we have nothing else to which we can compare our own existence; but it seems evident that if you wish to end up as a moderately advanced, thinking society, you need to be at the right end of a very long chain of outcomes involving reasonable periods of stability interspersed with just the right amount of stress and challenge (ice ages appear to be especially helpful in this regard) and marked by a total absence of real cataclysm. As we shall see in the pages ahead, we are very lucky to find ourselves in that position.

And on that note, let us now turn briefly to the elements that made us.

There are ninety-four naturally occurring elements on Earth, plus a further twenty-four that have been created artificially, but some of these we can immediately put to one side – as, in fact, chemists themselves tend to do. Not a few of our earthly chemicals are surprisingly little known. Astatine, for instance, is largely unstudied. It has a name and a place on the periodic table (next door to Marie Curie's polonium), but little else. The problem isn't scientific indifference, but rarity. At any one time, it is thought, there may be no more than one gram of astatine scattered through Earth's crust. That's the weight of a paper clip. Several other elements – francium, actinium, neptunium and promethium, for instance – also are only found in traces. Altogether, only about thirty of the naturally occurring elements are widespread on Earth, and barely half a dozen are of central importance to life.

As you might expect, oxygen is our most abundant element, accounting for just under 50 per cent of the Earth's crust, but after that the relative abundances are often surprising. Who would guess, for instance, that silicon is the second most common element on the planet, or that titanium is tenth? Abundance has little to do with their familiarity or utility to us. Many of the more obscure elements are actually more common than the better-known ones. There is more cerium on the Earth than copper, more neodymium and lanthanum than cobalt or nitrogen. Tin barely makes it into the top fifty, eclipsed by such relative obscurities as praseodymium, samarium, gadolinium and dysprosium.

Abundance also has little to do with ease of detection. Aluminium/aluminum is the third most common of our elements, accounting for nearly a tenth of everything that's underneath your feet, but its existence wasn't even suspected until it was discovered in the nineteenth century by Humphry Davy, and for a long time after that it was treated as rare and precious. Congress nearly put a shiny lining of aluminium foil atop the Washington Monument to show what a classy and prosperous nation America had become, and the French imperial family in the same period discarded the state silver dinner service and replaced it with an aluminium one. The fashion was cutting-edge even if the knives weren't.

Nor does abundance necessarily relate to importance. Carbon is only the fifteenth most common element, accounting for a very modest 0.048 per cent of Earth's crust, but we would be lost without it. What sets the carbon atom apart is that it is shamelessly promiscuous. It is the party animal of the atomic world, latching on to many other atoms (including itself) and holding tight, forming molecular conga lines of hearty robustness – the very trick of nature necessary to build proteins and DNA. As Paul Davies has written: 'If it wasn't for carbon, life as we know it would be impossible. Probably any sort of life would be impossible.' Yet carbon is not all that plentiful even in us who so vitally depend on it. Of every 200 atoms in your body, 126 are hydrogen, 51 are oxygen, and just 19 are carbon.*

Other elements are critical not for creating life but for sustaining it. We need iron to manufacture haemoglobin, and without it we would die. Cobalt is necessary for the creation of vitamin B12. Potassium and a very little sodium are literally good for your nerves. Molybdenum, manganese and vanadium help to keep your enzymes purring. Zinc – bless it – oxidizes alcohol.

---

* Of the remaining four, three are nitrogen and the remaining atom is divided among all the other elements.

We have evolved to utilize or tolerate these things – we could hardly be here otherwise – but even then we live within narrow ranges of acceptance. Selenium is vital to all of us, but take in just a little too much and it will be the last thing you ever do. The degree to which organisms require or tolerate certain elements is a relic of their evolution. Sheep and cattle now graze side by side, but actually have very different mineral requirements. Modern cattle need quite a lot of copper because they evolved in parts of Europe and Africa where copper was abundant. Sheep, on the other hand, evolved in copper-poor areas of Asia Minor.

As a rule, and not surprisingly, our tolerance for elements is directly proportionate to their abundance in the Earth's crust. We have evolved to expect, and in some cases actually need, the tiny amounts of rare elements that accumulate in the flesh or fibre that we eat. But step up the doses, in some cases by only a tiny amount, and we can soon cross a threshold. Much of this is only imperfectly understood. No one knows, for example, whether a tiny amount of arsenic is necessary for our well-being or not. Some authorities say it is; some not. All that is certain is that too much of it will kill you.

The properties of the elements can become more curious still when they are combined. Oxygen and hydrogen, for instance, are two of the most combustion-friendly elements around, but put them together and they make incombustible water.* Odder still in combination are sodium, one of the most unstable of all elements, and chlorine, one of the most toxic. Drop a small lump of pure sodium into ordinary water and it

---

* Oxygen itself is not combustible; it merely facilitates the combustion of other things. This is just as well, for if oxygen were combustible, each time you lit a match all the air around you would burst into flame. Hydrogen gas, on the other hand, *is* extremely combustible, as the dirigible *Hindenburg* demonstrated on 6 May 1937, in Lakehurst, New Jersey, when the hydrogen that provided its lift burst explosively into flame, killing thirty-six people.

will explode with enough force to kill. Chlorine is even more notoriously hazardous. Though useful in small concentrations for killing microorganisms (it's chlorine you smell in bleach), in larger volumes it is lethal. Chlorine was the element of choice for many of the poison gases of the First World War. And, as many a sore-eyed swimmer will attest, even in exceedingly dilute form the human body doesn't appreciate it. Yet put these two nasty elements together and what do you get? Sodium chloride – common table salt.

By and large, if an element doesn't naturally find its way into our systems – if it isn't soluble in water, say – we tend to be intolerant of it. Lead poisons us because we were never exposed to it until we began to fashion it into food vessels and pipes for plumbing. (Not incidentally, lead's symbol is Pb for the Latin *plumbum*, the source word for our modern *plumbing*.) The Romans also flavoured their wine with lead, which may be part of the reason they are not the force they used to be. As we have seen elsewhere, our own performance with lead (not to mention mercury, cadmium and all the other industrial pollutants with which we routinely dose ourselves) does not leave us a great deal of room for smirking. When elements don't occur naturally on Earth, we have evolved no tolerance for them and so they tend to be extremely toxic to us, as with plutonium. Our tolerance for plutonium is zero: there is no level at which it is not going to make you want to lie down.

I have brought you a long way to make a small point: a big part of the reason that Earth seems so miraculously accommodating is that we evolved to suit its conditions. What we marvel at is not that it is suitable to life but that it is suitable to *our* life – and hardly surprising really. It may be that many of the things that make it so splendid to us – well-proportioned Sun, doting Moon, sociable carbon, more molten magma than you can shake a stick at and all the rest – seem splendid simply because they are what we were born to count on. No one can altogether say.

Other worlds may harbour beings thankful for their silvery lakes of mercury and drifting clouds of ammonia. They may be delighted that their planet doesn't shake them silly with its grinding plates or spew messy gobs of lava over the landscape, but rather exists in a permanent non-tectonic tranquillity. Any visitors to the Earth from afar would almost certainly, at the very least, be bemused to find us living in an atmosphere composed of nitrogen, a gas sulkily disinclined to react with anything, and oxygen, which is so partial to combustion that we must place fire stations throughout our cities to protect ourselves from its livelier effects. But even if our visitors were oxygen-breathing bipeds with shopping malls and a fondness for action movies, it is unlikely that they would find the Earth ideal. We couldn't even give them lunch because all our foods contain traces of manganese, selenium, zinc and other elemental particles at least some of which would be poisonous to them. To them the Earth might not seem a wondrously congenial place at all.

The physicist Richard Feynman used to make a joke about *a posteriori* conclusions – reasoning from known facts back to possible causes. 'You know, the most amazing thing happened to me tonight,' he would say. 'I saw a car with the licence plate ARW 357. Can you imagine? Of all the millions of licence plates in the state, what was the chance that I would see that particular one tonight? Amazing!' His point, of course, is that it is easy to make any banal situation seem extraordinary if you treat it as fateful.

So it is possible that the events and conditions that led to the rise of life on the Earth are not quite as extraordinary as we like to think. Still, they were extraordinary enough, and one thing is certain: they will have to do until we find some better.

17

INTO THE TROPOSPHERE

Thank goodness for the atmosphere. It keeps us warm. Without it, Earth would be a lifeless ball of ice with an average temperature of minus 50 degrees Celsius. In addition, the atmosphere absorbs or deflects incoming swarms of cosmic rays, charged particles, ultraviolet rays and the like. Altogether, the gaseous padding of the atmosphere is equivalent to a 4.5-metre thickness of protective concrete, and without it these invisible visitors from space would slice through us like tiny daggers. Even raindrops would pound us senseless if it weren't for the atmosphere's slowing drag.

The most striking thing about our atmosphere is that there isn't very much of it. It extends upwards for about 190 kilometres, which might seem reasonably bounteous when viewed from ground level, but if you shrank the Earth to the size of a standard desktop globe it would only be about the thickness of a couple of coats of varnish.

For scientific convenience, the atmosphere is divided into four unequal layers: troposphere, stratosphere, mesosphere and ionosphere (now often called the thermosphere). The troposphere is the part that's dear to us. It alone contains enough warmth and oxygen to allow us to function, though even it swiftly becomes uncongenial to life as you climb up through it. From ground level to its highest point, the troposphere (or 'turning sphere') is about 16 kilometres thick at the equator and no more than 10 or 11 kilometres high in the temperate latitudes where most of us live. Eighty per cent of the atmosphere's mass, virtually all the water and thus virtually all the weather are contained within this thin and wispy layer. There really isn't much between you and oblivion.

Beyond the troposphere is the stratosphere. When you see the top of a storm cloud flattening out into the classic anvil shape, you are looking at the boundary between the troposphere and the stratosphere. This invisible ceiling is known as the tropopause and was discovered in 1902 by a Frenchman using a balloon, Léon-Philippe Teisserenc de Bort. *Pause* in this sense doesn't mean to stop momentarily but to cease altogether; it's from the same Greek root as *menopause*. Even at the troposphere's greatest extent, the tropopause is not very distant. A fast lift of the sort used in modern skyscrapers would get you there in about twenty minutes, though you would be well advised not to make the trip. Such a rapid ascent without pressurization would, at the very least, result in severe cerebral and pulmonary oedemas, a dangerous excess of fluids in the body's tissues. When the doors opened at the viewing platform, anyone inside would almost certainly be dead or dying. Even a more measured ascent would be accompanied by a great deal of discomfort. The temperature 10 kilometres up can be minus 57 degrees Celsius and you would need, or at least very much appreciate, supplementary oxygen.

After you have left the troposphere the temperature soon warms up again, to about 4 degrees Celsius, thanks to the absorptive effects of ozone (something else de Bort discovered). It then plunges to as low as minus 90 degrees Celsius in the mesosphere before sky-rocketing to 1,500 degrees Celsius or more in the aptly named but very erratic thermosphere, where temperatures can vary by over 500 degrees from day to night – though it must be said that 'temperature' at such a height becomes a somewhat notional concept. Temperature is really just a measure of the activity of molecules. At sea level, air molecules are so thick that one molecule can move only the tiniest distance – about eight-millionths of a centimetre, to be precise – before banging into another. Because trillions of molecules are constantly colliding, a lot of heat gets exchanged. But at the height of the thermosphere, at 80 kilometres or more, the air is so thin that any two molecules will be miles apart and hardly ever come into contact. So although each molecule is very warm, there are few interactions between them and thus little heat transference. This is good news for satellites and spaceships, because if the exchange of heat were more efficient any man-made object orbiting at that level would burst into flame.

Even so, spaceships have to take care in the outer atmosphere, particularly on return trips to Earth, as the space shuttle *Columbia* demonstrated all too tragically in February 2003 when it fell apart on re-entry over Texas, killing all seven astronauts onboard. Although the atmosphere is very thin, if a craft comes in at too steep an angle – more than about 6 degrees – or too swiftly it can strike enough molecules to generate drag of an exceedingly combustible nature. Conversely, if an incoming vehicle hits the thermosphere at too shallow an angle, it could well bounce back into space, like a pebble skipped across water.

But you needn't venture to the edge of the atmosphere to be reminded of what hopelessly ground-hugging beings we are. As anyone who has spent time in a lofty city will know, you don't have to rise too many hundreds of metres from sea level before your body begins to protest. Even experienced mountaineers, with the benefits of fitness, training and bottled oxygen, quickly become vulnerable at height to confusion, nausea, exhaustion, frostbite, hypothermia, migraine, loss of appetite and a great many other stumbling dysfunctions. In a hundred emphatic ways the human body reminds its owner that it wasn't designed to operate so far above sea level.

'Even under the most favourable circumstances,' the climber

INTO THE TROPOSPHERE

Peter Habeler has written of conditions atop Everest, 'every step at that altitude demands a colossal effort of will. You must force yourself to make every movement, reach for every handhold. You are perpetually threatened by a leaden, deadly fatigue.' In *The Other Side of Everest*, the British mountaineer and filmmaker Matt Dickinson records how Howard Somervell, on a 1924 British expedition up Everest, 'found himself choking to death after a piece of infected flesh came loose and blocked his windpipe'. With a supreme effort Somervell managed to cough up the obstruction. It turned out to be 'the entire mucous lining of his larynx'.

Bodily distress is notorious above 7,500 metres — the area known to climbers as the Death Zone — but many people become severely debilitated, even dangerously ill, at heights of no more than 4,500 metres or so. Susceptibility has little to do with fitness. Grannies sometimes caper about in lofty situations while their fitter offspring are reduced to helpless, groaning heaps until conveyed to lower altitudes.

The absolute limit of human tolerance for continuous living appears to be about 5,500 metres, but even people conditioned to living at altitude could not tolerate such heights for long. Frances Ashcroft, in *Life at the Extremes*, notes that there are Andean sulphur mines at 5,800 metres, but that the miners prefer to descend 460 metres each evening and climb back up the following day rather than live continuously at that elevation. People who habitually live at altitude have often spent thousands of years developing disproportionately large chests and lungs, and increasing their density of oxygen-bearing red blood cells by almost a third, though there are limits to how much thickening with red cells the blood supply can stand before it becomes too thick to flow smoothly. Moreover, above 5,500 metres even the most well-adapted women cannot provide a growing foetus with enough oxygen to bring it to its full term.

In the 1780s, when people began to make experimental balloon ascents in Europe, something that surprised them was how chilly it got as they rose. The temperature drops about 2 degrees centigrade with every 1,000 feet you climb. Logic would seem to indicate that the closer you get to a source of heat, the warmer you should feel. Part of the explanation is that you are not really getting nearer the Sun in any meaningful sense. The Sun is 93 million miles away. To move a few hundred metres closer to it is like taking one step closer to a bushfire in Australia and expecting to smell smoke when you are standing in Ohio.

The answer again takes us back to the question of the density of molecules in the atmosphere. Sunlight energizes atoms. It increases the rate at which they jiggle and jounce, and in their enlivened state they crash into one another, releasing heat. When you feel the sun warm on your back on a summer's day, it's really excited atoms you feel. The higher you climb, the fewer molecules there are, and so the fewer collisions between them.

Air is deceptive stuff. Even at sea level, we tend to think of the air as being ethereal and all but weightless. In fact, it has plenty of bulk, and that bulk often exerts itself. As a marine scientist named Wyville Thomson wrote more than a century ago: 'We sometimes find when we get up in the morning, by a rise of an inch in the barometer, that nearly half a ton has been quietly piled upon us during the night, but we experience no inconvenience, rather a feeling of exhilaration and buoyancy, since it requires a little less exertion to move our bodies in the denser medium.' The reason you don't feel crushed under that extra half-ton of pressure is the same reason your body would not be crushed deep beneath the sea: it is made mostly of incompressible fluids, which push back, equalizing the pressures within and without.

But get air in motion, as with a hurricane or even a stiff breeze, and you will quickly be reminded that it has very considerable mass. Altogether there are about 5,200 million million tonnes of air around us – 25 million tonnes for every square mile of the planet – a not inconsequential volume. When you get millions of tonnes of atmosphere rushing past at 50 or 60 kilometres an hour, it's hardly a surprise that tree-limbs snap and roof tiles go flying. As Anthony Smith notes, a typical weather front may consist of 750 million tonnes of cold air pinned beneath a billion tonnes of warmer air. Hardly a wonder that the result is at times meteorologically exciting.

Certainly there is no shortage of energy in the world above our heads. One thunderstorm, it has been calculated, can contain an amount of energy equivalent to four days' use of electricity for the whole United States. In the right conditions, storm clouds can rise to heights of 10 to 15 kilometres and contain updraughts and downdraughts of over 150 kilometres an hour. These are often side by side, which is why pilots don't want to fly through them. In all the internal turmoil, particles within the cloud pick up electrical charges. For reasons not entirely understood, the lighter particles tend to become positively charged and to be wafted by air currents to the top of the cloud. The heavier particles linger at the base, accumulating negative charges. These negatively charged particles have a powerful urge to rush to the positively charged Earth, and good luck to anything that gets in their way. A bolt of lightning travels at 435,000 kilometres an hour and can heat the air around it to a decidedly crisp 28,000 degrees Celsius, several times hotter than the surface of the Sun. At any one moment 1,800 thunderstorms are in progress around the globe – some 40,000 a day. Day and night across the planet, every second about 100 lightning bolts hit the ground. The sky is a lively place.

Much of our knowledge of what goes on up there is surprisingly recent. Jet streams, usually located about 9,000–10,000 metres up, can bowl along at up to nearly 300 kilometres an hour and vastly influence weather systems over whole continents, yet their existence wasn't suspected until pilots began to fly into them during the Second World War. Even now a great deal of atmospheric phenomena is barely understood. A form of wave motion popularly known as clear-air turbulence sometimes enlivens aeroplane flights. These events are not associated with cloud structures or anything else that can be detected visually. They are just pockets of startling turbulence in the middle of tranquil skies. Because of their unexpectedness they often cause injuries. In one incident in May 2024, a Singapore Airlines flight from London to Singapore encountered sudden severe turbulence over Myanmar. The plane plunged several hundred feet and was shaken with such violence that over 100 passengers and crew – nearly half of those aboard – were injured, many seriously, and one man died (albeit indirectly, of a heart attack). Such incidents are universally agreed to be growing more common, though the number is nowhere centrally recorded, and the chances of it happening on any given flight are low but unknown.

The process that moves air around in the atmosphere is the same process that drives the internal engine of the planet, namely convection. Moist, warm air from the equatorial regions rises until it hits the barrier of the tropopause and spreads out. As it travels away from the equator and cools, it sinks. When it hits bottom, some of the sinking air looks for an area of low pressure to fill and heads back for the equator, completing the circuit.

At the equator the convection process is generally stable and the weather predictably fair, but in temperate zones the patterns are far more seasonal, localized and random, which results in an endless battle between systems of high-pressure and low-pressure air. Low-pressure systems are created by rising air, which conveys water molecules into the sky, forming clouds and eventually rain. Warm air can hold more moisture than cool air, which is why tropical and summer storms tend to be the heaviest. Thus low areas tend to be associated with cloud and rain, and highs generally spell sunshine and fair weather. When two such systems meet, it often becomes manifest in the clouds. For instance, stratus clouds – those unlovable, featureless sprawls that give us our overcast skies – happen when moisture-bearing updraughts lack the oomph to break through a level of more stable air above and instead spread out, like smoke hitting a ceiling. Indeed, if you watch a smoker sometime, you can get a very good idea of how things work by watching how smoke rises from a cigarette in a still room. At first it goes straight up (this is called a laminar flow if you need to impress anyone), and then it spreads out in a diffused, wavy layer. The greatest supercomputer in the world, taking measurements in the most carefully controlled environment, cannot accurately predict precisely what forms these ripplings will take, so you can imagine the difficulties that confront meteorologists when they try to forecast such motions in a spinning, windy, large-scale world.

What we do know is that because heat from the Sun is unevenly distributed, differences in air pressure arise on the planet. Air can't abide this, so it rushes around trying to equalize things everywhere. Wind is simply the air's way of trying to keep things in balance. Air always flows from areas of high pressure to areas of low pressure (as you would expect; think of anything with air under pressure – a balloon or an air tank or an aeroplane with a missing window – and think how insistently that pressured air wants to go somewhere else), and the greater the discrepancy in pressures, the faster the wind blows.

Incidentally, wind speeds, like most things that accumulate, grow exponentially, so a wind blowing at 300 miles an hour is not simply ten times stronger than a wind blowing at 30 miles an hour, but 100 times stronger – and hence that much more destructive. Introduce several million tonnes of air to this accelerator effect and the result can be exceedingly energetic. A tropical hurricane can release in twenty-four hours as much energy as a rich, medium-sized nation like Britain or France uses in a year.

The impulse of the atmosphere to seek equilibrium was first suspected by Edmond Halley – the man who was everywhere – and elaborated upon in the eighteenth century by his fellow

Briton George Hadley, who saw that rising and falling columns of air tended to produce 'cells' (known ever since as 'Hadley cells'). Though a lawyer by profession, Hadley had a keen interest in the weather (he was, after all, English) and also suggested a link between his cells, the Earth's spin and the apparent deflections of air that give us our trade winds. However, it was an engineering professor at the École Polytechnique in Paris, Gustave-Gaspard de Coriolis, who worked out the details of these interactions in 1835, and thus we call it the Coriolis effect. (Coriolis's other distinction at the school was to introduce water coolers, which were once commonly known as Corios, apparently.) The Earth revolves at a brisk 1,000 miles an hour at the equator, though as you move towards the poles the speed slopes off considerably, to about 650 miles an hour in London, for instance. The reason for this is self-evident when you think about it. If you are on the equator the spinning Earth has to carry you quite a distance – about 40,000 kilometres – to get you back to the same spot, whereas if you stand beside the North Pole you may need to travel only a few metres to complete a revolution; yet in both cases it takes twenty-four hours to get you back to where you began. Therefore, it follows that the closer you get to the equator the faster you must be spinning.

The Coriolis effect explains why anything moving through the air in a straight line laterally to the Earth's spin will, given enough distance, seem to curve to the right in the northern hemisphere and to the left in the southern as the Earth revolves beneath it. The standard way to envision this is to imagine yourself at the centre of a large carousel and tossing a ball to someone positioned on the edge. By the time the ball gets to the perimeter, the target person has moved on and the ball passes behind him. From his perspective, it looks as if it has curved away from him. That is the Coriolis effect and it is what gives weather systems their curl and sends hurricanes spinning off like tops. The Coriolis effect is also why naval guns firing artillery shells have to adjust to left or right; a shell fired 15 miles would otherwise deviate by about 100 yards and plop harmlessly into the sea.

Considering the practical and psychological importance of the weather to nearly everyone, meteorology didn't really get going as a science until shortly before the beginning of the nineteenth century (though the term *meteorology* itself had been around since 1626, when it was coined by a T. Granger in a book of logic).

Part of the problem was that successful meteorology requires the precise measurements of temperatures, and thermometers for a long time proved more difficult to make than you might expect. An accurate reading was dependent on getting a very even bore in a glass tube, and that wasn't easy to do. The first person to solve the problem was Daniel Gabriel Fahrenheit, a Dutch maker of instruments, who produced an accurate thermometer in 1717. However, for reasons unknown he calibrated the instrument in a way that put freezing at 32 degrees and boiling at 212 degrees. From the outset this numeric eccentricity bothered some people and in 1742 Anders Celsius, a Swedish astronomer, came up with a competing scale. In proof of the proposition that inventors seldom get matters entirely right, Celsius made boiling point zero and freezing point 100 on his scale, but that was soon reversed.

The person most frequently identified as the father of modern meteorology was an English pharmacist named Luke Howard, who came to prominence at the beginning of the nineteenth century. Howard is chiefly remembered now for giving cloud types their names in 1803. Although he was an active and respected member of the Linnaean Society and employed Linnaean principles in his new scheme, Howard chose the rather more obscure Askesian Society as the forum in which to announce his new scheme of classification. (The Askesian Society, you may just recall from an earlier chapter, was the body whose members were unusually devoted to the pleasures of nitrous oxide, so we can only hope they treated Howard's presentation with the sober attention it deserved. It is a point on which Howard scholars are curiously silent.)

Howard divided clouds into three groups: stratus for the layered clouds, cumulus for the fluffy ones (the word means heaped in Latin) and cirrus (meaning curled) for the high, thin, feathery formations that generally presage colder weather. To these he subsequently added a fourth term, nimbus (from the Latin for cloud), for a rain cloud. The beauty of Howard's system was that the basic components could be freely recombined to describe every shape and size of passing cloud – stratocumulus, cirrostratus, cumulonimbus, and so on. It was an immediate hit, and not just in England. Goethe was so taken with the system that he dedicated four poems to Howard.

Howard's system has been much added to over the years, so much so that the encyclopedic if little-read *International Cloud Atlas* runs to two volumes, but interestingly virtually all the post-Howard cloud types – mammatus, pileus, nebulosis, spissatus, floccus and mediocris are a sampling – have never caught on with anyone outside meteorology and not terribly much within it, I'm told. Incidentally, the first, much thinner edition of that atlas, produced in 1896, divided clouds into ten basic types, of which the plumpest and most cushiony-looking was number nine, cumulonimbus.* That seems to have been the source of the expression 'to be on cloud nine'.

---

* If you have ever been struck by how beautifully crisp and well defined the edges of cumulus clouds tend to be, while other clouds are more blurry, the explanation is that there is a pronounced boundary between the moist interior of a cumulus cloud and the dry air beyond it. Any water molecule that strays beyond the edge of the cloud is immediately zapped by the dry air beyond, allowing the cloud to keep its fine edge. Much higher cirrus clouds are composed of ice and the zone between the edge of the cloud and the air beyond it not so clearly delineated, which is why they tend to be blurry at the edges.

INTO THE TROPOSPHERE

For all the heft and fury of the occasional anvil-headed storm cloud, the average cloud is actually a benign and surprisingly insubstantial thing. A fluffy summer cumulus several hundred metres to a side may contain no more than 100–150 litres of water – 'about enough to fill a bathtub', as James Trefil has noted. You can get some sense of the immaterial quality of clouds by strolling through fog – which is, after all, nothing more than a cloud that lacks the will to fly. To quote Trefil again: 'If you walk 100 yards through a typical fog, you will come into contact with only about half a cubic inch of water – not enough to give you a decent drink.' In consequence clouds are not great reservoirs of water. Only about 0.035 per cent of the Earth's fresh water is floating around above us at any moment.

Depending on where it falls, the prognosis for a water molecule varies widely. If it lands in fertile soil it will be soaked up by plants or re-evaporated directly within hours or days. If it finds its way down to the groundwater, however, it may not see sunlight again for many years – thousands if it gets really deep. When you look at a lake, you are looking at a collection of molecules that have been there on average for about a decade. In the ocean the residence time is thought to be more like 100 years. Altogether, about 60 per cent of water molecules in a rainfall are returned to the atmosphere within a day or two. Once evaporated, they spend no more than a week or so – Stephen Drury says twelve days – in the sky before falling again as rain.

Evaporation is a swift process, as you can easily gauge by the fate of a puddle on a summer's day. Even something as large as the Mediterranean would dry out in 1,000 years if it were not continually replenished. Such an event occurred a little under 6 million years ago and provoked what is known to science as the Messinian Salinity Crisis. What happened was that continental movement closed the Strait of Gibraltar. As the Mediterranean dried, its evaporated contents fell as fresh-water rain into other seas, mildly diluting their saltiness – indeed, making them just dilute enough to freeze over larger areas than normal. The enlarged area of ice bounced back more of the Sun's heat and pushed Earth into an ice age. So, at least, the theory goes.

What is certainly true, as far as we can tell, is that a little change in the Earth's dynamics can have repercussions beyond our imagining. Such an event, as we shall see further on, may even have created us.

The real powerhouse of the planet's surface behaviour are the oceans. Indeed, meteorologists increasingly treat oceans and atmosphere as a single system, which is why we must give them a little of our attention here. Water is marvellous at holding and transporting heat – unimaginably vast quantities of it. Every day, the Gulf Stream carries an amount of thermal energy to Europe many times greater than all the energy produced and consumed by humans on Earth in a year, which is why Britain and Ireland have such mild winters compared with Canada and Russia. But water also warms slowly, which is why lakes and swimming pools are cold even on the hottest days. For that reason there tends to be a lag in the official, astronomical start of a season and the actual feeling that that season has started. So spring may officially start in the northern hemisphere in March, but it doesn't feel like it in most places until April at the very earliest.

The oceans are not one uniform mass of water. Their differences in temperature, salinity, depth, density and so on have huge effects on how they move heat around, which in turn affects climate. The Atlantic, for instance, is saltier than the Pacific, and a good thing too. The saltier a water is the denser it is, and dense water sinks. Without its extra burden of salt the Atlantic currents would proceed up to the Arctic, warming the North Pole but depriving Europe of all that kindly warmth. The main agent of heat transfer on Earth is what is known as thermohaline circulation, which originates in slow, deep currents far below the surface — a process first detected by the scientist-adventurer Count von Rumford in 1797.* What happens is that surface waters, as they get to the vicinity of Europe, grow dense and sink to great depths and begin a slow trip back to the southern hemisphere. When they reach Antarctica, they are caught up in the Antarctic Circumpolar Current, where they are driven onward into the Pacific. The process is very slow — it can take 1,500 years for water to travel from the North Atlantic to the mid-Pacific — but the volumes of heat and water they move are very considerable and the influence on the climate is enormous.

(As for the question of how anyone could possibly figure out how long it takes a drop of water to get from one ocean to another, the answer is that scientists can measure compounds in the water like chlorofluorocarbons and work out how long it has been since they were last in the air. By comparing a lot of measurements from different depths and locations, they can reasonably chart the water's movement.) Thermohaline circulation not only moves heat around, but also helps to stir up nutrients as the currents rise and fall, making greater volumes of the ocean habitable for fish and other marine creatures. Unfortunately, it appears the circulation may also be very sensitive to change. According to computer simulations, even a modest dilution of the ocean's salt content — from increased

---

* The term means a number of things to different people, it appears. In November 2002, Carl Wunsch of MIT published a report in *Science*, 'What Is the Thermohaline Circulation?', in which he noted that the expression has been used in leading journals to signify at least seven different phenomena (circulation at the abyssal level, circulation driven by differences in density or buoyancy, 'meridional overturning circulation of mass' and so on) — though all are to do with ocean circulations and the transfer of heat, the cautiously vague and embracing sense in which I have employed it here.

melting of the Greenland ice sheet, for instance – could disrupt the cycle disastrously.

The seas do one other great favour for us. They soak up tremendous volumes of carbon and provide a means for it to be safely locked away. One of the oddities of our solar system is that the Sun burns about 25 per cent more brightly now than when the solar system was young. This should have resulted in a much warmer Earth. Indeed, as the English geologist Aubrey Manning has put it, 'This colossal change should have had an absolutely catastrophic effect on the Earth and yet it appears that our world has hardly been affected.'

So what keeps the planet stable and cool? Life does. Trillions upon trillions of tiny marine organisms that most of us have never heard of – foraminiferans and coccoliths and calcareous algae – capture atmospheric carbon, in the form of carbon dioxide, when it falls as rain and use it (in combination with other things) to make their tiny shells. By locking the carbon up in their shells they keep it from being re-evaporated into the atmosphere, where it would build up dangerously as a greenhouse gas. Eventually all the tiny foraminiferans and coccoliths and so on die and fall to the bottom of the sea, where they are compressed into limestone. It is remarkable, when you behold an extraordinary natural feature like the White Cliffs of Dover in England, to reflect that it is made up almost entirely of tiny deceased marine organisms, but even more remarkable when you realize how much carbon they cumulatively sequester. A six-inch cube of Dover chalk will contain well over 1,000 litres of compressed carbon dioxide that would otherwise be doing us no good at all. Eventually much of that limestone will end up feeding volcanoes and the carbon will return to the atmosphere and fall to the Earth in rain, which is why the whole is called the long-term carbon cycle. The process takes a very long time – about half a million years for a typical carbon atom – but in the absence of any other disturbance it works remarkably well at keeping the climate stable.

INTO THE TROPOSPHERE

Unfortunately, human beings have a careless predilection for disrupting this cycle by putting lots of extra carbon dioxide into the atmosphere whether the foraminiferans are ready for it or not. When this book first came out, the amount of carbon dioxide in the atmosphere was 360 parts per million (up from a pre-industrial level of 280 parts per million). Today, two decades later, it has jumped to 420 parts per million and is rising fast – indeed, at a rate about 100 times faster than the fastest known natural increase, as at the end of the last ice age.

Every year now, the amount of carbon dioxide we expel into the atmosphere globally goes up by about 40 billion metric tons, enough to cover nearly the whole of the western United States (specifically, Arizona, California, Colorado, New Mexico, Nevada, Texas and most of Wyoming) under a dense, poisonous haze 8.2 metres deep. That is, let me stress, our *annual* contribution, so next year there will be 40 billion tonnes more and the year after that another 40 billion (but probably even more) and so on indefinitely.

Another way to look at it is to imagine a cube 3 miles across on each side filled to the very top with viscous black gunk. That is the amount of fossil fuels we inhabitants of Earth get through every year. By the middle of this century, at the rate we are going, we will need a cube 9 miles across on each side to contain our annual consumption.

Those global averages, it should be noted, disguise a lot of variability. We in the privileged West consume vastly more per person than in less prosperous parts of the globe. The average European, for instance, takes just two and a half days to produce the same volume of carbon emissions as a Tanzanian creates in a year. The average American needs just twenty-eight hours to do the same.

As I have noted in print elsewhere, one day – and don't expect it to be a distant day – many of those 6 billion or so less well-off people are bound to demand to have what we have already, and to get it as easily as we got it, and that will require more resources than this planet can easily, or even conceivably, yield. The greatest possible irony would be if in our endless quest to fill our lives with comfort and happiness we created a world that had neither.

The good news is that Earth is marvellously resilient, and it is almost certain that eventually the carbon cycle would reassert itself and return the world to a situation of stability and happiness. The last time this happened, it took a mere 60,000 years.

18

THE BOUNDING MAIN

Imagine trying to live in a world dominated by dihydrogen oxide, a compound that has no taste or smell and is so variable in its properties that it is generally benign but at other times swiftly lethal. Depending on its state, it can scald you or freeze you. In the presence of certain organic molecules it can form carbonic acids so nasty that they can strip the leaves from trees and eat the faces off statuary. In bulk, when agitated, it can strike with a fury that no human edifice could withstand. Even for those who have learned to live with it, it is an often murderous substance. We call it water.

Water is everywhere. A potato is 80 per cent water, a cow 74 per cent, a bacterium 75 per cent. A tomato, at 95 per cent, is little *but* water. Even humans are 65 per cent water, making us more liquid than solid by a margin of almost two to one. Water is strange stuff. It is formless and transparent, and yet we long to be beside it. It has no taste and yet we love the taste of it. We will travel great distances and pay small fortunes to see it in sunshine. And even though we know it is dangerous and drowns tens of thousands of people every year, we can't wait to frolic in it.

Because water is so ubiquitous we tend to overlook what an extraordinary substance it is. Almost nothing about it can be used to make reliable predictions about the properties of other liquids, and vice versa. If you knew nothing of water and based your assumptions on the behaviour of compounds most chemically akin to it — hydrogen selenide or hydrogen sulphide, notably — you would expect it to boil at minus 93 degrees Celsius and to be a gas at room temperature.

Most liquids when chilled contract by about 10 per cent. Water does too, but only down to a point. Once it is within whispering distance of freezing, it begins — perversely, beguilingly, extremely improbably — to expand. By the time it is solid, it is almost a tenth more voluminous than it was before. Because it expands, ice floats on water — 'an utterly bizarre property', according to John Gribbin. If it lacked this splendid waywardness, ice would sink, and lakes and oceans would freeze from the bottom up. Without surface ice to hold heat in, the water's warmth would radiate away, leaving it even chillier and creating yet more ice. Soon even the oceans would freeze and almost certainly stay that way for a very long time, probably for ever — hardly the conditions to nurture life. Thankfully for us, water seems unaware of the rules of chemistry or laws of physics. Everyone knows that water's chemical formula is H_2O, which means that it consists of one largish oxygen atom with two smaller hydrogen atoms attached to it. The hydrogen atoms cling fiercely to their oxygen host, but also make casual bonds with other water molecules. The nature of a water molecule means that it engages in a kind of dance with other water molecules, briefly pairing and then moving on, like the ever-changing partners in a quadrille, to use Robert Kunzig's nice phrase. A glass of water may not appear terribly lively, but every molecule in it is changing partners billions of times a second. That's why water molecules stick together to form bodies like puddles and lakes, but not so tightly that they can't be easily separated as when, for instance, you dive into a pool of them. At any given moment only 15 per cent of them are actually touching.

THE BOUNDING MAIN

In one sense the bond is very strong – it is why water molecules can flow uphill when siphoned and why water droplets on a car bonnet show such a singular determination to bead with their partners. It is also why water has surface tension. The molecules at the surface are attracted more powerfully to the like molecules beneath and beside them than to the air molecules above. This creates a sort of membrane strong enough to support insects and skipping stones. It is what gives the sting to a bellyflop.

I hardly need point out that we would be lost without it. Deprived of water, the human body rapidly falls apart. Within days, the lips vanish 'as if amputated, the gums blacken, the nose withers to half its length, and the skin so contracts around the eyes as to prevent blinking', according to one account. Water is so vital to us that it is easy to overlook that all but the smallest fraction of the water on Earth is poisonous to us – deadly poisonous – because of the salts within it.

We need salt to live, but only in very small amounts, and sea water contains way more – about seventy times more – salt than we can safely metabolize. A typical litre of sea water will contain only about 2.5 teaspoons of common salt – the kind we sprinkle on food – but much larger amounts of other elements, compounds and other dissolved solids, which are collectively known as salts. The proportions of these salts and minerals in our tissues are uncannily similar to those in sea water – we sweat and cry sea water, as Margulis and Sagan have put it – but curiously we cannot tolerate them as an input. Take a lot of salt into your body and your metabolism very quickly goes into crisis. From every cell, water molecules rush off like so many volunteer firemen to try to dilute and carry off the sudden intake of salt. This leaves the cells dangerously short of the water they need to carry out their normal functions. They become, in a word, dehydrated. In extreme situations, dehydration will lead to seizures, unconsciousness and brain damage. Meanwhile, the overworked blood cells carry the salt to the kidneys, which eventually become overwhelmed and shut down. Without functioning kidneys you die. That is why we don't drink sea water.

There are 1.3 billion cubic kilometres of water on Earth and that is all we're ever going to get. The system is closed: practically speaking, nothing can be added or subtracted. The water you drink has been around doing its job since the Earth was young. By 3.8 billion years ago, the oceans had (at least more or less) achieved their present volumes.

The water realm is known as the hydrosphere and it is overwhelmingly oceanic. Ninety-seven per cent of all the water on Earth is in the seas, the greater part of it in the Pacific, which is bigger than all the land masses put together. Altogether the Pacific holds just over half of all the ocean water (51.6 per cent); the Atlantic has 23.6 per cent and the Indian Ocean 21.2 per cent, leaving just 3.6 per cent to be accounted for by all the other seas. The average depth of the ocean is 3.86 kilometres, with the Pacific on average about 300 metres deeper than the Atlantic and Indian Oceans. Sixty per cent of the planet's surface is ocean more than 1.6 kilometres deep. As Philip Ball notes, we would better call our planet not Earth but Water.

Of the 3 per cent of Earth's water that is fresh, most exists as ice sheets. Only the tiniest amount – 0.036 per cent – is found in lakes, rivers and reservoirs, and an even smaller part – just 0.001 per cent – exists in clouds or as vapour. Nearly 90 per cent of the planet's ice is in Antarctica and most of the rest is in Greenland. Go to the South Pole and you will be standing on over 2 miles of ice, at the North Pole just 15 feet of it. Antarctica alone has 6 million cubic miles of ice – enough to raise the oceans by a height of 200 feet if it all melted. But if all the water in the atmosphere fell as rain, evenly everywhere, the oceans would deepen by only a couple of centimetres.

Sea level, incidentally, is an almost entirely notional concept. Seas are not level at all. Tides, winds, the Coriolis force and other effects alter water levels considerably from one ocean to another and even within oceans. The Pacific is about a foot and a half higher along its western edge – a consequence of the centrifugal force created by the Earth's spin. Just as when you pull on a tub of water the water tends to flow towards the other end, as if reluctant to come with you, so the eastward spin of Earth piles water up against the ocean's western margins.

Considering the age-old importance of the seas to us, it is striking how long it took the world to take a scientific interest in them. Until well into the nineteenth century most of what was known about the oceans was based on what washed ashore or came up in fishing nets, and nearly all that was written was based more on anecdote and supposition than on physical evidence. In the 1830s, the British naturalist Edward Forbes surveyed ocean beds throughout the Atlantic and Mediterranean and declared that there was no life at all in the seas below 600 metres. It seemed a reasonable assumption. There was no light at that depth, so no plant life, and the pressures of water at such depths were known to be extreme. So it came as something of a surprise when, in 1860, one of the first transatlantic telegraph cables was hauled up for repairs from more than 3 kilometres down and found to be thickly encrusted with corals, clams and other living detritus.

The first really organized investigation of the seas didn't come until 1872, when a joint expedition set up by the British Museum, the Royal Society and the British government set forth from Portsmouth on a former warship called HMS *Challenger*. For three and a half years they sailed the world, sampling waters, netting fish and hauling a dredge through sediments. It was evidently dreary work. Out of a complement of 240 scientists and crew, one in four jumped ship and eight more died or went mad – 'driven to distraction by the mind-numbing routine of years of dredging', in the words of the historian Samantha Weinberg. But they sailed across almost 70,000 nautical miles of sea, collected over 4,700 new species of marine organisms, gathered enough information to create a fifty-volume report (which took nineteen years to put together), and gave the world the name of a new scientific discipline: *oceanography*. They also discovered, by means of depth measurements, that there appeared to be submerged mountains in the mid-Atlantic, prompting some excited observers to speculate that they had found the lost continent of Atlantis.

Because the institutional world mostly ignored the seas, it fell to devoted – and very occasional – amateurs to tell us what was down there. Modern deep-water exploration begins with Charles William Beebe and Otis Barton in 1930. Although they were equal partners, the more colourful Beebe has always received far more written attention. Born in 1877 into a well-to-do family in New York City, Beebe studied zoology at Columbia University, then took a job as a birdkeeper at the New York Zoological Society. Tiring of that, he decided to adopt the life of an adventurer and for the next quarter-century travelled extensively through Asia and South America with a succession of attractive female assistants whose jobs were inventively described as 'historian and technicist' or 'assistant in fish problems'. He supported these endeavours with a series of popular books with titles like *Edge of the Jungle* and *Jungle Days*, though he also produced some respectable books on wildlife and ornithology.

In the mid-1920s, on a trip to the Galápagos Islands, he discovered 'the delights of dangling', as he described deep-sea diving. Soon afterwards he teamed up with Barton, who came from an even wealthier family, had also attended Columbia and also longed for adventure. Although Beebe nearly always gets the credit, it was in fact Barton who designed the first bathysphere (from the Greek word for 'deep') and funded the $12,000 cost of its construction. It was a tiny and necessarily robust chamber, made of cast iron 1.5 inches thick and with two small portholes containing quartz blocks 3 inches thick. It held two men, but only if they were prepared to become extremely well acquainted. Even by the standards of the age, the technology was unsophisticated. The sphere had no manoeuvrability – it simply hung on the end of a long cable – and only the most primitive breathing system: to neutralize their own carbon dioxide they set out open cans of soda lime, and to absorb moisture they opened a small tub of calcium chloride, over which they sometimes waved palm fronds to encourage chemical reactions.

But the nameless little bathysphere did the job it was intended to do. On the first dive, in June 1930 in the Bahamas, Barton and Beebe set a world record by descending to 183 metres. By 1934 they had pushed the record to over 900 metres, where it would stay until after the Second World War. Barton was confident the device was safe to a depth of about 1,400 metres, though the strain on every bolt and rivet was audibly evident with every fathom they descended. At any depth, it was brave and risky work. At 900 metres, their little porthole was subjected to 19 tons of pressure per square inch. Should they pass the structure's limits of tolerance, death at such a depth would have been instantaneous, as Beebe never failed to observe in his many books, articles and radio broadcasts. Their main concern, however, was that the shipboard winch, straining to hold on to a metal ball and 2 tons of steel cable, would snap and send the two men plunging to the sea floor. In such an event, nothing could have saved them.

The one thing their descents didn't produce was a great deal of worthwhile science. Although they encountered many creatures that had not been seen before, the limits of visibility and the fact that neither of the intrepid aquanauts was a trained oceanographer meant they often weren't able to describe their findings in the kind of detail that real scientists craved. The sphere didn't carry an external light, merely a 250-watt bulb they could hold up to the window, but the water below 150 metres was practically impenetrable anyway, and they were peering into it through 3 inches of quartz, so anything they hoped to view would have to be nearly as interested in them as they were in it. About all they could report, in consequence, was that there were a lot of strange things down there. On one dive in 1934, Beebe was startled to spy a giant serpent 'more than twenty feet long and very wide'. It passed too swiftly to be more than a shadow. Whatever it was, nothing like it has been seen by anyone since. Because of such vagueness, their reports were generally ignored by academics.

After their record-breaking descent of 1934, Beebe lost interest in diving and moved on to other adventures, but Barton persevered. To his credit, Beebe always told anyone who asked that Barton was the real brains behind the enterprise, but Barton seemed unable to step from the shadows. He, too, wrote thrilling accounts of their underwater adventures and even starred in a Hollywood movie called *Titans of the Deep*, featuring a bathysphere and many exciting and largely fictionalized encounters with aggressive giant squid and the like. He even advertised Camel cigarettes ('They don't give me jittery nerves'). In 1948 he increased the depth record by 50 per cent, with a dive to 1,370 metres in the Pacific Ocean near California, but the world seemed determined to overlook him. One newspaper reviewer of *Titans of the Deep* actually thought the star of the film was Beebe. Nowadays, Barton is lucky to get a mention.

At all events, he was about to be comprehensively eclipsed by a father and son team from Switzerland, Auguste and Jacques Piccard, who were designing a new type of probe called a bathyscaphe (meaning 'deep boat'). Christened *Trieste*, after the Italian city in which it was built, the new device manoeuvred independently, though it did little more than just go up and down. On one of its early dives, in early 1954, it descended to below 4,000 metres, nearly three times Barton's record-breaking dive of six years earlier. But deep-sea dives required a great deal of costly support and the Piccards were gradually going broke.

In 1958, they did a deal with the US Navy which gave the

Navy ownership but left them in control. Now flush with funds, the Piccards upgraded the vessel so that it could withstand the pressures of a really deep dive – and at the deepest point in the ocean those pressures rise to a crushing 16,000 pounds per square inch, from just 15 pounds per square inch at the surface. Now the *Trieste* was ready for the greatest dive challenge ever attempted – to send two men into the Mariana Trench.*

If it were on the surface, the Mariana Trench would be the most celebrated landmark on Earth. Massively dwarfing the Grand Canyon, it runs for 1,500 miles through the western Pacific near Guam, forming a canyon about 40 miles wide and up to 7 miles deep at its deepest point. You could drop Mount Everest into the Mariana Trench at this point and still have 2 kilometres of water above it. Ships sailing over it are as far above the seabed as a plane flying between continents is above land.

The expedition had a crew of two: Jacques Piccard and Don Walsh, a young US Navy lieutenant. Walsh was a submarine officer, but with no experience of great depths. 'The last sub that I had served on had a maximum operating depth of 300 feet,' he recalled cheerfully in a reminiscence for *Scientific American* in 2014. The *Trieste* was his first command.

The two men squeezed into a container not much larger than a standard phone booth, suspended beneath two giant gas tanks. The descent took five hours at a rate of about 1 metre per second, and must have been hair-raising as the chamber around them groaned and strained under the immense and ever-increasing pressures as they fell through the pitch-black sea to the distant ocean floor. It was decidedly chilly,

---

* The Mariana Trench is named after the nearby Mariana Islands (in Spanish *Las Marianas*, which is why they are also sometimes called the Marianas Islands). Not incidentally, the Mariana Trench was discovered by Harry Hess with his fathometer.

too – about the same temperature as in a domestic refrigerator, Walsh recalled. When at last they touched the bottom, they kicked up so much sediment that they weren't able to see much of anything through their tiny porthole and couldn't take even a single photograph to commemorate the experience. Walsh believed he glimpsed a flatfish of some sort scuttle away just in the moment before their settling, but that was probably an illusion. The deepest fish ever found so far is a tiny snailfish of an undescribed species discovered off the coast of Japan at a depth of 8,336 metres in 2023, which is to say a considerably smaller creature at a considerably shallower depth.

The *Trieste*'s instruments indicated a depth of 11,521 metres, but that was later revised to 10,916 metres.* The vessel had no lateral manoeuvrability, so all they could do was sit and peer into the cloud of sediment outside – 'like looking into a bowl of milk,' Walsh recalled afterwards. After twenty minutes they began their ascent. Upon reaching the surface, they discovered that the Plexiglas window had cracked under the strain. They were very lucky to survive.

The event was widely celebrated and made the cover of *Life* magazine. Walsh and Piccard were summoned to the White House to meet the President and treated everywhere as heroes. Everyone assumed that a new era of deep-sea exploration had begun. In fact, more than half a century would pass before anyone attempted such a feat again. The problem wasn't technical, but rather came in the form of a US Navy vice-admiral named Hyman G. Rickover, a man with a lively temperament, forceful views and, most pertinently, control of the departmental chequebook. He thought underwater exploration a waste of resources and pointed out that the Navy was not a research institute. The nation, moreover, was about to become fully

---

* In 2021, using the latest instruments, the greatest depth of the Mariana Trench, an area known as the Challenger Deep, was updated to 10,935 metres, plus or minus 6 metres.

preoccupied with space travel and the quest to send a man to the Moon, which made deep sea investigations seem unimportant and rather old-fashioned. But the decisive consideration is that the *Trieste* descent didn't actually achieve much. As a Navy official explained years later: 'We didn't learn a hell of a lot from it, other than that we could do it. Why do it again?'

When underwater researchers realized that the Navy had no intention of pursuing a promised exploration programme, there was a pained outcry. Partly to placate its critics, the Navy provided funding for a more advanced submersible, to be operated by the Woods Hole Oceanographic Institution of Massachusetts. Called *Alvin*, in somewhat contracted honour of the oceanographer Allyn C. Vine, it would be a fully manoeuvrable mini-submarine, though unable to go more than half as deep as *Trieste*. (Initially, it could descend to 4,500 metres, which was later increased to 6,500.)

There was just one problem: the designers couldn't find anyone willing to build it. According to William J. Broad in *The Universe Below*: 'No big company like General Dynamics, which made submarines for the Navy, wanted to take on a project disparaged by both the Bureau of Ships and Admiral Rickover, the gods of naval patronage.' Eventually, not to say improbably, *Alvin* was constructed by General Mills, the food company, at a factory where it made the machines to produce breakfast cereals.

The *Alvin* is still going strong (and has been many times upgraded – and indeed effectively rebuilt) after more than sixty years. It has undertaken 5,200 missions, spent more than 36,000 hours submerged and has added greatly to scientific knowledge, most notably in 1977 when it made one of the most important and startling biological discoveries of the twentieth century. In that year *Alvin* found teeming colonies of large organisms living on and around deep-sea vents off the Galápagos Islands – tube worms over 3 metres long, clams 30 centimetres wide, shrimps and mussels in profusion, wriggling spaghetti worms. They all owed their existence to vast colonies of bacteria that were deriving *their* energy and sustenance from hydrogen sulphides – compounds profoundly toxic to surface creatures – that were pouring steadily from the vents. It was a world independent of sunlight, oxygen or anything else normally associated with life. This was a living system based not on photosynthesis but on chemosynthesis, an arrangement that biologists would have dismissed as preposterous had anyone been imaginative enough to suggest it.

Huge amounts of heat and energy are released by these vents. Two dozen of them together will produce as much energy as a large power station and the range of temperatures around them is enormous. The temperature at the point of outflow can be as much as 400 degrees Celsius, while a couple of metres away the water may be only two or three degrees above freezing. A type of worm called alvinellids were found living right on the margins, with the water temperature 78 degrees Celsius warmer at their heads than at their tails. Before this it had been thought that no complex organisms could survive in water warmer than about 54 degrees Celsius, and here was one that was surviving warmer temperatures than that *and* extreme cold to boot. The discovery transformed our understanding of the requirements for life.

It also answered one of the great puzzles of oceanography – something that many of us didn't realize *was* a puzzle – namely, why the oceans don't grow saltier with time. At the risk of stating the obvious, there is a lot of salt in the sea – enough to bury every bit of land on the planet to a depth of about 150 metres. It had been known for centuries that rivers carry minerals to the sea and that these minerals combine with ions in the ocean water to form salts. So far, no problem. But what was puzzling was that the salinity levels of the sea were stable. Millions of gallons of fresh water evaporate from the ocean daily, leaving all their salts behind, so logically the seas ought to grow more salty with the passing years, but they don't. Something takes an amount of salt out of the water equivalent to the amount being put in. For a very long time, no one could figure out what could be responsible for this. *Alvin*'s discovery of the deep-sea vents provided the answer. Geophysicists realized that the vents were acting much like the filters in a fish tank. As water is taken down into the Earth's crust, salts are stripped from it, and eventually clean water is blown out again through the chimney stacks. The process is not swift – it can take up to 10 million years to clean an ocean – but if you are not in a hurry it is marvellously efficient.

As for the Mariana Trench, it remained unvisited for fifty-two years until March 2012, when the Canadian film director James Cameron spent three hours on the bottom, at a recorded depth of 10,908 metres in a vessel called *Deepsea Challenger*. Seven years after that, an American businessman and adventurer named Victor Vescovo descended to a slightly greater reported depth of 10,927 metres in a submersible called DSV *Limiting Factor*. Such ventures don't come cheap – Vescovo self-funded his vessel at a reported cost of $50 million – which is precisely why they don't happen often.

Famously, when Vescovo looked out his porthole he saw a plastic bag and some candy wrappers – a painful reminder that the ocean bottom may seem remote, but it is still very much a part of our world.

Most ocean science these days is done by unmanned robotic submersibles or by surface vessels using sonar,* but we are still far short of having comprehensive knowledge of Earth's most dominant feature. As recently as 2017, just 6 per cent of the Earth's ocean floors had been mapped. Things have improved

---

* Unmanned submersibles are vulnerable too. In 2008, the Woods Hole Oceanographic Institution proudly unveiled its $8 million remotely operated vehicle called *Nereus*. Six years later, *Nereus* imploded while exploring the depths of the Pacific Ocean near New Zealand. Soon afterwards Woods Hole announced that the craft would not be replaced.

since – by 2022, the total was above 20 per cent – but there is still a long way to go. A project called Seabed 2030, run by two nonprofits, aims to have the entire sea floor mapped by 2030, but at this stage that is more a noble hope than a realistic target. It remains a truism that we have better maps of Mars than of our own sea floors.

Perhaps nothing speaks more clearly of our psychological distance from the ocean depths than that the main expressed goal for oceanographers during International Geophysical Year, 1957/8, was to study 'the use of ocean depths for the dumping of radioactive wastes'. This wasn't a secret assignment, you understand, but a proud public boast. In fact, though it wasn't much publicized, by 1957/8 the dumping of radioactive wastes had already been going on, with a certain appalling vigour, for over a decade. Since 1946, the United States had been ferrying 55-gallon drums of radioactive gunk out to the Fallarone Islands, some 50 kilometres off the California coast near San Francisco, where it simply threw them overboard.

It was all quite extraordinarily sloppy. Most of the drums were exactly the sort you see rusting behind petrol stations or standing outside factories, with no protective linings of any type. When they failed to sink, which was usually, Navy gunners riddled them with bullets to let water in (and, of course, plutonium, uranium and strontium out). Before this dumping was halted in the 1990s, the United States had dumped many hundreds of thousands of drums into about fifty ocean sites – almost 50,000 of them in the Fallarones alone.

As Tim Flannery has noted: 'Until the practice was banned in 1993, 142,000 tons of radioactive waste was dumped in the North Atlantic alone.' Russia, he added, dumped seventeen 'entire nuclear reactors into the Arctic Sea'. China, Japan and nearly all the nations of Europe added tons more.

And what effect might all this have had on life in the ocean depths? Well, little, we hope, but we actually have no idea.

To this day we remain astoundingly, sumptuously, radiantly ignorant of life beneath the seas.

Altogether the oceans cover some 71 per cent of the planet's surface, but in volume terms they account for 99 per cent of its habitable space – though the great bulk of that space is lightless and practically unknown. Only the top 200 metres of the oceans receive enough sunlight to support the kind of aquatic life we are familiar with. Between 200 and 1,000 metres below the surface is a layer of water called the mesopelagic, which few have heard of but which we have come increasingly to realize is vitally important to our very existence.

Every evening when darkness falls, trillions upon trillions of tiny creatures known collectively as zooplankton – an estimated 10 billion tons of them globally – rise up from the ocean depths to feed on their plant counterparts, phytoplankton. It is the greatest daily migration on the planet, and was only discovered during the Second World War when submarines' sonar signals were scattered and confused by the migrating swarms twice a day. In the process of eating, they capture lots and lots of carbon and convey it to the depths, where it eventually falls to the bottom and is safely sequestered.

The mesopelagic is also home to masses of little fish called bristlemouths. These exist in such numbers that it is thought they may be the most numerous vertebrates on the planet, and, by weight, may constitute more than 90 per cent of all the fish in the sea. They are no good as a human food, but they have great potential to be ground up and turned into fishmeal. The problem is that the bristlemouths play a vital role in taking carbon (in the form of ingested algae) from shallow waters down to the depths, where it can also sink to the bottom and become permanently trapped within sediments. Norway in 2017 issued forty-six licences permitting fishing in mesopelagic waters. This is probably not a process we should be interfering with on an industrial scale.

As to what else is down there, we have only barely begun

to take note. The sea holds over 90 per cent of the biodiversity of the planet, and yet most of it is unknown. According to one estimate, there could be as many as 30 million species of animals living in the sea, most still undiscovered. Even the most substantial ocean creatures are often remarkably little known to us – including the most mighty of them all, the great blue whale, a creature of such leviathan proportions that (to quote David Attenborough) its 'tongue weighs as much as an elephant, its heart is the size of a car and some of its blood vessels are so wide that you could swim down them'. It is by far the most gargantuan beast on Earth, stretching up to 110 feet in length and weighing as much as 330,000 pounds. Even the most cumbrous of dinosaurs were not that big. Yet the lives of blue whales are largely a mystery to us. Much of the time we have no idea where they are – where they go to breed and what routes they follow to get there, why some of them migrate great distances seasonally while others stay put. We are not sure to what age they live or how long one of their pregnancies lasts. What little we know of them comes largely from eavesdropping on their songs, but even these are a mystery. Blue whales will sometimes break off a song, then pick it up again at exactly the same spot six months later. Sometimes they strike up with a new song, which no member can have heard before but which each somehow understands. How they do this and why are not at all well understood. And these are animals that must routinely come to the surface to breathe.

For animals that need never surface, obscurity can be even more tantalizing. Consider our knowledge of the fabled giant squid and its cousin the colossal squid (which, name notwithstanding, is smaller than the giant squid). Though nothing on the scale of the blue whale, these are decidedly substantial creatures. They have eyes the size of soccer balls and trailing tentacles that can reach lengths of 13 metres in the case of the giant squid and a little less for the colossal squid. If you dumped one in a typical domestic swimming pool, there wouldn't be much room for anything else. Yet until 2004 no scientist – no person, as far as we know – had ever seen either type of squid alive. In that year, Japanese scientists briefly caught a living giant squid near the remote Ogasawara Islands in the western Pacific and filmed it before it escaped back to the depths. A second giant squid was filmed by researchers in 2010, and the colossal squid was finally caught on camera for the first time in 2025 – but only briefly, and, it must be said, was a juvenile only about 12 inches long.

And that's it. Everything else we know about these magnificent animals is from carcasses washed up on beaches or caught in nets. Giant squid must exist in large numbers because they form a central part of the sperm whale's diet, and sperm whales take a lot of feeding, but that is as much as we know. As for colossal squid, they inhabit the cold waters around Antarctica and the southern oceans, but almost nothing is known about their abundance, longevity, feeding habits or lifestyle.*

The first hint of how truly abundant life is in the deep seas didn't come until as recently as the 1960s with the invention of the epibenthic sled – a dredging device that captures organisms not just on and near the sea floor but also buried in the sediments beneath. In a single one-hour trawl along the continental shelf, at a depth of about 1.5 kilometres, Woods Hole oceanographers Howard Sandler and Robert Hessler netted over 25,000 creatures – worms, starfish, sea cucumbers and the like – representing 365 species. Even at a depth of nearly 5 kilometres, they found some 3,700 creatures representing almost 200 species of organism. But the dredge could capture only those things that were too slow or stupid to get out of the way. In the late 1960s a marine biologist named John Isaacs had the idea of lowering a

---

* The indigestible parts of giant squid, in particular their beaks, accumulate in sperm whales' stomachs into the substance known as ambergris, which is used as a fixative in perfumes. The next time you spray on Chanel N°5 (assuming you do), you may wish to reflect that you are dousing yourself in distillate of little-known sea monster.

camera with bait attached to it and found still more, in particular dense swarms of writhing hagfish, a primitive eel-like creature, as well as darting shoals of grenadier fish. Where a good food source is suddenly available – for instance, when a whale dies and sinks to the bottom – as many as 390 species of marine creature have been found dining off it. Intriguingly, many of these creatures were found to have come from vents up to 1,600 kilometres away. These included such types as mussels and clams, which are hardly known as great travellers. It is now thought that the larvae of certain organisms may drift through the water until, by some unknown chemical means, they detect that they have arrived at a food opportunity and fall on to it.

So why, if the seas are so vast, do we so easily overtax them? Well, to begin with, the world's seas are not uniformly bounteous. Altogether less than a tenth of the ocean is considered naturally productive. Most aquatic species like to be in shallow waters, where there are warmth and light and an abundance of organic matter to prime the food chain. Coral reefs, for instance, constitute well under 1 per cent of the ocean's space but are home to about 25 per cent of its fish. Elsewhere, the oceans aren't nearly so rich. Take Australia. With 36,735 kilometres of coastline and over 23 million square kilometres of territorial waters, it has more sea lapping its shores than any other country, yet, as Tim Flannery notes, it doesn't even make it into the top fifty among fishing nations. Indeed, Australia is a large net importer of seafood. This is because much of Australia's water is, like much of Australia itself, essentially desert. Because the soil is poor, it produces practically no nutrients in its run-offs. (A notable exception is the Great Barrier Reef off Queensland, which is sumptuously fecund, though increasingly stressed by warming sea temperatures, resulting in bleaching.)

Even where life thrives it is often extremely sensitive to disturbance. In the 1970s, fishermen from Australia and, to a lesser extent, New Zealand discovered shoals of a little-known

THE BOUNDING MAIN

fish living at a depth of about 800 metres on their continental shelves. They were known as orange roughy, they were delicious and they existed in huge numbers. In no time at all, fishing fleets were hauling in 40,000 tonnes of roughy a year. Then marine biologists made some alarming discoveries. Roughy are extremely long-lived and slow-maturing. Some may be 150 years old; any roughy you have eaten may well have been born when Victoria was Queen. Roughy have adopted this exceedingly unhurried lifestyle because the waters they live in are so resource-poor. In such waters, some fish spawn just once in a lifetime. Clearly these are populations that cannot stand a great deal of disturbance. Unfortunately, by the time this was realized the stocks had been severely depleted. Even with good management it will be decades before the populations recover, if they ever do.

Elsewhere, however, the misuse of the oceans has been more wanton than inadvertent. Many fishermen 'fin' sharks – that is, slice their fins off, then dump them back into the water to die. But sharks are also killed in great numbers for their oils and cartilage (which is made into homeopathic medicines that have absolutely no actual curative effects). Altogether as many as 100 million sharks are taken from the seas each year. Globally their numbers have fallen by 70 per cent since the 1970s.

Trawlers are sometimes now as big as cruise ships and haul behind them nets big enough to hold a dozen jumbo jets. Some even use spotter planes to locate shoals of fish from the air. The World Wildlife Fund estimates that as much as 40 per cent of every net hauled from the sea consists of 'by-catch' – fish that can't be landed because they are too small or of the wrong type or caught in the wrong season. As one observer told *The Economist*: 'We're still in the Dark Ages. We just drop a net down and see what comes up.' Perhaps as much as 40 million tonnes of such unwanted marine life is dumped back in the sea each year, mostly in the form of corpses. That's nearly double the estimated total when this book was first published.

Overfishing is a global phenomenon. At least two-thirds of species in the North Sea, by many estimates, are being taken at levels that cannot be sustained. In Grimsby, in northern England, which up to the mid-twentieth century was the largest fishing port in the world – let me repeat that: the world – in 1950 they landed over 100,000 tonnes of cod. Today it's less than 300. Grimsby's total fish haul in the same period has dropped from 200,000 tonnes to just 658. Across the Atlantic, halibut once abounded in such numbers off New England that individual boats could land 20,000 pounds of it in a day. Now halibut is all but extinct in many areas. The US National Oceanic and Atmospheric Administration, which makes the rules for commercial fishing in American waters, describes halibut stocks as 'at a very low level' and acknowledges that its 'overfishing status is unknown', yet still allows 60,000 pounds of halibut to be legally landed every year.

The staggering extent to which we have abused the bounty of the seas is not easy to grasp. In the late fifteenth century, the explorer John Cabot found cod in such incredible densities on North America's eastern banks that sailors scooped them up in baskets. Stocks were thought to be inexhaustible. Of course they were anything but. By 1960, the number of spawning cod in the North Atlantic had fallen to a worryingly low 1.6 million tonnes, but by 1990 this had crashed to a scant 22,000 tonnes. In commercial terms, the cod were extinct. 'Fishermen', wrote Mark Kurlansky in his fascinating history, *Cod*, 'had caught them all.' The fish of fish fillets or fish fingers was originally cod, but then was successively replaced by haddock, redfish and Pacific pollock. In the end, Kurlansky noted drily, 'fish' is 'whatever is left'.

In 2008, the Canadian Department of Fisheries and Oceans conceded that cod stocks had gone past a tipping point and would never recover. The problem, it appears, is that seals are eating whatever cod they can find, keeping stocks from ever recovering and ensuring their extinction in Canadian waters by

2050. (Cod stocks elsewhere, it should be noted, have been at least partially restored.)

A separate, but growing, issue is what is darkly referred to as 'ocean road kill' – the inadvertent killing of whales, dolphins, turtles and other large marine creatures by passing ships. The world's fleet of large cargo ships has doubled in twenty years, to more than 100,000 vessels today, with often dire consequences for animals that get in their way. No one can say precisely how many marine animals die in this way, but the number is known to be substantial. An experiment in the Bay of Fundy in Canada in which large ships were rerouted by just 4 nautical miles reduced the number of strikes on right whales by 90 per cent, but such helpful efforts are rare.

The simple fact is that in every possible way we are actively or benignly failing the world's oceans. As of 2023, not one of the United Nations' seventeen sustainable development goals for the oceans will be met by the target deadline of 2030, and some are actually moving in reverse. (The goals were set up in 2015.) Among the newer threats is mining. One area of particular interest to mining companies is a vast and little-explored sprawl of abyssal plains in the Pacific known as the Clarion-Clipperton Zone, where the seabed is extensively scattered with potato-sized, rock-like nodules containing nickel, cobalt, copper, titanium and the rare earth elements that are most urgently needed for making the batteries, solar cells, generators, magnets and other super-engineered infrastructure that will help us battle climate change.

It is easy to see the temptation. These nodules exist in their trillions and are just sitting on the seabed like so many Easter eggs, ready to be collected by trundling submersible harvesters or suction devices. Best of all, each nodule is almost 100 per cent commercially useful. The average grade of copper ore at the surface, by comparison, is just 0.5 per cent, which means that 99.5 per cent of everything dug up at a copper mine is waste, resulting in massive scarring of the earth to get a comparatively meagre pay-off. Ocean mining, it is argued, would be much less disruptive.

With this in mind, the International Seabed Authority, the body endowed with responsibility for the management of the seabed beyond national jurisdictions, has granted permission for experimental dredging at sixteen sites in the Clarion-Clipperton Zone, and other sites around the world are also being eagerly investigated. In June 2023, Norway gave out $18.5 billion worth of ocean-drilling permits and said it would open 280,000 square kilometres of its coastal waters to deep-sea mining.

Hardly any conservationists welcome these moves. It is impossible to gather nodules, they argue, without stirring up sediments, and no one knows what the consequences of that would be. Such studies as exist are not encouraging. As the Pew Charitable Trusts has noted: 'Scientific monitoring of experimental dredge sites in deep-sea sediment has shown that decades after a site is disturbed, few if any communities of organisms have recovered.'

In 2023, researchers at the Natural History Museum in London did a survey of marine life in the Clarion-Clipperton Zone and recorded 5,142 entirely new species. These were mostly shrimp, crabs, nematodes and worms – the bottom-dwelling creatures most at risk – of which all but six appear to be unique to that region of the ocean. Muriel Rabone, a researcher at the museum, suggested that there were probably some 6,000 to 8,000 more species in the zone yet to be identified. Altogether, 90 per cent of all creatures in the zone were not known to science before 2023.

In a scathing editorial, the journal *Nature* described deep-sea mining as 'a nascent industry that risks wreaking havoc on poorly understood ocean ecosystems'. Among much else, the editorial noted that the world's seabeds sequester massive volumes of carbon, which mining could easily stir up and inadvertently release back into the atmosphere. It has also been noted that the rate of recycling for rare earth elements is less than 1 per cent.

Perhaps above all, no one knows what role, if any, the nodules may play in helping to maintain the ecosystems in which they reside. A study in 2024 by the Scottish Association for Marine Science proposed that the nodules engage in chemical processes that produce significant amounts of oxygen – 'dark oxygen', as it has been dubbed – without which life in the sunless depths would not be possible. That finding has been disputed – most vehemently, as you would imagine, by mining companies, but also by some academics. One thing is certain: it takes millions of years for each nodule to form, so once they are removed, they won't come back in a hurry.

We are, in short, remarkably ignorant of the dynamics that rule life in the sea. While marine life is poorer than it ought to be in areas that have been overfished, in some naturally impoverished waters there is far more life than there ought to be. The southern oceans around Antarctica produce only about 3 per cent of the world's phytoplankton – far too little, it would seem, to support a complex ecosystem, and yet they do. Crab-eater seals are not a species of animal that most of us have heard of, but they may actually be the second most numerous large species of animal on Earth, after humans. As many as 15 million of them may live on the pack ice around Antarctica. There are also perhaps 2 million Weddell seals, at least half a million Emperor penguins, and maybe as many as 4 million Adélie penguins. The food chain is thus hopelessly top-heavy, but somehow it works. Remarkably, no one knows how.

All this is a very roundabout way of making the point that the world's oceans and the life they support remain tantalizingly little known to us. But then, as we shall see in the pages remaining to us, once you start talking about life on Earth, there is a very great deal we've yet to understand – not least, how it got going in the first place.

19

THE RISE OF LIFE

In 1953 Stanley Miller, a graduate student at the University of Chicago, took two flasks – one containing a little water to represent a primeval ocean, the other holding a mixture of methane, ammonia and hydrogen sulphide gases to represent the Earth's early atmosphere – connected them with rubber tubes and introduced some electrical sparks as a stand-in for lightning. After a few days, the water in the flasks had turned green and yellow in a hearty broth of amino acids, fatty acids, sugars and other organic compounds. 'If God didn't do it this way,' observed Miller's delighted supervisor, the Nobel laureate Harold Urey, 'He missed a good bet.'

Press reports of the time made it sound as if about all that was needed now was for somebody to give the flasks a good shake and life would crawl out. As time has shown, it wasn't nearly so simple. Despite seventy years of further study, we are no nearer to synthesizing life today than we were in 1953 – and much further away from thinking we can. Scientists are now pretty certain that the early atmosphere was nothing like as primed for development as Miller and Urey's gaseous stew, but rather was a much less reactive blend of nitrogen and carbon dioxide. Repeating Miller's experiments with these more challenging inputs has so far produced only one fairly primitive amino acid. At all events, creating amino acids is not really the problem. The problem is proteins.

Proteins are what you get when you string amino acids together, and we need a lot of them. No one really knows, but there may be as many as a million types of protein in the human body, and each one is a little miracle. By all the laws of probability proteins shouldn't exist. To make a protein you need to assemble amino acids (which I am obliged by long tradition to refer to here as 'the building blocks of life') in a particular order, in much the same way that you assemble letters in a particular order to spell a word. The problem is that words in the amino-acid alphabet are often exceedingly long. To spell 'collagen', the name of a common type of protein, you need to arrange eight letters in the right order. To *make* collagen, you need to arrange 1,055 amino acids in precisely the right sequence. But – and here's an obvious but crucial point – *you don't* make it. It makes itself, spontaneously, without direction, and this is where the unlikelihoods come in.

The chances of a 1,055-sequence molecule like collagen spontaneously self-assembling are, frankly, nil. It just isn't going to happen. To grasp what a long shot its existence is, visualize a standard Las Vegas slot machine but broadened greatly – to about 27 metres, to be precise – to accommodate 1,055 spinning wheels instead of the usual three or four, and with twenty symbols on each wheel (one for each common amino acid).* How long would you have to pull the handle before all 1,055 symbols came up in the right order? Effectively, for ever. Even if you reduced the number of spinning wheels to 200, which is actually a more typical number of

---

* Some 500 animo acids occur naturally on Earth, but just twenty-two make proteins and only twenty of those are necessary to produce us and most other living things. The twenty-second, called pyrrolysine, was discovered in 2002 by researchers at Ohio State University and is found only in some archaea (a basic form of life that we will discuss a little further on in the story) and bacteria.

amino acids for a protein, the odds against all 200 coming up in a prescribed sequence are 1 in 10^{260} (that is a 1 followed by 260 zeros). That in itself is a larger number than all the atoms in the universe.

Proteins, in short, are complex entities. Haemoglobin is only 146 amino acids long, a runt by protein standards, yet even it offers 10^{190} possible amino-acid combinations, which is why it took the Cambridge University chemist Max Perutz twenty-three years – a career, more or less – to unravel it. For random events to produce even a single protein would seem a stunning improbability – like a whirlwind spinning through a junkyard and leaving behind a fully assembled jumbo jet, in the colourful simile of the astronomer Fred Hoyle.

Yet we are talking about several hundred thousand types of protein, perhaps a million, each unique and each, as far as we know, vital to the maintenance of a sound and happy you. And it goes on from there. To be of use, a protein must not only assemble amino acids in the right sequence, it must then engage in a kind of chemical origami and fold itself into a very specific shape. Even having achieved this structural complexity, a protein is no good to you if it can't reproduce itself, and proteins can't. For this you need DNA. DNA is a whiz at replicating – it can make a copy of itself in seconds – but can do virtually nothing else. So we have a paradoxical situation. Proteins can't exist without DNA and DNA has no purpose without proteins. Are we to assume, then, that they arose simultaneously with the purpose of supporting each other? If so: wow.

And there is more still. DNA, proteins and the other components of life couldn't prosper without some sort of membrane to contain them. Pluck any atom from your body and it is no more alive than is a grain of sand. It is only when they come together within the nurturing refuge of a cell that these diverse materials can take part in the amazing dance that we call life. Without the cell, they are nothing more than interesting chemicals. But without the chemicals, the cell has no purpose. As the

THE RISE OF LIFE

physicist Paul Davies has put it, 'If everything needs everything else, how did the community of molecules ever arise in the first place?' It is rather as if all the ingredients in your kitchen somehow got together and baked themselves into a cake – but a cake that could moreover divide when necessary to produce *more* cakes. It is little wonder that we call it the miracle of life. It is also little wonder that we have barely begun to understand it.

So what accounts for all this wondrous complexity? Well, one possibility is that perhaps it isn't actually quite so wondrous as at first it seems. Take those amazingly improbable proteins. The wonder we see in their assembly comes in assuming that they arrived on the scene fully formed. But what if the protein chains didn't assemble all at once? What if, in the great slot machine of creation, some of the wheels could be held, as a gambler might hold a number of promising cherries? What if, in other words, proteins didn't suddenly burst into being, but *evolved*?

In short, there must have been some kind of cumulative selection process that allowed amino acids to assemble in chunks. Perhaps two or three amino acids linked up for some simple purpose and then after a time bumped into some other similar small cluster and in so doing 'discovered' some additional improvement.

Chemical reactions of the sort associated with life are actually something of a commonplace. It may be beyond us to cook them up in a lab, à la Stanley Miller and Harold Urey, but the universe does it readily enough. Lots of molecules in nature get together to form long chains called polymers. Sugars constantly assemble to form starches. Crystals can do a number of lifelike things – replicate, respond to environmental stimuli, take on a patterned complexity. They've never achieved life itself, of course, but they demonstrate repeatedly that complexity is a natural, spontaneous, entirely reliable event. There may or may not be a great deal of life in the universe at large, but there is no shortage of ordered self-assembly, in everything from the transfixing symmetry of snowflakes to the comely rings of Saturn.

So powerful is this natural impulse to assemble that many scientists now believe that life may be more inevitable than we think – that it is, in the words of the Belgian biochemist and Nobel laureate Christian de Duve, 'an obligatory manifestation of matter, bound to arise wherever conditions are appropriate'. De Duve thought it likely that such conditions would be encountered perhaps a million times in every galaxy.

Certainly there is nothing remotely exotic in the chemicals that animate us. If you wished to create another living object, whether a goldfish or a head of lettuce or a human being, you would need really only four principal elements, carbon, hydrogen, oxygen and nitrogen, plus small amounts of a few others, principally sulphur, phosphorus, calcium and iron. These are the same things you would find in a pile of dirt. It cannot be said too often: the only extraordinary thing about the elements that make us is that they make us.

The bottom line is that life is amazing and gratifying, perhaps even miraculous, but hardly impossible – as we repeatedly attest with our own modest existences. To be sure, many of the fine details of life's beginnings remain pretty imponderable. Every scenario you have ever read concerning the conditions necessary for life involves water – from the 'warm little pond' where Darwin supposed life began to the bubbling sea vents that are now the most popular candidates for life's beginnings – but all this overlooks the fact that to turn monomers into polymers (which is to say, to begin to create proteins) involves a type of reaction known to biology as 'dehydration linkages'. As one leading biology text puts it, with perhaps just a tiny hint of discomfort, 'Researchers agree that such reactions would not have been energetically favorable in the primitive sea, or indeed in any aqueous medium, because of the mass action law.' It is a little like putting sugar in a glass of water and having it become a cube. It shouldn't happen, but somehow in nature it does. The

THE RISE OF LIFE

actual chemistry of all this is a little arcane for our purposes here, but it is enough to know that if you make monomers wet they don't turn into polymers – except when creating life on the Earth. How and why it happens then and not otherwise is one of biology's great unanswered questions.

One of the biggest surprises in the earth sciences in recent decades was discovering just how early in Earth's history life arose. Well into the 1950s, it was thought that life was less than 600 million years old. By the 1970s, a few adventurous souls felt that maybe it went back 2.5 billion years. But the present date of 3.85 billion years is stunningly early. The Earth's surface had only recently (geologically speaking) become *solid*.

'We can only infer from this rapidity that it is not "difficult" for life of bacterial grade to evolve on planets with appropriate conditions,' Stephen Jay Gould observed in the *New York Times* in 1996. Or as he put it elsewhere, it is hard to avoid the conclusion that 'life, arising as soon as it could, was chemically destined to be.'

The age in which life arose was the Archaean (from the Greek for 'beginning'), which followed the brutal and well-named Hadean (after 'Hades'), when Earth was repeatedly bombarded by rocks from space. The Archaean ran from 4 billion to 2.5 billion years ago and was, on the face of it, a scarcely less congenial environment for life. Certainly, if you were to step from a time machine into that ancient Archaean world you would very swiftly scamper back inside, for there was no more oxygen to breathe on the Earth back then than there is on Mars today. It was also full of noxious vapours from hydrochloric and sulphuric acids powerful enough to eat through clothing and blister skin. The world was a darker place, too; the Sun's luminosity was only 75 to 80 per cent what it is now. What little you could see would be illumined only briefly by bright and frequent lightning flashes. In short, it was the Earth, but an Earth we wouldn't recognize as our own.

Yet somehow in this harsh and unimaginably distant past,

life arose. Some little bag of chemicals absorbed some nutrients, gently pulsed, had a brief existence. This much may have happened before, perhaps many times. But one ancestral packet did something additional and extraordinary: it cleaved itself and produced an heir. A tiny bundle of genetic material passed from one living entity to another, and has never stopped moving since. It was the moment of creation for us all.

As far as we can tell, it happened just once, which means that every living thing you see around you – every bird, insect, blade of grass, human being, cow, tree, unassuming nubbin of lichen – descends from that one magical moment. That is the most extraordinary fact in biology, perhaps the most extraordinary fact we know. Everything that has ever lived, plant or animal, dates its beginnings from a single primordial twitch. Biologists sometimes call it the Big Birth.

The College of All Souls of the Faithful Departed – universally known more simply as All Souls – stands along the High Street in Oxford, and is perhaps the most gloriously eccentric, and exclusive, of Oxford University's ancient institutions, which is of course saying something. It has all the trappings of an Oxford college – gracious buildings, immaculate quadrangle, magnificent chapel and library – but no students. Instead it is home to something in the region of 120 Fellows who are mostly engaged in research and whose common denominators appear to be relaxed congeniality and enormous intellects.

The entrance examination for All Souls is said to be the hardest in the world. Here, for purposes of illustration, are some questions from the history paper in 2024:

How important were the Achaemenids as a template for Sasanian power?

Is Abbasid silver a sufficient explanation for the 'Viking Age'?

Did Wang Anshi's reforms destroy the Northern Song?

Altogether, candidates at All Souls may choose from seven specialized subject categories – classical studies, law, history,

English literature, economics, politics and philosophy – from which they must answer six exceedingly challenging questions in well-formed essays, followed by two 'general' papers of three hours each. Each year about 150 brave candidates sit the exam, from which one to three (and sometimes none) are offered Fellowships. Cleverness of a very high order is the only way in.

It was at All Souls, on a morning in early autumn 2024, that I met Dr Ross Anderson, a friendly, enthusiastic, strikingly youthful (in his early thirties, but looking even younger) scientist and academic who spends his working days tackling perhaps the biggest question there is: how did we get here? We met in a very comfortable, club-like common room filled with the kind of furniture you don't sit on so much as sink into. Anderson is a senior researcher in natural history, and his particular interest is the complex organisms known as eukaryotes. He grew up near Nottingham in the village of Gotham (from which New York City, very improbably, got its nickname).* Anderson had never been to America and neither of his parents had attended university, but he was encouraged to apply for – and subsequently won – a scholarship to Harvard (followed by a masters degree and PhD at Yale). 'I went intending to do physics,' he told me, 'but in my first year I took a geology class on the history of Earth and I was captivated. It completely changed the direction I went academically.'

Working out when life began, and how it then made the extraordinary, gloriously improbable leap into the kind of

---

* Clearly the nickname for New York City has nothing to do with science or the rise of life, but having raised the matter, I feel I should perhaps explain. The village in Nottinghamshire – which is in fact pronounced 'goat-um,' not 'goth-um' – was once famous for a legend that its inhabitants feigned lunacy to discourage some officials who proposed to build a road through the village at the locals' expense. Confused and frustrated, the officials gave up and put the road elsewhere. In 1807, the author Washington Irving applied the name to his native city as a way of stating that his fellow New Yorkers seemed deranged but actually were cunning.

complexity that would eventually lead to us is what preoccupies his working days. It is, to put it mildly, a big challenge, not least because so few fossils survive from the very ancient past. Even at the best of times becoming a fossil is a tricky business. First, a living thing must die in the right place. In practical terms, the deceased must become buried in sediment, where it can leave an impression, like a leaf in wet mud, or decompose without exposure to oxygen, permitting the molecules within it to be replaced by dissolved minerals, creating a petrified copy of the original. Then, as the sediments in which the fossil lies are carelessly pressed and folded and pushed about by Earth's processes, the fossil must somehow maintain an identifiable shape. Finally, but above all, after tens or dozens or hundreds of millions of years hidden away, it must be found and recognized as something worth keeping. That's for fossils generally, but for ancient life the situation is even more complicated. Earth's early organisms were tiny and soft. When their brief existence was over they vanished swiftly and utterly, like a popped soap bubble, but on a scale mostly measured in microns. That scientists have found any at all is testament to their patience and ingenuity, but it is also an area fraught with controversy.

'The oldest fossils we can be absolutely confident of are ones from South Africa that are about 3.2 billion years old,' Anderson told me. 'They match fossils that we find in younger rocks and are preserved in the same way, so there's a pretty high degree of confidence. But then it starts to get murky. The easiest way to tell if something is a fossil is when you have a complex morphology that cannot be created by physical processes alone. That's pretty obvious when you are looking at multicellular organisms. It's less obvious when you go really far back in time and you are looking at single-celled organisms because then you've got a filament or rod or a sphere, and there are physical processes which can make accumulations of carbon that look like that. The further you go back, the harder it gets

THE RISE OF LIFE

to tease apart the distinctions that indicate whether a thing was ever living or not.'

That's where the controversy comes in. There are many claimed examples of fossils from Canada, Greenland and elsewhere that go back almost 4 billion years. 'All of those are somewhat speculative,' Anderson says. 'There are some reasons to believe that they are fossils and some reasons to conclude that they are not,' he adds diplomatically. 'But there's no certainty either way. Fortunately, we have other, indirect evidence that there *was* life early on, from traces of microbes made in rocks called stromatolites and classic isotopic signatures, but whether or not they are actual fossils is often a matter of debate.'

Anniversaries were few and far between in the Archaean world. For 2 billion years, microbial organisms were the only forms of life. They lived, they reproduced, they swarmed, but they didn't show any particular inclination to move on to another, more challenging level of existence. And then, rather magically, at some unknown point between about 1.6 billion and 2.7 billion years ago, an entirely new type of cell arose called a eukaryote (meaning 'truly nucleated'), as contrasted with the old type, known as a prokaryote ('pre-nucleated'). Prokaryotes were little more than simple 'bags of chemicals', in the words of geologist Stephen Drury. Eukaryotes were bigger – eventually as much as 10,000 times bigger – and far more sophisticated. They contained a nucleus and little bodies called *organelles* (from a Greek word meaning 'little tools'), and could hold as much as 1,000 times more DNA than their predecessors.

The process is thought to have started when some blundering or adventuresome microbe either invaded or was captured by some other microbe and it turned out that this suited them both. The captive microbe became, it is thought, a mitochondrion. Mitochondria are very tiny – you could pack a billion into the space occupied by a grain of sand – but vitally important to our existence.

We rely on them still – indeed, we couldn't live for two minutes without them – yet even after hundreds of millions of years mitochondria behave as if they think things might not work out between us. They maintain their own DNA, RNA and ribosomes. They reproduce at a different time from their host cells. They look like bacteria, divide like bacteria and sometimes respond to antibiotics in the way bacteria do. They don't even speak the same genetic language as the cell in which they live. In short, they keep their bags packed. It is like having a stranger in your house, but one who has been there for the better part of a billion years.

This mitochondrial invasion (or endosymbiotic event, as biologists term it) made complex life possible. It would be hard to exaggerate what a revolution this was. The great Harvard zoologist Ernst Mayr thought it 'the most important and dramatic event in the history of life' – more important even than the moment of creation itself. Thanks to that small but profound act of cellular reorganization Earth went from being a teeming smear of microbes to a place of unimaginable richness and glory as these new eukaryotes developed the last great feat of early life: the creation of the vast, resourceful and staggeringly diverse kingdom known as Animalia.

That is the evolutionary moment that captivates Anderson. 'How did this modern planet with so much animal diversity come to be? I think that is the most exciting question in science. For billons of years Earth was unicellular, and then it filled with an abundance of animal life. How and why did that happen?'

The problem is that scientists can't say precisely when this pivotal moment occurred. Paleobiologists have two ways of determining the ages of ancient organisms – by dating fossils based on the ages of the rocks in which they are embedded or, more notionally, by counting genetic changes by means of what is known as the molecular clock.

'If you have two animals,' Anderson explained to me, 'you can determine how distinct their DNA is – how many mutations they've undergone over time. That gives you the genetic distance which separates them. Then you can make models of how fast mutations happen, which gives you a speed. If you have distance and speed, you can calculate how long ago they departed from a common ancestor. That's the molecular clock. When you do that for all animal life, you commonly get an age of origin somewhere around 700 to 800 million years ago – and occasionally younger.'

Unfortunately, that is stubbornly at odds with the fossil record, which goes back only about 575 million years. Before that there is no bona fide sign of animal life. After it, there came a surge of complex life – the famous Ediacaran and Cambrian outbursts – that left a wondrous abundance of fossils in scattered locations all around the world, many of them soft-bodied yet beautifully preserved.

Anderson's ambition was to resolve this discrepancy. He wondered why some ancient rocks preserved delicate animal fossils while others of the same age were barren. Anderson and his team discovered that the key fossils had formed in very particular mudstones containing the clays berthierine and kaolinite; these have antibacterial properties that could act as a shield against decay. (This was a lot harder than it sounds; it involved infrared and X-ray spectroscopy, X-ray diffraction and the use of the UK's national synchrotron.)

The challenge now is to study as many rocks of this type and age as they can find – there are known outcrops in Canada, Norway and Russia – and see if they contain fossils. When we met, Ross had just returned from an expedition with scientists from Oxford, Dartmouth, Williams College and the University of Washington to the remote Mackenzie Mountains of western Canada, where he had collected about 100 kilograms of 800 million- to 900 million-year-old rocks. All these had to be helicoptered out and shipped to Oxford to be cut into smaller pieces, a little at a time, and laboriously examined for signs of early life.

After our talk at All Souls, Ross took me to the nearby Earth Sciences Building where a DPhil student named George Wedlake was hunched over a microscope. I asked him what he was doing. 'Well, we start by putting a fragment of rock in hydrofluoric acid,' he explained. 'That dissolves the minerals that make up the rock, but it leaves the organic matter, which ends up as a sludge at the bottom of the container. It's quite a protracted process because the material has to be thoroughly washed several times to rinse away the acid, which is quite dangerous. Then we examine the organic material a little at a time through the microscope. If I find anything interesting, I suck it up with a pipette and put it aside for further study. Then I repeat the process.'

'Until you have got through a hundred kilograms of rocks?' I said.

He smiled. 'We don't look at every bit of every rock, but we do look at a *lot*,' he agreed.

I could hardly imagine a more tedious undertaking, but then George invited me to peer down the microscope at the kind of things he finds – an organism called *Palaeastrum*. 'It's one of the very first complex multicellular organisms known,' he said.

To the untrained eye it was just a cluster of translucent circular blobs, some quite dark, others pale. It was a bit like looking at quinoa – but quinoa from the earliest moments of complex life. It was an unexpectedly entrancing experience to peer into an eyepiece and find something that felt so miraculously ancestral. This particular specimen may have been just a random fragment of prehistoric algae, but somewhere in the ancient past there lived something very like this that eventually, over hundreds of millions of years of chemical fidgets and blundering encounters, became us. What a thought.

Suddenly I could appreciate why people like George and

Ross are prepared to devote years of their lives to slogging across remote and forbidding landscapes collecting rocks and staring for endless days into microscopes in the hope of finding some tiny thing that lived for perhaps less than a day more than half a billion years ago on a planet we would scarcely recognize as our own.

One reason life took so long to grow complex was that the world had to wait until the simpler organisms had oxygenated the atmosphere sufficiently in what is sometimes called the Great Oxidation Event. 'Animals could not summon up the energy to work,' as Richard Fortey put it. The driver of this massive change was a simple blue-green algae called cyanobacteria, the same stuff you find today forming a scummy green film on garden ponds. Cyanobacteria developed a trick that ultimately proved very beneficial to us: it learned to feed off the hydrogen that exists in spectacular abundance in sea water. In so doing it produced oxygen as a waste product – yet another of the happy accidents that eventually made us possible.

As cyanobacteria proliferated the world began to fill with O_2, to the consternation of those organisms that found it poisonous – which in those days was all of them. In an anaerobic (or non-oxygen-using) world, oxygen is extremely poisonous. Our white blood cells actually use oxygen to kill invading bacteria. That oxygen is fundamentally toxic often comes as a surprise to those of us who find it so convivial to our well-being, but that is only because we have evolved to exploit it. To other things it is a terror. It is what turns butter rancid and makes iron rust. Even we can tolerate it only up to a point. The oxygen level in our cells is only about a tenth the level found in the atmosphere. Oxygen was a more efficient way to produce energy, and it vanquished competitor organisms. Some retreated into the oozy, anaerobic world of bogs and lake bottoms. Others did likewise but then later (much later) migrated to the digestive tracts of beings like you and me.

Quite a number of these primeval entities are alive inside your body right now, helping to digest your food, but abhorring even the tiniest hint of O_2. Untold numbers of others failed to adapt and died.

At first, the extra oxygen the cyanobacteria produced didn't accumulate in the atmosphere but combined with iron to form ferric oxides, which sank to the bottom of primitive seas. For millions of years, the world literally rusted – a phenomenon vividly recorded in the banded iron deposits that provide so much of the world's iron ore today. It took some 2 billion years of such tiny exertions – nearly 40 per cent of Earth's total history – to raise the level of oxygen in the atmosphere to 20 per cent, the level that sustains us now.

The world was finally ready for us. But before we get too excited about that it is worth remembering that that Earth, as we are about to see, still belongs to the very small.

20

SMALL WORLD

It's perhaps not a good idea to take too personal an interest in your microbes. Louis Pasteur, the great French chemist and bacteriologist, became so preoccupied with his that he took to peering critically with a magnifying glass at every dish placed before him, a habit that presumably did not win him many repeat invitations to dinner.

In fact, there is no point in trying to hide from your bacteria, for they are on and around you always, in numbers you can't conceive of. If you are in good health and averagely diligent about hygiene, you will have a herd of about 1 trillion bacteria grazing on your fleshy plains – about 100,000 of them on every square centimetre of skin. They are there to dine off the 10 billion or so flakes of skin you shed every day, plus all the tasty oils and fortifying minerals that seep out from every pore and fissure. You are for them the ultimate buffet, with the convenience of warmth and constant mobility thrown in. By way of thanks, they give you B.O.

And those are just the bacteria that inhabit your skin. There are billions more tucked away in your gut and nasal passages, clinging to your hair and eyelashes, swimming over the surface of your eyes, drilling through the enamel of your teeth. Your digestive system alone is host to at least 400 types of microbes.

Some deal with sugars, some with starches, some attack other bacteria. A surprising number, like the ubiquitous intestinal spirochetes, have no detectable function at all. They just seem to like to be with you. Every human body consists of about 37.2 trillion cells but is host to up to 50 trillion bacterial cells, making you considerably more bacterial than human. Looked at genetically, the difference is starker still: you have 20,000 genes of your own, but perhaps as many as 20 million bacterial genes, so from that perspective you are roughly 99 per cent bacterial and not quite 1 per cent you. However you look at it, bacteria are a very big part of us. From the bacteria's point of view, of course, we are a rather small part of them.

Because we humans are big and clever enough to produce and use antibiotics and disinfectants, it is easy to convince ourselves that we have banished bacteria to the fringes of existence. Don't you believe it. Bacteria may not build cities or have interesting social lives, but they will be here when the Sun explodes. This is their planet, and we are on it only because they allow us to be.

Bacteria, never forget, got along for billions of years without us. We couldn't survive a day without them. They process our wastes and make them usable again. They purify our water and keep our soils productive. Bacteria synthesize vitamins in our gut, convert the things we eat into useful sugars and polysaccharides, and go to war on alien microbes that slip down our gullet.

We depend totally on bacteria to pluck nitrogen from the air and convert it into useful nucleotides and amino acids for us. It is a prodigious and gratifying feat. As Margulis and Sagan noted in *Microcosmos*, to do the same thing industrially (as when making fertilizers) manufacturers must heat the source materials to 500 degrees centigrade and squeeze them to 300 times normal pressures. Bacteria do the same thing all the time without fuss, and thank goodness, for no larger organism could survive without the nitrogen they pass on. Above all, microbes provide us with the air we breathe and keep the atmosphere stable; cyanobacteria supply the greater part of the planet's breathable oxygen.

And they are amazingly prolific. The more frantic among them can yield a new generation in less than ten minutes; *Clostridium perfringens*, the disagreeable little organism that causes gangrene, can reproduce in nine minutes and then begin at once to split again. At such a rate, a single bacterium could theoretically produce more offspring in two days than there are protons in the universe. 'Given an adequate supply of nutrients, a single bacterial cell can generate 280,000 billion individuals in a single day,' according to the Belgian biochemist and Nobel laureate Christian de Duve. In the same period, a human cell can just about manage a single division.

About once every million divisions, they produce a mutant. Usually this is bad luck for the mutant – for an organism, change is always risky – but just occasionally the new bacterium is endowed with some accidental advantage, such as the ability to elude or shrug off an attack of antibiotics. With this ability to evolve rapidly goes another, even scarier advantage. Bacteria share information. Any bacterium can take pieces of genetic coding from any other. Any adaptive change that occurs in one area of the bacterial universe can spread to any other. It's rather as if a human could go to an insect to get the necessary genetic coding to sprout wings or walk on ceilings. It means that from a genetic point of view bacteria have become a single super-organism – tiny, dispersed, but invincible.

They will live and thrive on almost anything you spill, dribble or shake loose. Just give them a little moisture – as when you run a damp cloth over a counter – and they will bloom as if created from nothing, though many others, residing deep in the Earth, don't even need moisture. They will eat wood, the glue in wallpaper, the metals in hardened paint. Some bacteria break down chemical materials from which, as far as we can tell, they gain no benefit at all.

Scientists in Australia found microbes known as *Thiobacillus concretivorans* which lived in – indeed, could not live without – concentrations of sulphuric acid strong enough to dissolve metal. In short, there are few environments in which bacteria aren't prepared to live. 'They are finding now that when they push probes into ocean vents so hot that the probes actually start to melt, there are bacteria even there,' Victoria Bennett of the Australian National University in Canberra told me.

In the 1920s two scientists at the University of Chicago, Edson Bastin and Frank Greer, announced that they had isolated from oil wells strains of bacteria that had been living at depths of 600 metres. The notion was dismissed as fundamentally preposterous – there was nothing to live *on* at 600 metres – and for fifty years it was assumed that their samples had been contaminated with surface microbes. We now know that there are a lot of microbes living deep within the Earth, many of which have nothing at all to do with the conventionally organic world. They eat rocks or, rather, the stuff that's in rocks – iron, sulphur, manganese and so on. And they breathe odd things too – iron, chromium, cobalt, even uranium. Such processes may be instrumental in concentrating gold, copper and other precious metals, and possibly deposits of oil and natural gas. It has even been suggested that their tireless nibblings created the Earth's crust.

Some scientists now think that there could be as much as 100 trillion tonnes of bacteria living beneath our feet in what are known as subsurface lithoautotrophic microbial ecosystems – SLiME for short. Thomas Gold of Cornell University once estimated that if you took all the bacteria out of the Earth's interior and dumped them on the surface, they would cover the planet to a depth of 15 metres – the height of a four-storey building. If the estimates are correct, there could be more life under the Earth than on top of it.

At depth, microbes shrink in size and become extremely sluggish. The liveliest of them may divide no more than once a century, some no more than perhaps once in 500 years. As *The*

Economist has put it: 'The key to long life, it seems, is not to do too much.' When things are really tough, bacteria are prepared to shut down all systems and wait for better times. In 1997 scientists successfully activated some anthrax spores that had lain dormant for eighty years in a museum display in Trondheim, Norway. Other microorganisms have leaped back to life after being released from a 118-year-old can of meat and a 166-year-old bottle of beer. In 1996, scientists at the Russian Academy of Science claimed to have revived bacteria frozen in Siberian permafrost for 3 million years.

It is a remarkable fact that well into the space age, most school textbooks divided the world of the living into just two categories – plant and animal. Microorganisms hardly featured. Amoebas and similar single-celled organisms were treated as proto-animals and algae as proto-plants. Bacteria were usually lumped in with plants, too, even though everyone knew they didn't belong there. As far back as the late nineteenth century the German naturalist Ernst Haeckel had suggested that bacteria deserved to be placed in a separate kingdom, which he called Monera, but the idea didn't begin to catch on among biologists until the 1960s, and then only among some of them.

Many organisms in the visible world were also poorly served by the traditional division. Fungi, the group that includes mushrooms, moulds, mildews, yeasts and puffballs, were nearly always treated as botanical objects, though in fact almost nothing about them – how they reproduce and respire, how they build themselves – matches anything in the plant world. Structurally, they have more in common with animals in that they build their cells from chitin, a material that gives them their distinctive texture. The same substance is used to make the shells of insects and the claws of mammals, though it isn't nearly so tasty in a stag beetle as in a Portobello mushroom. Above all, unlike all plants, fungi don't photosynthesize, so they have no chlorophyll and thus are not green. Instead they grow directly on their food source, which can be almost anything. Fungi will eat the sulphur off a concrete wall or the decaying matter between your toes – two things no plant will do. Almost the only plant-like quality they have is that they root.

Even less comfortably susceptible to categorization was the peculiar group of organisms formally called myxomycetes but more commonly known as slime moulds. The name no doubt has much to do with their obscurity. An appellation that sounded a little more dynamic – 'ambulant self-activating protoplasm', say – and less like the stuff you find when you reach deep into a clogged drain would almost certainly have earned these extraordinary entities a more immediate share of the attention they deserve, for slime moulds are, make no mistake, among the most interesting organisms in nature. When times are good, they exist as one-celled individuals, much like amoebas. But when conditions grow tough, they crawl to a central gathering place and become, almost miraculously, a slug. The slug is not a thing of beauty and it doesn't go terribly far – usually just from the bottom of a pile of leaf litter to the top, where it is in a slightly more exposed position – but for millions of years this may well have been the niftiest trick in the universe.

And it doesn't stop there. Having hauled itself up to a more favourable locale, the slime mould transforms itself yet again, taking on the form of a plant. By some curious orderly process the cells reconfigure, like the members of a tiny marching band, to make a stalk atop of which forms a bulb known as a fruiting body. Inside the fruiting body are millions of spores which, at the appropriate moment, are released to the wind to blow away to become single-celled organisms that can start the process again.

For years, slime moulds were claimed as protozoa by zoologists and as fungi by mycologists, though most people could see they didn't really belong anywhere. When genetic testing arrived, people in lab coats were surprised to find that slime moulds were so distinctive and peculiar that they weren't directly related to anything else in nature, and sometimes not even to each other.

In 1969, in an attempt to bring some order to the growing inadequacies of classification, an ecologist from Cornell named R. H. Whittaker unveiled in the journal *Science* a proposal to divide life into five principal branches – kingdoms, as they are known – called Animalia, Plantae, Fungi, Protista and Monera. Protista was a modification of an earlier term, *Protoctista*, which had been suggested a century earlier by a Scottish biologist named John Hogg and was meant to describe any organisms that were neither plant nor animal.

Though Whittaker's new scheme was a great improvement, Protista remained ill defined. Some taxonomists preserved the term for large unicellular organisms, but others treated it as the kind of odd-sock drawer of biology, putting into it anything that didn't fit anywhere else. It included (depending on which text you consulted) slime moulds, amoebas, even seaweed, among much else. By one calculation it contained as many as 200,000 different species of organism all told. That's a lot of odd socks.

Ironically, just as Whittaker's five-kingdom classification was beginning to find its way into textbooks, an unassuming academic at the University of Illinois was groping his way towards a discovery that would challenge everything. His name was Carl Woese (rhymes with rose) and since the mid-1960s – or about as early as it was possible to do so – he had been quietly studying genetic sequences in bacteria. In the early days, this was an exceedingly painstaking process. Work on a single bacterium could easily consume a year. At that time, according to Woese, only about 500 species of bacteria were formally recorded, which is fewer than the number of species you have in your mouth. Today the number is over 40,000 (up from just 5,000 twenty years ago), but still vastly short of the hundreds of millions of species thought to exist.

It isn't simple indifference that keeps the total low. Bacteria can be exasperatingly difficult to isolate and study. Only about 1 per cent will grow in culture. Considering how wildly adaptable they are in nature, it is an odd fact that the one place they seem not to wish to live is a petri dish. Plop them on a bed of agar and pamper them as you will, and most will just lie there, declining every inducement to bloom. Any bacterium that thrives in a lab is by definition exceptional, and yet these were, almost exclusively, the organisms studied by microbiologists. It was, said Woese, 'like learning about animals from visiting zoos'.

Genes, however, allowed Woese to approach microorganisms from another angle. As he worked, Woese realized that there were more fundamental divisions in the microbial world than anyone suspected. A lot of little organisms that looked like bacteria and behaved like bacteria were actually something else altogether – something that had branched off from bacteria a long time ago. Woese called these organisms archaebacteria, later shortened to archaea.

It has to be said that the attributes that distinguish archaea from bacteria are not the sort that would quicken the pulse of any but a biologist. They are mostly differences in their lipids and an absence of something called peptidoglycan. But in practice they make a world of difference. Archaea are more different from bacteria than you and I are from a crab or spider. Singlehandedly, Woese had discovered an unsuspected division of life so fundamental that it stood above the level of kingdom at the apogee of the Universal Tree of Life, as it is rather reverentially known.

In 1976 he startled the world – or at least the little bit of it that was paying attention – by redrawing the Tree of Life to incorporate not five main divisions, but twenty-three. These he grouped under three new principal categories – Bacteria, Archaea and Eukarya (sometimes spelled Eucarya) – which he called domains. The new arrangement was as follows:

- *Bacteria*: cyanobacteria, purple bacteria, gram-positive bacteria, green non-sulphur bacteria, flavobacteria and thermotogales
- *Archaea*: halophilicarchaeans, methanosarcina, methanobacterium, methanococcus, thermoceler, thermoproteus and pyrodictium
- *Eukarya*: diplomads, microsporidia, trichomonads, flagellates, entameba, slime moulds, ciliates, plants, fungi and animals

Woese's new divisions did not take the biological world by storm. Some dismissed his system as much too heavily weighted towards the microbial. Many just ignored it. Woese, according to Frances Ashcroft, 'felt bitterly disappointed'. But slowly his new scheme began to catch on among microbiologists. Botanists and zoologists were much slower to appreciate its virtues. It's not hard to see why. In Woese's model, the worlds of botany and zoology are relegated to a few twigs on the outermost branch of the eukaryotic limb. Everything else belongs to unicellular beings.

'These folks were brought up to classify in terms of gross morphological similarities and differences,' Woese told an interviewer in 1996. 'The idea of doing so in terms of molecular sequence is a bit hard for many of them to swallow.' In short, if they couldn't see a difference with their own eyes, they didn't like it. And so they persisted with the more conventional five-kingdom division – an arrangement that Woese called 'not very useful' in his milder moments and 'positively misleading' much of the rest of the time. 'Biology, like physics before it,' Woese wrote, 'has moved to a level where the objects of interest and their interactions often cannot be perceived through direct observation.'

In 1998 the great and ancient Harvard zoologist Ernst Mayr (who then was in his ninety-fourth year and still going strong: he died aged 100 in 2005; Woese followed in 2012, aged eighty-four) stirred the pot further by declaring that there should be just two prime divisions of life – 'empires' he called them. In a paper published in the *Proceedings of the National Academy of Sciences*, Mayr said that Woese's findings were interesting but ultimately misguided, noting that 'Woese was not trained as a biologist and quite naturally does not have an extensive familiarity with the principles of classification,' which is perhaps as close as one distinguished scientist can come to saying of another that he doesn't know what he is talking about.

The specifics of Mayr's criticisms are highly technical – they involved issues of meiotic sexuality, Hennigian cladification and controversial interpretations of the genome of *Methanobacterium thermoautrophicum*, among rather a lot else – but essentially he maintained that Woese's arrangement unbalanced the Tree of Life. Mayr noted that the bacterial realm, as it then stood, consisted of no more than a few thousand species while the archaean had a mere 175 named specimens, with perhaps a few thousand more to be found – 'but hardly more than that'.

By contrast, the eukaryotic realm – that is, the complicated organisms with nucleated cells, like us – numbered already in the millions of species. For the sake of 'the principle of balance', Mayr argued for combining the simple bacterial organisms in a single category, Prokaryota, while placing the more complex and 'highly evolved' remainder in the empire Eukaryota, which would stand alongside as an equal. Put another way, he wanted to keep things much as they were. This division between simple cells and complex cells 'is where the great break is in the living world'.

If Woese's new arrangement teaches us anything it is that life really is various and that most of that variety is small, unicellular and unfamiliar. It is a natural human impulse to think of evolution as a long chain of improvements, of a never-ending advance towards largeness and complexity – in a word, towards us. We flatter ourselves. Most of the real diversity in evolution has been small-scale. We large things are just flukes – an interesting side branch. Of the twenty-three main divisions of life, only three – plants, animals and fungi – are large enough to be seen by the human eye, and even they contain species that are microscopic. Indeed, according to Woese, if you totalled up all the biomass of the planet – every living thing, plants included – microbes would account for at least 80 per cent of all there is, perhaps more. The world belongs to the very small – and has done for a very long time.

So why, you are bound to ask at some point in your life, do microbes so often want to hurt us? What possible satisfaction could there be to a microbe in having us grow feverish or chilled, or disfigured with sores, or above all deceased? A dead host, after all, is hardly going to provide long-term hospitality.

To begin with, it is worth remembering that most microorganisms are neutral or even beneficial to human well-being. The most rampantly infectious organism on Earth, a bacterium called *Wolbachia*, doesn't hurt humans at all – or, come to that, any other vertebrates – but if you are a shrimp or worm or fruit fly it can make you wish you had never been born. Altogether, only about one microbe in 1,000 is a pathogen for humans, according to the *National Geographic* – though, knowing what some of them can do, we could be forgiven for thinking that that is quite enough. Even if most of them are benign, microbes are still the number-three killer in the Western world – and even many that don't kill us make us deeply rue their existence.

Making a host unwell has certain benefits for the microbe. The symptoms of an illness often help to spread the disease. Vomiting, sneezing and diarrhoea are excellent methods of getting out of one host and into position for boarding another. The most effective strategy of all is to enlist the help of a mobile third party. Infectious organisms love mosquitoes because the mosquito's sting delivers them directly into a bloodstream where they can get straight to work before the victim's defence mechanisms can figure out what's hit them. This is why so many grade-A diseases – malaria, yellow fever, dengue fever, encephalitis and 100 or so other less celebrated but often rapacious maladies – begin with a mosquito bite.

It is a mistake, however, to consider the matter too carefully from the position of logic because microorganisms clearly are not calculating entities. They don't care what they do to you any more than you care what distress you cause when you slaughter them by the millions with a soapy shower or a swipe of deodorant. The only time your continuing well-being is of consequence to a pathogen is when it kills you too well. If they eliminate you before they can move on, then they may well die out themselves. History, Jared Diamond notes, is full of diseases that 'once caused terrifying epidemics and then disappeared as mysteriously as they had come'. He cites the robust but mercifully transient English sweating sickness, which raged from 1485 to 1552, killing tens of thousands as it went before burning itself out. Too much efficiency is not a good thing for any infectious organism.

A great deal of sickness arises not because of what the organism has done to you but because of what your body is trying to do to the organism. In its quest to rid the body of pathogens, the immune system sometimes destroys cells or damages critical tissues, so often when you are unwell what you are feeling is not the pathogens but your own immune responses. Anyway, getting sick is a sensible response to infection. Sick people retire to their beds and thus are less of a threat to the wider community.

Because there are so many things out there with the potential to hurt you, your body holds lots of different varieties of defensive white blood cells – some 10 million types in all, each designed to identify and destroy a particular sort of invader. It would be impossibly inefficient to maintain 10 million separate standing armies, so each variety of white blood cell keeps only a few scouts on active duty.

When an infectious agent – what's known as an antigen – invades, relevant scouts identify the attacker and put out a call for reinforcements of the right type. While your body is manufacturing these forces, you are likely to feel wretched. The onset of recovery begins when the troops finally swing into action.

White cells are merciless and will hunt down and kill every last pathogen they can find. To avoid extinction, attackers have evolved two elemental strategies. Either they strike quickly and move on to a new host, as with common infectious illnesses like flu, or they disguise themselves so that the white cells fail to spot them, as with HIV, the virus responsible for AIDS, which can sit harmlessly and unnoticed in the nuclei of cells for years before springing into action.

One of the odder aspects of infection is that microbes that normally do no harm at all sometimes get into the wrong parts of the body and 'go kind of crazy', in the words of Dr Bryan Marsh, an infectious diseases specialist at Dartmouth–Hitchcock Medical Center in Lebanon, New Hampshire. 'It happens all the time with car accidents when people suffer internal injuries. Microbes that are normally benign in the gut get into other parts of the body – the bloodstream, for instance – and cause terrible havoc.'

Perhaps the scariest, most out-of-control bacterial disorder is a disease called necrotizing fasciitis, in which bacteria essentially eat the victim from the inside out, devouring internal tissue and leaving behind a pulpy, noxious residue. Patients often come in with comparatively mild complaints – a skin rash and fever, typically – but then dramatically deteriorate. When they are opened up it is often found that they are simply being consumed. The only treatment is what is known as 'radical excisional surgery' – cutting out every bit of infected area. Seventy per cent of victims die; many of the rest are left terribly disfigured. The source of the infection is a mundane family of bacteria called Group A Streptococcus, which normally do no more than cause strep throat. Very occasionally, for reasons unknown, some of these bacteria get through the lining of the throat and into the body proper, where they wreak the most devastating havoc. They are completely resistant to antibiotics. About 1,000 cases a year occur in the United States. The number rose significantly during the Covid epidemic, but has since mercifully receded.

Precisely the same thing happens with meningitis. At least 10 per cent of young adults, and perhaps 30 per cent of teenagers, carry the deadly meningococcal bacterium, but it lives quite harmlessly in the throat. Just occasionally – in about one young person in 100,000 – it gets into the bloodstream and makes them very ill indeed. In the worst cases, death can come in twelve hours. That's shockingly quick. 'You can have a person who's in perfect health at breakfast and dead by evening,' Marsh told me.

We would have much more success with bacteria if we weren't so profligate with our best weapon against them: antibiotics. Remarkably, by one estimate some 66 per cent of the antibiotics used in the developed world, and no less than 80 per cent in the United States, are given to farm animals, often routinely in stock feed, simply to promote growth or as a precaution against infection. Such applications give bacteria every opportunity to evolve a resistance to them. It is an opportunity that they have enthusiastically seized.

In 1952, penicillin was fully effective against all strains of staphylococcus bacteria, to such an extent that by the early 1960s the US Surgeon-General, William Stewart, felt confident enough to declare: 'The time has come to close the book on infectious diseases. We have basically wiped out infection in the United States.' Even as he spoke, however, some 90 per cent of those strains were in the process of developing immunity to penicillin. Soon one of these new strains, called methicillin-resistant staphylococcus aureus, began to show up in hospitals. Only one type of antibiotic, vancomycin, remained effective against it, but in 1997 a hospital in Tokyo reported the appearance of a strain that could resist even that. Within months it had spread to six other Japanese hospitals. All over, the microbes are beginning to win the war again: in US hospitals alone, some 72,000 people (up from 14,000 when this book first came out) die each year from infections they pick up there. As James Surowiecki noted in a *New Yorker* article, given a choice between developing antibiotics that people will take every day for two weeks and antidepressants that people will take every day for ever, drug companies not surprisingly opt for the latter.

Our carelessness is all the more alarming since the discovery that many other ailments may be bacterial in origin. The process began in 1983 when Barry Marshall, a doctor in Perth, Western Australia, found that many stomach cancers and most stomach ulcers are caused by a bacterium called *Helicobacter pylori*. Even though his findings were easily tested, the notion was so radical that years would pass before they were generally accepted. America's National Institutes of Health, for instance, didn't officially endorse the idea until 1994. 'Hundreds, even thousands of people must have died from ulcers who wouldn't have,' Marshall told a reporter from *Forbes* in 1999.

Since then, further research has shown that there is or may well be a bacterial component in all kinds of other disorders – heart disease, asthma, arthritis, multiple sclerosis, several types of mental disorders, many cancers, even, it has been suggested (in *Science* no less), obesity. The day may not be far off when we desperately require an effective antibiotic and haven't got one to call on.

It may come as a slight comfort to know that bacteria can themselves get sick. They are sometimes infected by bacteriophages (or simply phages), a type of virus. A virus is a strange and unlovely entity – 'a piece of nucleic acid surrounded by bad news' in the memorable phrase of the Nobel laureate Peter Medawar. Smaller and simpler than bacteria, viruses aren't themselves alive. In isolation they are inert and harmless.

But introduce them into a suitable host and they burst into busyness – into life. Virus numbers are constantly changing as new species emerge or old ones are discovered, but as of early 2025 the number of species recognized by the International Committee on Taxonomy of Viruses was 14,691. Of these, only about 270 cause diseases in humans, though that may be small comfort to anyone suffering from rabies, yellow fever, Ebola, polio, measles, flu, Covid or even the common cold.

Viruses prosper by hijacking the genetic material of a living cell and using it to produce more virus. They reproduce in a fanatical manner, then burst out in search of more cells to invade. Not being living organisms themselves, they can afford to be very simple. Many, including HIV, have ten genes or fewer, whereas even the simplest bacteria require several thousand. They are also very tiny, much too small to be seen with a conventional microscope. It wasn't until 1943 and the invention of the electron microscope that science got its first look at them. But they can do immense damage. Smallpox in the twentieth century alone killed an estimated 300 million people.

They also have an unnerving capacity to burst upon the world in some new and startling form and then to vanish again as quickly as they came. In 1916, in one such case, people in Europe and America began to come down with a strange sleeping sickness, which became known as encephalitis lethargica. Victims would go to sleep and not wake up. They could be roused without great difficulty to take food or go to the lavatory, and would answer questions sensibly – they knew who and where they were – though their manner was always apathetic. However, the moment they were permitted to rest they would at once sink back into deepest slumber and remain in that state for as long as they were left. Some went on in this manner for months before dying. A very few survived and regained consciousness but not their former liveliness. They existed in a state of profound apathy, 'like extinct volcanoes', in the words of one doctor. In ten years the disease killed some

5 million people and then quietly went away. It didn't get much lasting attention because in the meantime an even worse epidemic swept across the world.

It is sometimes called the Great Swine Flu epidemic and sometimes the Great Spanish Flu epidemic, but in either case it was ferocious. The First World War killed 21 million people in four years; swine flu did the same in its first four months. Almost 80 per cent of American casualties in the First World War came not from enemy fire, but from flu. In some units the mortality rate was as high as 80 per cent.

Swine flu arose as a normal, non-lethal flu in the spring of 1918, but somehow, over the following months – no one knows how or where – it mutated into something more severe. A fifth of victims suffered only mild symptoms, but the rest became gravely ill and many died. Some succumbed within hours; others held on for a few days.

In the United States, the first deaths were recorded among sailors in Boston in late August 1918, but the epidemic quickly spread to all parts of the country. Schools closed, public entertainments were shut down, people everywhere wore masks. It did little good. Between autumn 1918 and spring the following year, 548,452 people died of the flu in America. The toll in Britain was 220,000, with similar numbers in France and Germany. No one knows the global toll, as records in the Developing World were often poor, but it was not less than 20 million and probably more like 50 million. Some estimates have put the global total as high as 100 million.

In an attempt to devise a vaccine, medical authorities conducted experiments on volunteers at a military prison on Deer Island in Boston Harbor. The prisoners were promised pardons if they survived a battery of tests. These tests were rigorous to say the least. First, the subjects were injected with infected lung tissue taken from the dead and then sprayed in the eyes, nose and mouth with infectious aerosols. If they still failed to succumb, they had their throats swabbed with discharges taken straight from the sick and dying. If all else failed, they were required to sit open-mouthed while a gravely ill victim was sat up slightly and made to cough into their faces.

Out of – somewhat amazingly – 300 men who volunteered, the doctors chose sixty-two for the tests. None contracted the flu – not one. The only person who did grow ill was the ward doctor, who swiftly died. The probable explanation for this is that the epidemic had passed through the prison a few weeks earlier and the volunteers, all of whom had survived that visitation, had a natural immunity. Much about the 1918 flu epidemic is understood poorly or not at all. One mystery is how it erupted suddenly, all over, in places separated by oceans, mountain ranges and other earthly impediments. A virus can survive for no more than a few hours outside a host body, so how could it appear in Madrid, Bombay and Philadelphia all in the same week? The probable answer is that it was incubated and spread by people who had only slight symptoms or none at all. Even in normal outbreaks, about 10 per cent of people in any given population (and as much as 50 per cent, according to one study) have the flu but are unaware of it because they experience no ill effects. And because they remain in circulation they tend to be the great spreaders of the disease.

Even more mysterious is that it was most devastating to people in the prime of life. Flu is normally hardest on infants and the elderly, but in the 1918 outbreak deaths were overwhelmingly among people in their twenties and thirties. Older people may have benefited from resistance gained from an earlier exposure to the same strain, but why the very young were similarly spared is unknown. The greatest mystery of all is why the 1918 flu was so ferociously deadly when most flus are not.

New and frightening viruses crop up all the time – as we saw only too clearly with Covid-19. We forget how quickly this swept into our lives. The first mention of it in the West was in an article in the *New York Times* on 8 January 2020, which noted that it had arisen in Wuhan, China, where so far it had sickened fifty-nine people. Fortunately, the article added, there was no evidence that the virus spread easily among people.

That was clearly a touch premature. Within a month, China had almost 50,000 recorded cases and Covid was spreading through the world in a truly alarming fashion. Italy went from having five cases in early February to over 170,000 in just six weeks. New York State had just one recorded case on 1 March but 200,000 six weeks later.

The cumulative death toll was even more shocking. On 1 January 2020, there was one known fatality globally from Covid. By 1 April that number had gone up to 44,000, by 1 May to 236,000, by 1 June to 388,000, by 1 July to 535,000. And so it went. The total number of Covid deaths in the world as of this writing is essentially unknowable because of underreporting by some countries, but estimates range from 7 million to more than 18 million.

The miracle is that other such diseases haven't gone rampant. Lassa fever, which wasn't first detected until 1969, in West Africa, is extremely virulent and little understood. In 1969, a doctor at a Yale University lab in New Haven, Connecticut, who was studying Lassa fever came down with it. He survived, but, more alarmingly, a technician in a nearby lab, with no direct exposure, also contracted the disease and died.

Happily the outbreak stopped there, but we can't count on always being so fortunate. Our lifestyles invite epidemics. Air travel makes it possible to spread infectious agents across the planet with amazing ease. An Ebola virus could begin the day in, say, Benin, and finish it in New York or Hamburg or Nairobi, or all three. It means also that medical authorities increasingly need to be acquainted with pretty much every malady that exists everywhere, but of course they are not. In 1990, a Nigerian living in Chicago was exposed to Lassa fever on a visit to his homeland, but didn't develop symptoms until he had returned to the United States. He died in a Chicago hospital without diagnosis and without anyone taking any special precautions in treating him, unaware that he had one of the most lethal and infectious diseases on the planet. Miraculously, no one else was infected. Clearly we can't assume that we will always be so lucky.

And on that sobering note, it's time to return to the world of the visibly living.

21

LIFE GOES ON

It isn't easy, as we have already seen, to become a fossil, but it is worth noting that this applies to big things as well as small ones. The fate of nearly all living organisms – over 99.9 per cent of them – is to compost down to nothingness. When your spark is gone, every molecule you own will be nibbled off you or sluiced away to be put to use in some other system. That's just the way it is. Even if you make it into the small pool of organisms, the less than 0.1 per cent, that don't get devoured, the chances of being fossilized are very small.

Only about one bone in a billion, it is thought, ever becomes fossilized. If that is so, it means that the complete fossil legacy of all the Americans alive today – that's 333 million people with 206 bones each – will be fewer than 100 bones, not enough to make even half a complete skeleton. That's not to say, of course, that any of these bones will ever actually be found. Bearing in mind that they can be buried anywhere within an area of slightly over 9.3 million square kilometres, little of which will ever be turned over, much less examined, it would be something of a miracle if they were. Fossils are in every sense vanishingly rare. Most of what has lived on Earth has left behind no record at all. It has been estimated that less than one species in 10,000 has made it into the fossil record. That in itself is a stunningly infinitesimal proportion. However, if you accept the common estimate that the Earth has produced 30 billion species of creature in its time, and Richard Leakey and Roger Lewin's statement (in *The Sixth Extinction*) that there are 250,000 species of creature in the fossil record, that reduces the proportion to just one in 120,000. Either way, what we possess is the merest sampling of all the life that the Earth has spawned.

Moreover, the record we do have is hopelessly skewed. Most land animals, of course, don't die in sediments. They drop in the open and are eaten or left to rot or weather down to nothing. The fossil record, consequently, is almost absurdly biased in favour of marine creatures. About 95 per cent of all the fossils we possess are of animals that once lived under water, mostly in shallow seas.

I mention all this to explain why on a grey day in February, many years ago, I went to the Natural History Museum in London to meet a cheerful, vaguely rumpled, very likeable palaeontologist named Richard Fortey.

Fortey, who died in 2025, knew an awful lot about an awful lot. He was the author of a wry, splendid book called *Life: An Unauthorised Biography*, which covers the whole pageant of animate creation. His first love throughout his working life was a type of marine creature called trilobites, which once teemed in Ordovician seas but haven't existed for a long time except in fossilized form. All trilobites shared a basic body plan of three parts, or lobes – head, tail, thorax – from which comes the name. Fortey found his first when he was a boy clambering over rocks at St David's Bay in Wales. He was hooked for life.

He took me to a gallery of tall metal cupboards. Each cupboard was filled with shallow drawers, and each drawer was filled with stony trilobites – 20,000 specimens in all.

'It seems like a big number,' he agreed, 'but you have to remember that millions upon millions of trilobites lived for millions upon millions of years in ancient seas, so twenty thousand isn't a huge number. And most of these are only partial specimens. Finding a complete trilobite fossil is still a big moment for a palaeontologist.'

Trilobites first appeared – fully formed, seemingly from nowhere – about 540 million years ago, near the start of the great outburst of complex life popularly known as the Cambrian explosion, and then vanished, along with a great deal else, in the great and still largely mysterious Permian extinction 3 million or so centuries later. As with all extinct creatures, there is a natural temptation to regard them as failures, but in fact they were among the most successful animals ever to live. They reigned for 300 million years – twice the span of dinosaurs, which were themselves among history's great survivors. Humans, Fortey pointed out, have survived so far for one-half of 1 per cent as long.

With so much time at their disposal, the trilobites proliferated prodigiously. Most remained small, about the size of modern beetles, but some grew to be as big as platters. Altogether they formed at least 5,000 genera and 60,000 species – though more turn up all the time.

Throughout the nineteenth century, trilobites were almost the only known forms of early complex life, and for that reason were assiduously collected and studied. The big mystery about them was their sudden appearance. Even now, as Fortey told me, it can be startling to go to the right formation of rocks and to work your way upwards through the aeons, finding no visible life at all, and then suddenly 'a whole *Profallotaspis* or *Elenellus* as big as a crab will pop into your waiting hands'. These were creatures with limbs, gills, nervous systems, probing antennae, 'a brain of sorts', in Fortey's words, and the strangest eyes ever seen. Made of calcite rods, the same stuff that forms limestone, they constituted the earliest visual systems known. More than this, the earliest trilobites didn't consist of just one venturesome species but dozens, and didn't appear in one or two locations but all over. Many thinking people in the nineteenth century saw this as proof of God's handiwork and refutation of

Darwin's evolutionary ideals. If evolution proceeded slowly, they asked, then how did he account for this sudden appearance of complex, fully formed creatures? The fact is, he couldn't.

And so matters seemed destined to remain for ever until one day in 1909, three months shy of the fiftieth anniversary of the publication of Darwin's *On the Origin of Species*, when a palaeontologist named Charles Doolittle Walcott made an extraordinary find in the Canadian Rockies.

Walcott was born in 1850 and grew up near Utica, New York, in a family of modest means, which became more modest still with the sudden death of his father when Charles was an infant. As a boy Walcott discovered that he had a knack for finding fossils, particularly trilobites, and built up a collection of sufficient distinction that it was bought by Louis Agassiz for his museum at Harvard for a small fortune – about £45,000 in today's money. Although he had barely a high school education and was self-taught in the sciences, Walcott became a leading authority on trilobites and was the first person to establish that they were arthropods, the group that includes modern insects and crustaceans.

In 1879 Walcott took a job as a field researcher with the newly formed US Geological Survey and served with such distinction that within fifteen years he had risen to be its head. In 1907 he was appointed secretary of the Smithsonian Institution, where he remained until his death in 1927. Despite his administrative obligations, he continued to do fieldwork and to write prolifically. 'His books fill a library shelf,' according to Fortey. Not incidentally, he was also a founding director of the National Advisory Committee for Aeronautics, which eventually became the National Aeronautics and Space Agency, or NASA, and thus can rightly be considered the grandfather of the space age.

But what he is remembered for now is an astute but lucky find in British Columbia, high above the little town of Field, in the late summer of 1909. The customary version of the story is that Walcott, accompanied by his wife, was riding on a mountain trail when his wife's horse slipped on loose stones. Dismounting to assist her, Walcott discovered that the horse had turned a slab of shale that contained fossil crustaceans of an especially ancient and unusual type. Snow was falling – winter comes early to the Canadian Rockies – so they didn't linger, but the next year at the first opportunity Walcott returned to the spot. Tracing the presumed route of the rocks' slide, he climbed 750 feet to near the mountain's summit. There, 8,000 feet above sea level, he found a shale outcrop, about the length of a city block, containing an unrivalled array of fossils from soon after the moment when complex life burst forth in dazzling profusion – the famous Cambrian explosion. Walcott had found, in effect, the holy grail of palaeontology. The outcrop became known as the Burgess Shale, from the name of the ridge on which it was found, and for a long time it provided 'our sole vista upon the inception of modern life in all its fullness', as the late Stephen Jay Gould recorded in his popular book *Wonderful Life*.

Gould, ever scrupulous, discovered from reading Walcott's diaries that the story of the Burgess Shale's discovery appears to have been somewhat embroidered – Walcott makes no mention of a slipping horse or falling snow – but there is no disputing that it was an extraordinary find.

It is almost impossible for us, whose time on Earth is limited to a breezy few decades, to appreciate how remote in time from us the Cambrian outburst was. If you could fly backwards into the past at the rate of one year per second, it would take you about half an hour to reach the time of Christ, and a little over three weeks to get back to the beginnings of human life. But it would take you twenty years to reach the dawn of the Cambrian period. It was, in other words, an extremely long time ago and the world was a very different place.

For one thing, 500 million years ago and more, when the Burgess Shale was formed, it wasn't at the top of a mountain but at the foot of one. Specifically, it was in a shallow ocean basin at the bottom of a steep cliff. The seas of that time teemed with life, but normally the animals left no record because they were soft-bodied and decayed upon dying. At Burgess, however, the cliff collapsed and the creatures below, entombed in a mudslide (rich in the berthierine clays mentioned in Chapter 19), were pressed like flowers in a book, their features preserved in wondrous detail.

In annual summer trips from 1910 to 1925 (by which time he was seventy-five years old), Walcott excavated tens of thousands of specimens (Gould says 80,000; the fact-checkers of *National Geographic* say 60,000), which he brought back to Washington for further study. In both sheer numbers and diversity the collection was unparalleled. Some of the Burgess fossils had shells; many others did not. Some of the creatures were sighted, others blind. The variety was enormous, consisting of 140 species, by one count. 'The Burgess Shale included a range of disparity in anatomical designs never again equaled, and not matched today by all the creatures in the world's oceans,' Gould wrote.

Unfortunately, according to Gould, Walcott failed to discern the significance of what he had found. 'Snatching defeat from the jaws of victory,' Gould wrote in another work, *Eight Little Piggies*, 'Walcott then proceeded to misinterpret these magnificent fossils in the deepest possible way.' He placed them into modern groups, making them ancestral to today's worms, jellyfish and other creatures, and thus failed to appreciate their distinctness. 'Under such an interpretation,' Gould sighed, 'life began in primordial simplicity and moved inexorably, predictably onward to more and better.'

Walcott died in 1927 and the Burgess fossils were largely forgotten. For nearly half a century they stayed shut away in drawers in the American Museum of Natural History in Washington, seldom consulted and never questioned. Then, in 1973, a graduate student from Cambridge University named Simon Conway Morris paid a visit to the collection. He was astonished by what he found. The fossils were far more varied and magnificent than Walcott had indicated in his writings. In taxonomy the category that describes the basic body plans of organisms is the phylum, and here, Conway Morris concluded, were drawer after drawer of such anatomical singularities – all amazingly and unaccountably unrecognized by the man who had found them.

With his supervisor, Harry Whittington, and fellow graduate student Derek Briggs, Conway Morris spent the next several years making a systematic revision of the entire collection, and cranking out one exciting monograph after another as discovery piled upon discovery. Many of the creatures employed body plans that were not simply unlike anything seen before or since, but were *bizarrely* different. One, *Opabinia*, had five eyes and a nozzle-like snout with claws on the end. Another, a disc-shaped being called *Peytoia*, looked almost comically like a circular pineapple slice. A third had evidently tottered about on rows of stilt-like legs, and was so odd that they named it *Hallucigenia*. There was so much unrecognized novelty in the collection that at one point, upon opening a new drawer, Conway Morris famously was heard to mutter, 'Oh fuck, not another phylum.'

The English team's revisions showed that the Cambrian had been a time of unparalleled innovation and experimentation in body designs. For almost 4 billion years life had dawdled along without any detectable ambitions in the direction of complexity, and then suddenly, in the space of just 5 million or 10 million years, it had created all the basic body designs still in use today. Name a creature, from a nematode worm to Zendaya, and they all use architecture first created in the Cambrian party.

What was most surprising, however, was that there were so many body designs that had failed to make the cut, so to speak, and left no descendants. Altogether, according to Gould, at least fifteen and perhaps as many as twenty of the Burgess animals belonged to no recognized phylum. (The number soon grew in some popular accounts to as many as 100 – far more than the Cambridge scientists ever actually claimed.) 'The history of life,' wrote Gould, 'is a story of massive removal followed by differentiation within a few surviving stocks, not the conventional tale of steadily increasing excellence, complexity, and diversity.' Evolutionary success, it appeared, was a lottery.

One creature that *did* manage to slip through, a small wormlike being called *Pikaia gracilens*, was found to have a primitive spinal column, making it the earliest known ancestor of all later vertebrates, including us. *Pikaia* were by no means abundant among the Burgess fossils, so goodness knows how close they may have come to extinction. Gould, in a famous quotation, left no doubt that he saw our lineal success as a fortunate fluke: 'Wind back the tape of life to the early days of the Burgess Shale; let it play again from an identical starting point, and the chance becomes vanishingly small that anything like human intelligence would grace the replay.'

Gould's *Wonderful Life* was published in 1989 to general critical acclaim and was a great commercial success. What wasn't generally known was that many scientists didn't agree with Gould's conclusions at all, and that it was all soon to get very ugly. In the context of the Cambrian, 'explosion' would soon have more to do with modern tempers than ancient physiological facts.

Nearly forty years after Walcott made his discovery in Canada, on the other side of the planet in Australia a young geologist named Reginald Sprigg found something even older and in its way just as remarkable.

In 1946 Sprigg, then just an assistant government geologist for the state of South Australia, was sent to make a survey of abandoned mines in the Ediacaran Hills of the Flinders Range, an expanse of baking outback some 500 kilometres north of Adelaide. The idea was to see if there were any old mines that might be profitably reworked using newer technologies, so he wasn't studying surface rocks at all, still less fossils. But one day, while eating his lunch, Sprigg idly overturned a hunk of sandstone and was surprised – to put it mildly – to see that the rock's surface was covered in delicate fossils, rather like the impressions leaves make in mud. These rocks predated the Cambrian explosion. He was looking at the dawn of visible life.

Sprigg submitted a paper to *Nature*, but it was turned down. He read it instead at the next annual meeting of the Australian and New Zealand Association for the Advancement of Science, but it failed to find favour with the association's head, who said the Ediacaran imprints were merely 'fortuitous inorganic markings' – patterns made by wind or rain or tides, but not living beings. His hopes not yet entirely crushed, Sprigg travelled to London and presented his findings to the 1948 International Geological Congress, but failed to excite either interest or belief. Finally, for want of a better outlet, he published his findings in the *Transactions of the Royal Society of South Australia*. Then he quit his government job and took up oil exploration.

Nine years later, in 1957, a schoolboy named Roger Mason, while walking through Charnwood Forest in the English Midlands, found a rock with a strange fossil in it, similar to a modern sea pen and exactly like some of the specimens Sprigg had found and been trying to tell everyone about ever since. The schoolboy alerted a palaeontologist at the University of Leicester, who identified it at once as Precambrian. Young Mason got his picture in the papers and was treated as a precocious hero. The specimen was named in his honour *Charnia masoni*.

Today some of Sprigg's original Ediacaran specimens, along with many others that have been found throughout the Flinders Range since that time, can be seen in a glass case in an upstairs room of the stout and lovely South Australian Museum in Adelaide, but they don't attract a great deal of attention. The delicately etched patterns are rather faint and not terribly arresting to the untrained eye. They are mostly small and disc-shaped, with occasional, vague trailing ribbons. Fortey has described them as 'soft-bodied oddities'.

There is still some debate about what these things were or how they lived. They had, as far as can be told, no mouth or anus with which to take in and discharge digestive materials, and no internal organs with which to process them along the way. 'In life,' Fortey wrote, 'most of them probably simply lay upon the surface of the sandy sediment, like soft, structure-less and inanimate flatfish,' though there is some evidence, in the form of fossil impressions, that at least some of them had some mobility. All the Ediacaran creatures were diploblastic, meaning they were built from two layers of tissue. With the exception of jellyfish, all animals today are triploblastic.

Some experts have suggested that they weren't animals at all, but more like plants or fungi. The distinctions between plant and animal are not always clear-cut even now. The modern sponge spends its life fixed to a single spot and has no eyes or brain or beating heart, and yet is an animal. 'When we go back to the Precambrian the differences between plants and animals were probably even less clear,' Richard Fortey has noted. 'There isn't any rule that says you have to be demonstrably one or the other.'

Many authorities have viewed Ediacaran life as a kind of failed experiment, a stab at complexity that didn't take, possibly because the sluggish Ediacaran organisms were devoured or outcompeted by the lither and more sophisticated animals of the Cambrian period. The feeling was that ultimately they weren't terribly important to the development of life on Earth. One suggestion was that there was a mass extermination at the Precambrian–Cambrian boundary and that most or all the Ediacaran creatures failed to move on to the next phase. The real business of complex life, in other words, started with the Cambrian explosion. That's how Gould saw it, in any case.

As for the revisions of the Burgess Shale fossils, almost at once people began to question the interpretations and, in particular, Gould's interpretation of the interpretations. 'From the first there were a number of scientists who doubted the account that Steve Gould had presented, however much they admired the manner of its delivery,' Fortey wrote in *Life*. That is putting it mildly.

'If only Stephen Gould could think as clearly as he writes!' barked Richard Dawkins in the opening line of a review (in the *Sunday Telegraph*) of *Wonderful Life*. Dawkins acknowledged that the book was 'unputdownable' and a 'literary tour-de-force', but accused Gould of engaging in a 'grandiloquent and near-disingenuous' misrepresentation of the facts by suggesting that the Burgess revisions had stunned the palaeontological community. 'The view that he is attacking – that evolution marches inexorably towards a pinnacle such as man – has not been believed for 50 years,' Dawkins fumed.

Strangest of all was the response of one of the heroes of *Wonderful Life*, Simon Conway Morris, who startled many in the palaeontological community by rounding abruptly on Gould in a book of his own, *The Crucible of Creation*. 'I have never encountered such spleen in a book by a professional,' Fortey wrote later. 'The casual reader of *The Crucible of Creation*, unaware of the history, would never gather that the author's views had once been close to (if not actually shared with) Gould's.'

As a consequence of the controversy, the early Cambrian fossils underwent a period of critical reappraisal. Fortey and Derek Briggs – one of the other principals in Gould's book – used a method known as cladistics to compare the various Burgess fossils. In simple terms, cladistics consists of organizing organisms on the basis of shared features. Fortey gave as an example the idea of comparing a shrew and an elephant. If you considered the elephant's large size and striking trunk you might conclude that it could have little in common with a tiny, sniffling shrew. But if you compared both of them with a lizard, you would see that the elephant and shrew were in fact built to much the same plan. In essence, Fortey argued, Gould saw elephants and shrews where he and Briggs saw mammals. The Burgess creatures, they believed, weren't as strange and various as they appeared at first sight. 'They were often no stranger than trilobites,' Fortey told me. 'It is just that we have had a century or so to get used to trilobites. Familiarity, you know, breeds familiarity.'

The upshot is that the Burgess fossils were found to be not so different after all. *Hallucigenia*, it turned out, had been reconstructed upside down. Its stilt-like legs were actually spikes along its back. *Peytoia*, the weird creature that looked like a pineapple slice, was found to be not a distinct creature but merely part of a larger animal called *Anomalocaris*. Many of the Burgess specimens were reassigned to living phyla – just where Walcott put them in the first place. *Hallucigenia* and some others are thought to be related to *Onychophora*, a group of caterpillar-like animals. Others have been reclassified as precursors of the modern annelids. In fact, says Fortey, 'there are relatively few Cambrian designs that are wholly novel. More often they turn out to be just interesting elaborations of well-established designs.' As he wrote in *Life*: 'None was as strange as a present day barnacle, nor as grotesque as a queen termite.'

So the Burgess Shale specimens weren't so spectacular after all. This made them, as Fortey has written, 'no less interesting, or odd, just more explicable'. Their weird body plans were just a kind of youthful exuberance – the evolutionary equivalent, as it were, of spiked hair and tongue studs. Eventually the forms settled into a staid and stable middle age.

But that still left the enduring question of where all these animals had come from – how they had suddenly appeared from nowhere. In fact, it turns out the Cambrian explosion was not quite so explosive as once supposed. The Cambrian animals, it is now thought, were probably there all along, but were just too small to see. Once again it was trilobites that provided the clue – in particular, that seemingly mystifying appearance of different types of trilobite in widely scattered locations around the globe, all at more or less the same time. On the face of it, the sudden appearance of lots of fully formed but varied creatures would seem to enhance the miraculousness of the Cambrian outburst, but in fact it did the opposite. It is one thing to have one well-formed creature like a trilobite burst forth in isolation – that really is a wonder – but to have many of them, all distinct but clearly related, turning up simultaneously in the fossil record in places as far apart as China and New York clearly suggests that we are missing a big part of their history. There could be no stronger evidence that they simply had to have a forebear – some grandfather species that started the line in a much earlier past.

'It isn't necessary to be big to be a perfectly functioning, complex organism,' Fortey noted. 'The sea swarms with tiny arthropods today that have left no fossil record.' As an example, he cited the little copepod, which numbers in the trillions in modern seas and clusters in shoals large enough to turn vast areas of the ocean black, and yet our total knowledge of its ancestry is a single specimen found in the body of an ancient fossilized fish.

'The Cambrian explosion, if that's the word for it, probably was more an increase in size than a sudden appearance of new body types,' Fortey wrote in *Life*. 'And it could have happened quite swiftly, so in that sense I suppose it was an explosion.' The idea is that, just as mammals bided their time for 100 million years until the dinosaurs cleared off and then seemingly burst forth in profusion all over the planet, so too perhaps the arthropods and other triploblasts waited in semi-microscopic anonymity for the dominant Ediacaran organisms to have their day. As Fortey has put it: 'We know that mammals increased in size quite dramatically after the dinosaurs went – though when I say quite abruptly I of course mean it in a geological sense. We're still talking millions of years.'

Incidentally, Reginald Sprigg did eventually get a measure of overdue credit. One of the main early genera, *Spriggina*, was named in his honour, as were several species, and the whole became known as the Ediacaran fauna after the hills through which he had searched. By this time, however, Sprigg's fossil-hunting days were long over. After leaving geology he founded a successful oil company and eventually retired to an estate in his beloved Flinders Range where he created a wildlife reserve. He died in 1994 a rich man.

22

GOODBYE TO ALL THAT

When you consider it from a human perspective, and clearly it would be difficult for us to do otherwise, life is an odd thing. It couldn't wait to get going, but then, having got going, it seemed in very little hurry to move on.

Consider the lichen. Lichens are just about the hardiest visible organisms on Earth, but among the least ambitious. They will grow happily enough in a sunny churchyard, but they particularly thrive in environments where no other organism would go – on blowy mountaintops and Arctic wastes, wherever there is little but rock and rain and cold, and almost no competition. In areas of Antarctica where virtually nothing else will grow, you can find vast expanses of lichen – 400 types of them – adhering devotedly to every wind-whipped rock.

For a long time, people couldn't understand how they did it. Because lichens grew on bare rock without evident nourishment or the production of seeds, many people – educated people – believed they were stones caught in the process of becoming plants. 'Spontaneously, inorganic stone becomes living plant!' rejoiced one observer, a Dr Hornschuch, in 1819.

Closer inspection showed that lichens were more interesting than magical. They are in fact a partnership between fungi and algae. The fungi excrete acids which dissolve the surface of the rock, freeing minerals that the algae convert into food sufficient to sustain both. It is not a very exciting arrangement, but it is a conspicuously successful one. The world has more than 20,000 species of lichens.

Like most things that thrive in harsh environments, lichens are slow-growing. It may take a lichen more than half a century to attain the dimensions of a shirt button. Those the size of dinner plates, writes David Attenborough, are therefore 'likely to be hundreds if not thousands of years old'. It would be hard to imagine a less fulfilling existence. 'They simply exist,' Attenborough adds, 'testifying to the moving fact that life even at its simplest level occurs, apparently, just for its own sake.'

It is easy to overlook this thought that life just is. As humans we are inclined to feel that life must have a point. We have plans and aspirations and desires. We want to take constant advantage of all the intoxicating existence we've been endowed with. But what's life to a lichen? Yet its impulse to exist, to be, is every bit as strong as ours – arguably even stronger. If I were told that I had to spend decades being a furry growth on a rock in the woods, I believe I would lose the will to go on. Lichens don't. Like virtually all living things, they will suffer any hardship, endure any insult, for a moment's additional existence. Life, in short, just wants to be. But – and here's an interesting point – for the most part it doesn't want to be much.

This is perhaps a little odd, because life has had plenty of time to develop ambitions. If you imagine the 4,500 million years of Earth's history compressed into a normal earthly day, then life begins very early, about 4 a.m., with the rise of the first simple, single-celled organisms, but then advances no further for the next sixteen hours. Not until almost 8.30 p.m., with the day five-sixths over, has the Earth anything to show the universe but a restless skin of microbes. Then, finally, the first sea plants appear, followed twenty minutes later by the enigmatic Ediacaran fauna first seen by Reginald Sprigg in Australia.

GOODBYE TO ALL THAT

At 9.04 p.m. trilobites swim on to the scene, followed more or less immediately by the shapely creatures of the Burgess Shale. Just before 10 p.m. plants begin to pop up on the land. Soon after, with less than two hours left in the day, the first land creatures follow. Thanks to ten minutes or so of balmy weather, by 10.24 the Earth is covered in the great carboniferous forests whose residues give us all our coal, and the first winged insects are evident. Dinosaurs plod on to the scene just before 11 p.m. and hold sway for about three-quarters of an hour. At twenty-one minutes to midnight they vanish and the age of mammals begins. Humans emerge one minute and seventeen seconds before midnight.

The whole of our recorded history, on this scale, would be no more than a few seconds, a single human lifetime barely an instant. Throughout this greatly speeded-up day, continents slide about and bang together at a clip that seems positively reckless. Mountains rise and melt away, ocean basins come and go, ice sheets advance and withdraw. And throughout the whole, about three times every minute, somewhere on the planet there is a flashbulb pop of light marking the impact of a Manson-sized meteor or larger. It's a wonder that anything at all can survive in such a pummelled and unsettled environment. In fact, not many things do for long.

Perhaps an even more effective way of grasping our extreme recentness as a part of this 4.5-billion-year-old picture is to stretch your arms to their fullest extent and imagine that width as the entire history of the Earth. On this scale, according to John McPhee in *Basin and Range*, the distance from the fingertips of one hand to the wrist of the other is Precambrian. All of complex life is in one hand, 'and in a single stroke with a medium-grained nail file you could eradicate human history'.

Fortunately, that moment hasn't happened, but the chances are good that it will. I don't wish to interject a note of gloom just at this point, but the fact is that there is one other extremely pertinent quality about life on Earth: it goes extinct. Quite regularly. For all the trouble they take to assemble and preserve themselves, species crumple and die remarkably routinely. And the more complex they get, the more quickly they appear to go extinct. Which is perhaps one reason why so much of life isn't terribly ambitious.

So any time life does something bold it is quite an event, and few occasions were more eventful than when life moved on to the next stage in our narrative and came out of the sea. Land was a formidable environment: hot, dry, bathed in intense ultraviolet radiation, lacking the buoyancy that makes movement in water comparatively effortless. To live on land, creatures had to undergo wholesale revisions of their anatomies. Hold a fish at each end and it sags in the middle, its backbone too weak to support it. To survive out of water, marine creatures needed to come up with new load-bearing internal architecture – not the sort of adjustment that happens overnight.

Above all and most obviously, any land creature would have to develop a way to take its oxygen directly from the air rather than filter it from water. These were not trivial challenges to overcome. On the other hand, there was a powerful incentive to leave the water: it was getting dangerous down there. The slow fusion of the continents into a single land mass, Pangaea, meant there was much less coastline than formerly and thus less coastal habitat. So competition was fierce. There was also an omnivorous and unsettling new type of predator on the scene, one so perfectly designed for attack that it has scarcely changed in all the long aeons since its emergence: the shark. Never would there be a more propitious time to find an alternative environment to water.

Plants began the process of land colonization about 450 million years ago, accompanied of necessity by tiny mites and other organisms which they needed to break down and recycle dead organic matter on their behalf. Larger animals took a little longer to emerge, but by about 400 million years ago they were venturing out of the water, too. Popular illustrations have encouraged us to envision the first venturesome land dwellers as a kind of ambitious fish — something like the modern mudskipper, which can hop from puddle to puddle during droughts — or even as a fully formed amphibian. In fact, the first visible mobile residents on dry land were probably much more like modern woodlice, sometimes also known as pillbugs or sow bugs. These are the little bugs (crustaceans, in fact) that are commonly thrown into confusion when you upturn a rock or log.

For those that learned to breathe oxygen from the air, times were good. Oxygen levels in the Devonian and Carboniferous periods, when terrestrial life first bloomed, were as high as 35 per cent (as opposed to nearer 20 per cent now). This allowed animals to grow remarkably large remarkably quickly.

And how, you may reasonably wonder, can scientists know what oxygen levels were like hundreds of millions of years ago? The answer lies in a slightly obscure but ingenious field known as isotope geochemistry. The long-ago seas of the Carboniferous and Devonian swarmed with tiny plankton which wrapped themselves inside tiny protective shells. Then, as now, the plankton created their shells by drawing oxygen from the atmosphere and combining it with other elements (carbon especially) to form durable compounds such as calcium carbonate. It's the same chemical trick that goes on in (and is discussed elsewhere in relation to) the long-term carbon cycle — a process that doesn't make for terribly exciting narrative but is vital for creating a habitable planet.

Eventually in this process all the tiny organisms die and drift to the bottom of the sea, where they are slowly compressed into limestone. Among the tiny atomic structures the plankton take to the grave with them are two very stable isotopes — oxygen-16 and oxygen-18. (If you have forgotten what an isotope is, it doesn't matter, though for the record it's an atom with an abnormal number of neutrons.) This is where the geochemists come in, for the isotopes accumulate at different rates depending on how much oxygen or carbon dioxide is in the atmosphere at the time of their creation. By comparing the ancient rates of deposition of the two isotopes, geochemists can read conditions in the ancient world – oxygen levels, air and ocean temperatures, the extent and timing of ice ages and much else. By combining their isotope findings with other fossil residues that indicate other conditions such as pollen levels and so on, scientists can, with considerable confidence, recreate entire landscapes that no human eye ever saw.

The principal reason oxygen levels were able to build so robustly throughout the period of early terrestrial life was that much of the world's landscape was dominated by giant tree ferns and vast swamps, which by their boggy nature disrupted the normal carbon-recycling process. Instead of completely rotting down, falling fronds and other dead vegetative matter accumulated in rich, wet sediments which were eventually squeezed into vast coal beds. The heady levels of oxygen clearly encouraged outsized growth.

The oldest indication of a surface animal yet found is a track left 350 million years ago by a millipede-like creature on a rock in Scotland. It was over a metre long. Before the era was out some millipedes would reach lengths more than double that.

With such creatures on the prowl, it is perhaps not surprising that insects in the period evolved a trick that could keep them safely out of tongueshot: they learned to fly. Some took to this new means of locomotion with such uncanny facility that they haven't changed their techniques in all the time since. Then, as now, dragonflies could cruise at over 50 kilometres an hour, instantly stop, hover, fly backwards, and lift far more, proportionately, than any flying machine humans have come up with. 'The U.S. Air Force', one commentator has written, 'has put them in wind tunnels to see how they do it, and despaired.' They, too, gorged on the rich air. In carboniferous forests dragonflies grew as big as ravens. Trees and other vegetation likewise attained outsized proportions.

Horsetails and tree ferns grew to heights of 15 metres, club mosses to 40 metres.

The first terrestrial vertebrates – which is to say, the first land animals from which we would derive – are something of a mystery. This is partly because of a shortage of relevant fossils, but partly also because of an idiosyncratic Swede named Erik Jarvik, whose odd interpretations and secretive manner held back progress on this question for almost half a century. Jarvik was part of a team of Scandinavian scholars who went to Greenland in the 1930s and 1940s looking for fossil fish. In particular they sought lobe-finned fish of the type that presumably were ancestral to us and all other walking creatures, known as tetrapods.

Tetrapods comprise all the modern animals that have, or once had, four limbs, each ending in a maximum of five fingers or toes, so it includes creatures like snakes, birds and humans that have lost or conspicuously modified their limb structures over the long span of evolutionary time. Dinosaurs, whales and even fish are also all tetrapods, which clearly suggests they come from a single common ancestor. The clue to this ancestor, it was assumed, would be found in the Devonian era, from about 400 million years ago. Before that time nothing walked on land. After that time lots of things did. Luckily the team found just such a creature, a metre-long animal called an *Ichthyostega*. The analysis of the fossil fell to Jarvik, who began his study in 1948 and kept at it for the next forty-eight years. Unfortunately, Jarvik refused to let anyone else study his tetrapod. The world's palaeontologists had to be content with two sketchy interim papers in which Jarvik noted that the creature had five fingers on each of four limbs, confirming its ancestral importance.

Jarvik died in 1998. After his death, other palaeontologists eagerly examined the specimen and found that Jarvik had severely miscounted the fingers and toes – there were actually eight on the forelimbs and seven on the rear – and failed to observe that the fish could not have walked, but rather dragged itself. The transition from fish to land animals, it turns out, was more complicated, various and drawn out than long thought.

But come we did, though reaching our present state of eminence has not, of course, always been straightforward. Since life on land began, it has consisted of four megadynasties, as they are sometimes called. The first consisted of primitive, plodding but sometimes fairly hefty amphibians and reptiles. The best-known animal of this age was the Dimetrodon, a sail-backed creature that is commonly confused with dinosaurs (including, I note, in a picture caption in the Carl Sagan book *Comet*). The Dimetrodon was in fact a synapsid. So, once upon a time, were we. Synapsids were one of the four main divisions of early reptilian life, the others being anapsids, euryapsids and diapsids. The names simply refer to the number and location of small holes found in the sides of their owners' skulls. Synapsids had one hole in their lower temples; diapsids had two; euryapsids had a single hole higher up.

Over time, each of these principal groupings split into further subdivisions, of which some prospered and some faltered. Anapsids gave rise to the turtles, which for a time, perhaps a touch improbably, appeared poised to predominate as the planet's most advanced and deadly species, before an evolutionary lurch let them settle for durability rather than dominance. The synapsids divided into four streams, only one of which survived beyond the Permian. Happily, that was the stream we belonged to, and it evolved into a family of protomammals known as therapsids. These formed Megadynasty 2.

Unfortunately for the therapsids, their cousins the diapsids were also productively evolving, in their case into dinosaurs (among other things), which gradually proved too much for the therapsids. Unable to compete head-to-head with these aggressive new creatures, the therapsids by and large vanished from the record. A very few, however, evolved into small, furry, burrowing beings that bided their time for a very long while as little mammals. The biggest of them grew no larger than a house cat and most were no bigger than mice. Eventually, this would prove their salvation, but they would have to wait nearly 150 million years for Megadynasty 3, the Age of Dinosaurs, to come to an abrupt end and make way for Megadynasty 4 and our own Age of Mammals.

Each of these massive transformations, as well as many smaller ones between and since, was dependent on that paradoxically important motor of progress: extinction. It is a curious fact that on Earth species death is, in the most literal sense, a way of life. No one knows how many species of organisms have existed since life began. Thirty billion is a commonly cited figure, but the number has been put as high as 4,000 billion. Whatever the actual total, 99.99 per cent of all species that have ever lived are no longer with us. 'To a first approximation,' as the late David Raup of the University of Chicago liked to say, 'all species are extinct.' For complex organisms, the average lifespan of a species is thought to be only about 4 million years.

Extinction is always bad news for the victims, of course, but it appears to be a good thing for a dynamic planet. 'The alternative to extinction is stagnation,' Ian Tattersall of the American Museum of Natural History once told me, 'and stagnation is seldom a good thing in any realm.' (I should perhaps note that we are speaking here of extinction as a natural, long-term process. Extinction brought about by human carelessness is another matter altogether.)

Crises in the Earth's history are invariably associated with dramatic leaps afterwards. The fall of the Ediacaran fauna was followed by the creative outburst of the Cambrian period. The Ordovician extinction of 440 million years ago cleared the oceans of a lot of immobile filter feeders and, somehow, created conditions that favoured darting fish and giant aquatic reptiles. These in turn were in an ideal position to send colonists on to dry land when another blowout in the late Devonian period gave life another sound shaking. And so it has gone at scattered intervals through history. If most of these events hadn't happened just as they did, just when they did, we almost certainly wouldn't be here now.

Earth has seen five major extinction episodes in its time – the Ordovician, Devonian, Permian, Triassic and Cretaceous, in that order – and many smaller ones. The Ordovician (443 million years ago) and Devonian (374 million) each wiped out about 80 to 85 per cent of species. The Triassic (200 million years ago) and Cretaceous (66 million years) each wiped out 70–75 per cent of species. But the real whopper was the Permian extinction of about 250 million years ago, which raised the curtain on the long age of the dinosaurs. In the Permian, at least 95 per cent of animals known from the fossil record checked out, never to return. Even about a third of insect species went – the only occasion on which they were lost en masse. It is as close as we have ever come to total obliteration.*

'It was, truly, a mass extinction, a carnage of a magnitude that had never troubled the Earth before,' said Richard Fortey. The Permian event was particularly devastating to sea creatures. Trilobites vanished altogether. Clams and sea urchins nearly went. Virtually all other marine organisms were staggered. Altogether, on land and in the water, it is thought that Earth lost 52 per cent of its families – that's the level above genus and below order on the grand scale of life (the subject of the next chapter) – and perhaps as many as 96 per cent of all its species. It would be a long time – as much as 80 million years by one reckoning – before species totals recovered.

Two points need to be kept in mind. First, these are all just informed guesses. Estimates for the number of animal species alive at the end of the Permian range from as low as 45,000

---

* The timings vary, often by a few million years, from source to source. Those cited here are from the Natural History Museum in London.

to as high as 240,000. If you don't know how many species were alive, you can hardly specify with conviction the proportion that perished. Moreover, we are talking about the death of species, not individuals. For individuals the death toll could be much higher – in many cases, practically total. The species that survived to the next phase of life's lottery almost certainly owe their existence to a few scarred and limping survivors.

In between the big kill-offs, there have also been many smaller, less well-known extinction episodes – the Hemphillian, Frasnian, Famennian, Rancholabrean and a dozen or so others – which were not so devastating to total species numbers, but often critically hit certain populations. Grazing animals, including horses, were nearly wiped out in the Hemphillian event about 5 million years ago. Horses declined to a single species, which appears so sporadically in the fossil record as to suggest that for a time it teetered on the brink of oblivion. Imagine a human history without horses, without grazing animals.

In nearly every case, for both big extinctions and more modest ones, we have bewilderingly little idea of what the cause was. Even after stripping out the more crackpot notions, there are still more theories for what caused the extinction events than there have been events. At least two dozen potential culprits have been identified as causes or prime contributors, including global warming, global cooling, changing sea levels, oxygen depletion of the seas (a condition known as anoxia), epidemics, giant leaks of methane gas from the sea floor, meteor and comet impacts, runaway hurricanes of a type known as hypercanes, huge volcanic upwellings and catastrophic solar flares.

This last is a particularly intriguing possibility. Nobody knows how big solar flares can get because we have only been watching them closely since the beginning of the space age, but the Sun is a mighty engine and its storms are commensurately enormous. A typical solar flare – something we wouldn't even notice on Earth – will release the energy equivalent of a billion hydrogen bombs and fling into space 100 billion tonnes or so of murderous high-energy particles. The magnetosphere and atmosphere between them normally swat these back into space or steer them safely towards the poles (where they produce the Earth's comely auroras), but it is thought that an unusually big blast, say 100 times the typical flare, could overwhelm our ethereal defences. The light show would be a glorious one, but it would almost certainly kill a very high proportion of all that basked in its glow. Moreover, and rather chillingly, according to Bruce Tsurutani of the NASA Jet Propulsion Laboratory, 'it would leave no trace in history'.

What all this leaves us with, as one researcher has put it, is 'tons of conjecture and very little evidence'. Cooling seems to be associated with at least three of the big extinction events – the Ordovician, Devonian and Permian – but beyond that little is generally accepted, including whether a particular episode happened swiftly or slowly. Scientists can't universally agree, for instance, whether the late Devonian extinction – the event that was followed by vertebrates moving on to the land – happened over millions of years or thousands of years or in one lively day.

One of the reasons it is so hard to produce convincing explanations for extinctions is that it is so very hard to exterminate life on a grand scale. As we have seen from the Manson impact, you can receive a ferocious blow and still stage a full, if presumably somewhat wobbly, recovery. So why, out of all the thousands of impacts Earth has endured, was the KT event of 65 million years ago, which put paid to the dinosaurs, so singularly devastating? Well, first, it was positively enormous. It struck with the force of 100 million megatonnes. Such an outburst is not easily imagined, but, as the American geologist James Lawrence Powell pointed out in *Night Comes to the Cretaceous* (1998), if you exploded one Hiroshima-sized bomb for every person alive on Earth at that time, you would still be about a billion bombs short of the size of the KT impact. Yet even that alone may not have been enough to wipe out 70 per cent of Earth's life, dinosaurs included.

The KT meteor had the additional advantage – advantage if you are a mammal, that is – that it landed in a shallow sea just 10 metres deep, at a time when oxygen levels were 10 per cent higher than at present and so the world was more combustible. Above all, the floor of the sea where it landed was made of rock rich in sulphur. The result was an impact that turned an area of sea floor the size of Belgium into aerosols of sulphuric acid. For months afterwards, the Earth was subjected to rains acid enough to burn skin.

In a sense, an even greater question than 'What wiped out 70 per cent of the species that were existing at the time?' is 'How did the remaining 30 per cent survive?' Why was the event so irremediably devastating to every single dinosaur that existed, while other reptiles, like snakes and crocodiles, passed through unimpeded? So far as we can tell, no species of toad, newt, salamander or other amphibian went extinct in North America. 'Why should these delicate creatures have emerged unscathed from such an unparalleled disaster?' asks Tim Flannery in his fascinating prehistory of America, *The Eternal Frontier*.

In the seas it was much the same story. All the ammonites vanished, but their cousins the nautiloids, who lived similar lifestyles, swam on. Among plankton, some species were practically wiped out – 92 per cent of foraminiferans, for instance – while other organisms like diatoms, designed to a similar plan and living alongside them, were comparatively unscathed. These are difficult inconsistencies. As Richard Fortey observed: 'Somehow it does not seem satisfying just to call them "lucky ones" and leave it at that.' If, as seems entirely likely, the event was followed by months of dark and choking smoke, then many of the insect survivors become difficult to account for. 'Some insects, like beetles,' Fortey noted, 'could live on wood or other things lying around. But what about those like bees that navigate by sunlight and need pollen? Explaining their survival isn't so easy.'

Above all, there are the corals. Corals require algae to survive and algae require sunlight, and both together require steady minimum temperatures. Much publicity has been given in the last few years to corals dying from changes in sea temperature of only a degree or so. If they are that vulnerable to small changes, how did they survive the long-impact winter? There are also many regional variations that are hard to explain. Extinctions seem to have been far less severe in the southern hemisphere than the northern. New Zealand in particular appears to have come through largely unscathed, and yet it had almost no burrowing creatures. Even its vegetation was overwhelmingly spared, and yet the scale of conflagration elsewhere suggests that devastation was global. In short, there is just a great deal we don't know.

Some animals absolutely prospered – including, a little surprisingly, the turtles once again. As Tim Flannery notes, the period immediately after the dinosaur extinction could well be known as the Age of Turtles. Sixteen species survived in North America and three more came into existence soon after.

Clearly, it helped to be at home in water. The KT impact wiped out almost 90 per cent of land-based species but only 10 per cent of those living in fresh water. Water obviously offered protection against heat and flame, but also presumably provided more sustenance in the lean period that followed. All the land-based animals that survived had a habit of retreating to a safer environment during times of danger – into water or underground – either of which would have provided considerable shelter against the ravages without. Animals that scavenged for a living would also have enjoyed an advantage. Lizards were, and are, largely impervious to the bacteria in rotting carcasses. Indeed, often they are positively drawn to them, and for a long while there were clearly a lot of putrid carcasses about.

It is often wrongly stated that only small animals survived the KT event. In fact, among the survivors were crocodiles, which were not just large but three times larger than they are today. But on the whole, it is true, most of the survivors were small and furtive. Indeed, with the world dark and hostile, it was a perfect time to be small, warm-blooded, nocturnal, flexible in diet and cautious by nature – the very qualities that distinguished our mammalian forebears. Had our evolution been more advanced, we would probably have been wiped out. Instead, mammals found themselves in a world to which they were as well suited as anything alive. However, it wasn't as if mammals swarmed forward to fill every niche. 'Evolution may abhor a vacuum,' wrote the palaeobiologist Steven M. Stanley, 'but it often takes a long time to fill it.' For perhaps as many as 10 million years mammals remained cautiously small. In the early Tertiary, if you were the size of a bobcat you could be king.

But once they got going, mammals expanded prodigiously – sometimes to an almost preposterous degree. For a time, there were guinea pigs the size of rhinos and rhinos the size of a two-storey house. Wherever there was a vacancy in the predatory chain, mammals rose (often literally) to fill it. Early members of the raccoon family migrated to South America, discovered a vacancy, and evolved into creatures the size and ferocity of bears. Birds, too, prospered disproportionately. For millions of years, a gigantic, flightless, carnivorous bird called *Titanis* was possibly the most ferocious creature in North America. Certainly it was the most daunting bird that ever lived. It stood 3 metres high, weighed over 350 kilograms and had a beak that could tear the head off pretty much anything that irked it. Its family survived in formidable fashion for 50 million years, yet until a skeleton was discovered in Florida in 1963 we had no idea that it had ever existed.

Which brings us to another reason for our uncertainty about extinctions: the paltriness of the fossil record. We have touched already on the unlikelihood of any set of bones becoming fossilized, but the record is actually worse than you might think. Consider dinosaurs. Museums give the impression that we have a global abundance of dinosaur fossils. In fact, overwhelmingly museum displays are artificial. The giant diplodocus that long dominated the entrance hall of the Natural History Museum in London and a celebrated tableau of a giant barosaurus defending her baby from attack by a toothy allosaurus in the entrance hall of the American Museum of Natural History in New York are all entirely fake. Every one of the several hundred bones involved was made of plaster. Visit almost any large natural history museum in the world – in Paris, Vienna, Frankfurt, Buenos Aires, Mexico City – and what will greet you are antique models, not ancient bones.

The fact is, we don't really know a great deal about the dinosaurs. For the whole of the age of dinosaurs, only about 700 'valid nonavian species' (in the words of the American Museum of Natural History in New York) have been identified, almost half of them known from a single specimen. That's only about a quarter of the number of mammal species alive now. Dinosaurs, bear in mind, ruled the Earth for roughly three times as long as mammals have, so either dinosaurs were remarkably unproductive of species or we have barely scratched the surface (to use an irresistibly apt cliché).

For millions of years through the age of dinosaurs not a single fossil has yet been found. Even for the period of the late Cretaceous – the most studied prehistoric period there is, thanks to our long interest in dinosaurs and their extinction – some three-quarters of all species that lived may yet be undiscovered. Animals bulkier than the diplodocus or more forbidding than tyrannosaurus may have roamed the Earth in their thousands and we may never know it. Until very recently, everything known about the dinosaurs of this period came from only about 300 specimens representing just sixteen species. The scantiness of the record led to the widespread belief that dinosaurs were already on their way out when the KT impact occurred.

In the late 1980s a palaeontologist from the Milwaukee Public Museum, Peter Sheehan, decided to conduct an experiment. Using 200 volunteers, he made a painstaking census of a well-defined, but also well-picked-over, area of the famous Hell Creek Formation in Montana. Sifting meticulously, the volunteers collected every last tooth and vertebra and chip of bone – everything that had been overlooked by previous diggers. The work took three years. When they had finished, they found that they had more than tripled – for the planet – the number of dinosaur fossils from the late Cretaceous. The survey established unequivocally that dinosaurs remained numerous right up to the time of the KT impact.

We are so used to the notion of our own inevitability as life's dominant species that it is hard to grasp that we are here only because of timely extraterrestrial bangs and other random flukes. The one thing we have in common with all other living things is that for nearly 4 billion years our ancestors have managed to slip through a series of closing doors every time we needed them to. Stephen Jay Gould expressed it succinctly in a well-known line: 'Humans are here today because our particular line never fractured – never once at any of the billion points that could have erased us from history.'

We started this chapter with three points: life wants to be; life doesn't always want to be much; life from time to time goes extinct. To this we may add a fourth: life goes on. And often, as we shall see, it goes on in ways that are decidedly amazing.

23

THE RICHNESS OF BEING

Here and there in the Natural History Museum in London, built into recesses along the underlit corridors or standing between glass cases of minerals and ostrich eggs and a century or so of other productive clutter, are secret doors – at least secret in the sense that there is nothing about them to attract the visitor's notice. Occasionally you might see someone with the distracted manner and interestingly wilful hair that mark the scholar emerge from one of the doors and hasten down a corridor, probably to disappear through another door a little further on, but this is a relatively rare event. For the most part the doors stay shut, giving no hint that beyond them exists another – a parallel – Natural History Museum as vast as, and in many ways more wonderful than, the one the public knows and adores.

The Natural History Museum contains some 80 million objects (of which just 40,000 are on display) from every realm of life and every corner of the planet, with another 100,000 or so added to the collection each year, but it is really only behind the scenes that you get a sense of what a treasure house this is. In cupboards and cabinets and long rooms full of close-packed shelves are kept tens of thousands of pickled animals in bottles, millions of insects pinned to squares of card, drawers of shiny molluscs, bones of dinosaurs, skulls of early humans, endless folders of neatly pressed plants. It is a little like wandering through Darwin's brain. The spirit room alone holds 15 miles of shelving containing jar upon jar of animals preserved in methylated spirit.

Back here are specimens collected by Joseph Banks in Australia, Alexander von Humboldt in Amazonia and Darwin on the *Beagle* voyage – and much else that is either very rare or historically important or both. Many people would love to get their hands on these things. A few actually have. In 1954 the museum acquired an outstanding ornithological collection from the estate of a devoted collector named Richard Meinertzhagen, author of *Birds of Arabia*, among other scholarly works. Meinertzhagen had been a faithful attendee of the museum for years, coming almost daily to take notes for the production of his books and monographs. When the crates arrived, the curators excitedly levered them open to see what they had been left and were surprised, to put it mildly, to discover that a very large number of specimens bore the museum's own labels. Mr Meinertzhagen, it turned out, had been helping himself to their collections for years. It explained his habit of wearing a large overcoat even during warm weather.

A few years later a charming old regular in the molluscs department – 'quite a distinguished gentleman', I was told – was caught inserting valued sea shells into the hollow legs of his Zimmer frame.

'I don't suppose there's anything in here that somebody somewhere doesn't covet,' Richard Fortey said with a thoughtful air as he gave me a tour of the beguiling world that is the behind-the-scenes part of the museum. We wandered through a confusion of departments where people sat at large tables doing intent, investigative things with arthropods and palm fronds and boxes of yellowed bones. Everywhere there was an air of unhurried thoroughness, of people being engaged in a gigantic endeavour that could never be completed and mustn't be rushed. In 1967, I had read, the museum issued its report on the John Murray Expedition, an Indian Ocean survey, forty-four years after the expedition had concluded. This is a world where things move at their own pace, including a tiny lift Fortey and I shared with a scholarly-looking elderly man with whom Fortey chatted genially and familiarly as we proceeded upwards at about the rate that sediments are laid down.

When the man departed, Fortey said to me: 'That was a very nice chap named Norman who's spent forty-two years studying one species of plant, St John's wort.'

'How do you spend forty-two years on one species of plant?' I asked.

'It's remarkable, isn't it?' Fortey agreed. He thought for a moment. 'He's very thorough, apparently.'

The lift door opened to reveal a bricked-over opening. Fortey looked confounded. 'That's very strange,' he said. 'That used to be Botany back there.' He punched a button for another floor and we found our way at length to Botany by means of back staircases and discreet trespass through yet more departments where investigators toiled lovingly over once-living objects. And so it was that I was introduced to Len Ellis and the quiet world of bryophytes – mosses to the rest of us.

When Emerson poetically noted that mosses favour the north sides of trees ('The moss upon the forest bark, was pole-star when the night was dark') he really meant lichens, for in the nineteenth century mosses and lichens weren't distinguished. True mosses aren't actually fussy about where they grow, so they are no good as natural compasses. In fact, mosses aren't actually much good for anything. 'Perhaps no great group of plants has so few uses, commercial or economic, as the mosses,' wrote Henry S. Conard, perhaps just a touch sadly, in *How to Know the Mosses and Liverworts*, published in 1956 and still to be found on some library shelves as almost the only attempt to popularize the subject. They are, however, prolific. Even with lichens removed, bryophytes is a busy realm, with over 25,000 recorded species. Thanks to its moist, temperate climate, Britain is a congenial place for bryophytes. It is home to 1,098 species, an impressive total for a modest land mass.*

'The real variety,' Len told me when we first met, 'is in the tropics. You can go out into a place like the rainforests of Malaysia and find new varieties with relative ease. I did that myself not long ago. I looked down and there was a species that had never been recorded.'

'So we don't know how many species are still to be discovered?'

'Oh, no. No idea.'

A quiet, spare man, Len discovered several new species and one new genus in a long career. He retired from the Natural History Museum in 2024, exactly fifty years and two weeks after he started there. I asked him, when we met again in early 2025, if that was a record for employment at the museum. 'Oh no,' he said at once. 'There are people here older than me who have been here longer.' Ellis himself still comes in two days a week, mostly to help identify mosses sent to him by puzzled curators from abroad.

You might not think there would be that many people in the world prepared to devote lifetimes to the study of something so inescapably low-key, but in fact moss people number in the hundreds and they feel very strongly about their subject.

'Oh, yes,' Ellis told me when we first met in 2003, 'the meetings can get very lively at times.'

---

* The category bryophytes usually, but not always, also includes liverworts and the much rarer hornworts. Globally there are 20,000 recorded species of mosses, 5,000 of liverworts, and a rather vague 'around 150' of hornworts, according to the British Bryological Society. The common feature of all is an absence of roots, which means they take in nutrients directly through their tiny leaves rather than sucking them up from below. Apart from their similar lifestyles, the three varieties are not otherwise closely related.

I asked him for an example of controversy.

'Well, here's one inflicted on us by one of your countrymen,' he said, smiling lightly, and opened a hefty reference work containing illustrations of mosses whose most notable characteristic to the uninstructed eye was their uncanny similarity one to another. 'That,' he said, tapping a moss, 'used to be one genus, *Drepanocladus*. Now it's been reorganized into three: *Drepanocladus*, *Warnstorfia* and *Hamatacoulis*.'

'And did that lead to blows?' I asked, perhaps a touch hopefully.

'Well, it made sense. It made perfect sense. But it meant a lot of reordering of collections and it put all the books out of date for a time, so there was a bit of, you know, grumbling.'

Mosses offer mysteries as well, he told me. One famous case – famous to moss people, anyway – came in 1883 when a well-known bryologist named Robert Braithwaite found a moss called *Trematodon ambiguus* growing in solitary magnificence on the slopes of Schiehallion, the Scottish mountain that was central to determining the masses of the Sun, Earth and other planets (see Chapter 4). He saved a specimen, now part of the Natural History Museum's collection, but *T. ambiguus* has never been seen again on Schiehallion or anywhere else in Britain. Similarly, a moss called *Hyophila stanfordensis*, discovered on the campus of Stanford University in California in the nineteenth century, was found growing beside a path in Cornwall in 1961 but had never been encountered anywhere in between. (It has since been found elsewhere in Britain.) 'It's now known as *Hennediella stanfordensis*,' Len told me. 'Another revision.'

How *T. ambiguus* and *H. stanfordensis* came to exist in such disparate locations and not in between is anybody's guess, though presumably part of the answer is that in some places they may just have been overlooked. The fact is that even to the most discerning experts bryophytes can be hard to spot and even harder to distinguish from their cousins.

These days the bryology department at the Natural History

Museum is run by Jo Wilbraham, a friendly and enthusiastic woman who looks after the algae, lichen and seaweed collections as well – some 2 million objects in all. The big change in the last twenty years in all these areas, she told me, has been advances in plant genetics. Plants have DNA just like the rest of us, and this can be extracted and studied to determine characteristics that might not be visible to the naked eye, even through a powerful microscope. 'It gives us a whole new tool for investigation and classification,' she said. As an example, she cited a common liverwort, *Conocephalum conicum*, which can be found along almost any moist bank in the British Isles, and often elsewhere. 'Until quite recently it was thought to be a single species,' she said, 'but molecular data have shown that it is actually several species. That was a big surprise.'

The golden age of moss-collecting in Britain was the nineteenth century, when legions of devoted hunters crawled (often literally) across the landscape. As with the early geologists, they were nearly all amateurs, and they came from every walk of life. Among the keenest and most productive collectors were William Holmes Burrell, who ran a chemist's shop; William Gardiner, an umbrella maker; William Phillips Hamilton, a prison governor; William Palgrave, mayor of Yarmouth; Hugh Dixon, a teacher of deaf children and a celebrated outdoor enthusiast (he climbed Skiddaw in the Lake District on his eightieth birthday), and more vicars and rectors than would bear listing here. Of special note was Eleanora Armitage, who had to fight her way into the ferociously male-dominated world of bryology but made many important discoveries in Herefordshire and abroad, and was additionally notable for being the great-granddaughter of Spencer Perceval, the only British Prime Minister killed by an assassin, and for wearing trousers.

It is thanks to them, and countless more like them, that the Natural History Museum's collection of bryophytes is one of the world's most comprehensive. All 800,000 of its specimens are pressed into large folded sheets of heavy paper, some very old and covered with spidery Victorian script. Some are almost certainly in the hand of Robert Brown, the great Victorian botanist, unveiler of Brownian motion and the nucleus of cells, who ran the museum's botany department from its founding in 1837 to his death in 1858. All the specimens are kept in lustrous old mahogany cabinets so strikingly fine that I remarked upon them.

'Oh, those were Sir Joseph Banks's, from his house in Soho Square,' Ellis said casually, as if identifying a recent purchase from Ikea. 'He had them built to hold his specimens from the *Endeavour* voyage.' He regarded the cabinets thoughtfully, as if for the first time in a long while. 'I don't know how *we* ended up with them in bryology,' he added.

This was an amazing disclosure. Joseph Banks was England's greatest botanist and the *Endeavour* voyage — that is, the one on which Captain Cook charted the 1769 transit of Venus and claimed Australia for the crown, among rather a lot else — was the greatest botanical expedition in history. Banks paid £10,000, about £600,000 in today's money, to take himself and a party of nine others — a naturalist, a secretary, three artists and four servants — on the three-year adventure around the world. Goodness knows what the bluff Captain Cook made of such a velvety and pampered assemblage, but he seems to have liked Banks well enough and could not but admire his talents in botany — a feeling shared by posterity.*

Never before or since has a botanical party enjoyed greater triumphs. Partly it was because the voyage took in so many new or little-known places — Tierra del Fuego, Tahiti, New Zealand, Australia, New Guinea — but mostly it was because Banks was such an astute and resourceful collector. Even when

---

* That was actually fairly restrained by Banks's standards. On a later, much shorter voyage to Iceland he brought seventeen servants, including his own chef and two horn players. On another occasion he endeavoured, unsuccessfully, to sneak a mistress onboard disguised as a male assistant.

unable to go ashore at Rio de Janeiro because of a quarantine, he sifted through a bale of fodder sent for the ship's livestock and made new discoveries. Nothing, it seems, escaped his notice. Altogether he brought back 30,000 plant specimens, including 1,400 not seen before – enough to increase by about a quarter the number of known plants in the world.

But Banks's grand cache was only part of the total haul in what was an almost absurdly acquisitive age. Plant-collecting in the eighteenth century became a kind of international mania. Glory and wealth alike awaited those who could find new species, and botanists and adventurers went to the most incredible lengths to satisfy the world's craving for horticultural novelty. Thomas Nuttall, the man who named the wisteria after Caspar Wistar, came to America as an uneducated printer but discovered a passion for plants and walked halfway across the country and back again, collecting hundreds of growing things never seen before. John Fraser, for whom is named the Fraser fir, spent years in the wilderness collecting on behalf of Catherine the Great and emerged at length to find that Russia had a new tsar who thought he was mad and refused to honour his contract. Fraser took everything to Chelsea, where he opened a nursery and made a handsome living selling rhododendrons, azaleas, magnolias, Virginia creepers, asters and other colonial exotica to a delighted English gentry. Less lucky was David Douglas, discoverer of the Douglas fir, who died on a collecting expedition in Hawaii in 1834 after falling into a pit dug to capture wild bulls; the pit was unfortunately occupied by a bull, which butted him to death.

Huge sums could be made with the right finds. John Lyon, an amateur botanist, spent two hard and dangerous years collecting specimens, but cleared almost £125,000 in today's money for his efforts. Many, however, just did it for the love of botany. Nuttall gave most of what he found to the Liverpool Botanic Gardens. Eventually he became director of Harvard's Botanic Garden and author of the encyclopedic

Genera of North American Plants (which he not only wrote but also largely typeset).

And that was just plants. There was also all the fauna of the new worlds – kangaroos, kiwis, raccoons, bobcats, mosquitoes and other curious forms beyond imagining. The volume of life on Earth was seemingly infinite, as Jonathan Swift noted in some famous lines:

> So, naturalists observe, a flea
> Hath smaller fleas that on him prey;
> And these have smaller still to bite 'em;
> And so proceed *ad infinitum*.

All this new information needed to be filed, ordered and compared with what was known. The world was desperate for a workable system of classification. Fortunately there was a man in Sweden who stood ready to provide it.

His name was Carl Linné (later changed, with permission, to the more aristocratic *von* Linné), but he is remembered now by the Latinized form Carolus Linnaeus. He was born in 1707 in the village of Råshult in southern Sweden, the son of a poor but ambitious Lutheran curate, and was such a sluggish student that his exasperated father apprenticed him (or, by some accounts, nearly apprenticed him) to a cobbler. Appalled at the prospect of spending a lifetime banging tacks into leather, young Linné begged for another chance, which was granted, and he never thereafter wavered from academic distinction. He studied medicine in Sweden and Holland, though his passion became the natural world. In the early 1730s, still in his twenties, he began to produce catalogues of the world's plant and animal species, using a system of his own devising, and gradually his fame grew.

Rarely has a man been more comfortable with his own greatness. He spent much of his leisure time penning long and flattering portraits of himself, declaring that there had never 'been a greater botanist or zoologist', and that his system of classification was 'the greatest achievement in the realm of science'. Modestly, he suggested that his gravestone should bear the inscription *Princeps Botanicorum*, 'Prince of Botanists'. It was never wise to question his generous self-assessments. Those who did so were apt to find they had weeds named after them. He rarely ventured into the field himself, but instead sent students to collect specimens for him. It was dangerous work. According to Linnaeus's biographer Gunnar Broberg, as many as one in three of them never returned.

Linnaeus's other striking quality was an abiding – at times, one might say, a feverish – preoccupation with sex. He was particularly struck by the similarity between certain bivalves and the female pudenda. To the parts of one species of clam he gave the names 'vulva', 'labia', 'pubes', 'anus' and 'hymen'. He grouped plants by the nature of their reproductive organs and endowed them with an arrestingly anthropomorphic amorousness. His descriptions of flowers and their behaviour are full of references to 'promiscuous intercourse', 'barren concubines' and 'the bridal bed'. In spring, he wrote in one oft-quoted passage,

> Love comes even to the plants. Males and females ... hold their nuptials ... showing by their sexual organs which are males, which females. The flowers' leaves serve as a bridal bed, which the Creator has so gloriously arranged, adorned with such noble bed curtains, and perfumed with so many soft scents that the bridegroom with his bride might there celebrate their nuptials with so much the greater solemnity. When the bed has thus been made ready, then is the time for the bridegroom to embrace his beloved bride and surrender himself to her.

He named one genus of plants *Clitoria*. Not surprisingly, many people thought him strange. But his system of classification was irresistible. Before Linnaeus, plants were given names that were expansively descriptive. The common ground cherry was called *Physalis amno ramosissime ramis angulosis glabris foliis dentoserratis*. Linnaeus lopped it back to *Physalis angulata*, which name it still uses. The plant world was equally disordered by inconsistencies of naming. A botanist could not be sure if *Rosa sylvestris alba cum rubore, folio glabro* was the same plant that others called *Rosa sylvestris inodora seu canina*. Linnaeus solved the puzzlement by calling it simply *Rosa canina*. To make these excisions useful and agreeable to all required much more than simply being decisive. It required an instinct – a genius, in fact – for spotting the salient qualities of a species.

The Linnaean system is so well established that we can hardly imagine an alternative, but before Linnaeus, systems of classification were often highly whimsical. Animals might be categorized by whether they were wild or domesticated, terrestrial or aquatic, large or small, even whether they were thought handsome and noble or of no consequence. Buffon arranged his animals by their utility to man. Anatomical considerations barely came into it. Linnaeus made it his life's work to rectify this deficiency by classifying all that was alive according to its physical attributes. Taxonomy – which is to say, the science of classification – has never looked back.

It all took time, of course. The first edition of his great *Systema Naturae* in 1735 was just fourteen pages long. But it grew and grew, until by the twelfth edition – the last that Linnaeus would live to see – it extended to three volumes and 2,300 pages. In the end he named or recorded some 13,000 species of plant and animal. Other works were more comprehensive – John Ray's three-volume *Historia Generalis Plantarum* in England, completed a generation earlier, covered no fewer than 18,625 species of plants alone – but what Linnaeus had that no one else could touch was consistency, order, simplicity and timeliness. Though his work dates from the 1730s, it didn't become widely known in England until the 1760s, just in time to make Linnaeus a kind of father figure to British naturalists. Nowhere was his system adopted with greater enthusiasm (which is why, for one thing, the Linnaean Society has its home in London and not Stockholm).

Linnaeus was not flawless. He made room for mythical beasts and 'monstrous humans' whose descriptions he gullibly accepted from seamen and other imaginative travellers. Among these were a wild man, *Homo ferus*, who walked on all fours and had not yet mastered the art of speech, and *Homo caudatus*, 'man with a tail'. But then it was, as we should not forget, an altogether more credulous age. Even the great Joseph Banks took a keen and believing interest in a series of reported sightings of mermaids off the Scottish coast at the end of the eighteenth century. For the most part, however, Linnaeus's lapses were offset by sound and often brilliant taxonomy. Among other accomplishments, he saw that whales belonged with cows, mice and other common terrestrial animals in the order quadrupedia (later changed to mammalia), which no one had done before.

In the beginning, Linnaeus intended to give each plant only a genus name and a number – *Convolvulus 1*, *Convolvulus 2* and so on – but he soon realized that that was unsatisfactory and hit on the binomial arrangement that remains at the heart of the system to this day. The intention originally was to use the binomial system for everything – rocks, minerals, diseases, winds, whatever existed in nature. Not everyone embraced the system warmly. Many were disturbed by its tendency towards indelicacy, which was slightly ironic as before Linnaeus the common names of many plants and animals had been heartily vulgar. The dandelion was long popularly known as the 'pissabed' because of its supposed diuretic properties, and other names in everyday use included *mare's fart*, *naked ladies*, *twitch-ballock*, *hound's piss*, *open arse* and *bum-towel*. One or two of these earthy appellations may unwittingly survive in English yet. The 'maidenhair' in maidenhair moss, for instance, does *not* refer to the hair on the maiden's head. At all events, it had long been felt that the natural sciences would be appreciably dignified by a dose of classical renaming, so there was a certain dismay in discovering that the self-appointed Prince of Botany had sprinkled his texts with such designations as *Clitoria*, *Fornicata* and *Vulva*.

Over the years many of these were quietly dropped (though not all: the common slipper limpet still answers on formal occasions to *Crepidula fornicata*) and many other refinements introduced as the needs of the natural sciences grew more specialized. In particular, the system was bolstered by the gradual introduction of additional hierarchies. *Genus* (plural *genera*) and *species* had been employed by naturalists for over 100 years before Linnaeus, and *order*, *class* and *family* in their biological senses all came into use in the 1750s and 1760s. But *phylum* wasn't coined until 1876 (by the German Ernst Haeckel), and *family* and *order* were treated as interchangeable until early in the twentieth century. For a time zoologists used *family* where botanists placed *order*, to the occasional confusion of nearly everyone.*

Linnaeus had divided the animal world into six categories: mammals, reptiles, birds, fishes, insects and 'vermes', or worms, for everything that didn't fit into the first five. From the outset it was evident that putting lobsters and shrimp into the same category as worms was unsatisfactory, and various new categories such as mollusca and crustacea were created. Unfortunately these new classifications were not uniformly applied from nation to nation. In an attempt to re-establish order, the British in 1842 proclaimed a new set of rules called the Stricklandian Code, but the French saw this as high-handed, and the Société Zoologique countered with its own conflicting code.

---

* To illustrate, humans are in the domain Eucarya, in the kingdom Animalia, in the phylum chordata, in the subphylum vertebrata, in the class mammalia, in the order primates, in the family Hominidae, in the genus *Homo*, in the species *sapiens*. (The convention, I'm informed, is to italicize genus and species names, but not those of higher divisions.) Some taxonomists employ further subdivisions: tribe, suborder, infraorder, parvorder and more.

Meanwhile, the American Ornithological Society, for obscure reasons, decided to use the 1758 edition of *Systema Naturae* as the basis for all its naming, rather than the 1766 edition used elsewhere, which meant that many American birds spent the nineteenth century logged in different genera from their avian cousins in Europe. Not until 1902, at an early meeting of the International Congress of Zoology, did naturalists begin at last to show a spirit of compromise and adopt a universal code.

Taxonomy is described sometimes as a science and sometimes as an art, but really it's a battleground. Even today there is more disorder in the system than most people realize. Take the category of the phylum, the division that describes the basic body plans of organisms. A few phyla are generally well known, such as molluscs (the home of clams and snails), arthropods (insects and crustaceans) and chordates (us and all other animals with a backbone or proto-backbone); thereafter, things move swiftly in the direction of obscurity. Among the obscure we might list gnathostomulida (marine worms), cnidaria (jellyfish, medusae, anemones and corals) and the delicate priapulida (or little 'penis worms'). Familiar or not, these are elemental divisions. Yet there is surprisingly little agreement on how many phyla there are or ought to be. Most biologists fix the total at about thirty, but some opt for a number in the low twenties while Edward O. Wilson in *The Diversity of Life* puts the number at a surprisingly robust eighty-nine. It depends on where you decide to make your divisions – whether you are a 'lumper' or a 'splitter', as they say in the biological world.

At the more workaday level of species, the possibilities for disagreements are even greater. Whether a species of grass should be called *Aegilops incurva*, *Aegilops incurvata* or *Aegilops ovata* may not be a matter that would stir many non-botanists to passion, but it can be a source of very lively heat in the right quarters. The problem is that there are 5,000 species of grass and many of them look awfully alike even to people who know grass. In consequence, some species have been found and named at least twenty times, and there are hardly any, it appears, that haven't been independently identified at least twice. The two-volume *Manual of the Grasses of the United States* (which was last revised in 1951 but amazingly is still considered a standard reference) devotes 200 closely typeset pages to sorting out all the synonymies, as the biological world refers to its inadvertent but quite common duplications. And that is just for the grasses of a single country.

The core of the problem is that scientists have never been able to agree on what exactly a species is. By one count, biologists use no fewer than sixteen different methodologies for defining what constitutes a species. Determining at what point an organism stops being part of a subspecies and becomes a species in its own right is as much a matter of preference and prejudice as of scientific exactitude. A family of freshwater shrimp called *Gammarus fossarum* comprises thirty-two separate species or up to 152, depending on whose categorizations you follow.

To deal with disagreements on the global stage, a body known as the International Association for Plant Taxonomy arbitrates on questions of priority and duplication. At intervals it hands down decrees, declaring that *Zauschneria californica* (a common plant in rock gardens) is to be known henceforth as *Epilobium canum*; or that *Aglaothamnion tenuissimum* may now be regarded as conspecific with *Aglaothamnion byssoides*, but not with *Aglaothamnion pseudobyssoides*. Normally these are small matters of tidying up that attract little notice, but when they touch on beloved garden plants, as they sometimes do, shrieks of outrage inevitably follow. In the late 1980s the common chrysanthemum was banished (on apparently sound scientific principles) from the genus of the same name and relegated to the comparatively drab and undesirable world of the genus *Dendranthema*.

Chrysanthemum breeders are a proud and numerous lot, and they protested to the real-if-improbable-sounding Committee on Spermatophyta. (There are also committees for

Pteridophyta, Bryophyta and Fungi, among others, all reporting to an executive called the Rapporteur-Général; this is truly an institution to cherish.) Although the rules of nomenclature are supposed to be rigidly applied, botanists are not indifferent to sentiment, and in 1995 the decision was reversed. Similar adjudications have saved petunias, euonymus and a popular species of amaryllis from demotion, but not many species of geraniums, which some years ago were transferred, amid howls, to the genus *Pelargonium*. The disputes are entertainingly surveyed in Charles Elliott's *The Potting-Shed Papers*.

Disputes and reorderings of much the same type can be found in all the other realms of the living, so keeping an overall tally is not nearly as straightforward a matter as you might suppose. In consequence, the rather amazing fact is that we don't have the faintest idea – 'not even to the nearest order of magnitude', in the words of Edward O. Wilson – of the number of things that live on our planet. Estimates range from 3 million to 200 million. And that's just for complex, multicellular life. For microorganisms, the number of species is thought to be at least several hundred million, possibly a trillion, but again these are just guesses.

Because there is no central registry of information, it can take years to work out whether a 'new' species really is new or has simply been hiding (so to speak) in a collection somewhere, either under a different name or just unrecognized. A review by twenty-two academics in 2020 concluded that some 3 billion biological specimens are held in museums and other collections around the world, and the great majority of these have never been formally identified, but simply collected and stored away for future study. According to the journal *Nature*, the 'lag time' between the discovery of a new species and its scientific description is twenty-one years on average, though it is not unusual for specimens to sit for a century or more before anyone gives them a close look.

Of the organisms that we *do* know about, more than 99 in

100 are only sketchily described – 'a scientific name, a handful of specimens in a museum, and a few scraps of description in scientific journals' is how Wilson describes the state of our knowledge. In *The Diversity of Life*, he estimated the number of known species of all types – plants, insects, microbes, algae, everything – at 1.4 million, but added that that was just a guess. Other authorities have put the number of known species slightly higher, at around 1.5 million to 1.8 million, but there is no central registry of these things, so nowhere to check numbers. In short, the remarkable position in which we find ourselves is that we don't actually know what we actually know.

In principle you ought to be able to go to experts in each area of specialization, ask how many species there are in their fields, then add the totals. Many people have in fact done so. The problem is that seldom do any two come up with matching figures. Some sources put the number of known types of fungi at 70,000, others at 100,000 – nearly half as many again. You can find confident assertions that the number of described earthworm species is 4,000 and equally confident assertions that the figure is 12,000. For insects, the numbers run from 750,000 to 950,000 species. These are, you understand, supposedly the *known* number of species. For plants, the commonly accepted numbers range from 248,000 to 265,000.

One of the most informed and comprehensive surveys is the *State of the World's Plants and Fungi* from the Royal Botanic Gardens at Kew in London. Its survey in 2023 concluded that there are 450,000 species of plant in the world (of which about 400,000 have been identified) and more than 2.5 million species of fungi (of which just 155,000 have been identified). 'At the current rate that scientists are naming species,' the report notes, 'it would take 750 to 1,000 years to fully describe the world's fungi.' Despite being nominally about plants and fungi, the report also helpfully calculates that the world contains an estimated 80,500 vertebrate animals (of which 74,420 have been formally identified) and 8.5 million invertebrates (of which a very precise 1,461,728 have been identified). Only about 10 per cent of those, it is worth noting, have been studied closely.

But, not surprisingly, other studies have come up with very different numbers. A survey by Dalhousie University in Nova Scotia, using computer modelling and patterns of distribution based on known data, estimated that there are about 8.7 million eukaryotic species globally (plus or minus a decidedly generous 1.3 million) of which about 6.5 million are based on land and 2.2 million are in the seas. It put the number of known and catalogued species at 1.2 million, which suggests that 86 per cent of species on the surface and 91 per cent in the sea are still to be discovered.

Putting things in order is not the easiest of tasks. In the early 1960s, Colin Groves of the Australian National University began a systematic survey of the 250-plus known species of primate. Oftentimes it turned out that the same species had been described more than once – sometimes several times – without any of the discoverers realizing that they were dealing with an animal that was already known to science. It took Groves four decades to untangle everything, and that was with a comparatively small group of easily distinguished, generally non-controversial creatures.

Working out the number of *individuals* in a population is even more problematic. One extraordinary example was in 1960 when a British entomologist named C. B. Williams, writing in the journal *The American Naturalist*, estimated the global population of individual insects at 10 quintillion (that is 10^{18}), based purely on the number found on one acre of farmland in Hertfordshire, England, multiplied by the total amount of surface area on Earth.

Williams was well aware that the number of insects per acre in Amazonia or Greenland would vary considerably from the 400 million found in Hertfordshire, but argued that because

England fell neatly midway between tropical munificence and desert paucity, it represented a reasonable average. He never pretended that his number was anything more than a very rough estimate, yet even today, more than sixty years later, if you do an internet search of how many individual insects there are on Earth, the answer that comes up again and again is 10 quintillion, even from the most respectable sources. That number wasn't correct then and is no doubt even less correct now, but no one can say for sure because the number of insects on Earth simply cannot be accurately computed.

Perhaps nothing is more telling about the state of taxonomy than that taxonomists cannot even tell you how many taxonomists there are in the world. The number globally may be as low as 10,000 or as high as 40,000. No one actually knows.

'It's not a biodiversity crisis, it's a taxonomist crisis!' barked Koen Maes, Belgian-born head of invertebrates at the Kenyan National Museum in Nairobi, whom I met on a visit to the country in the autumn of 2002, and that situation still applies more than twenty years later. There were no specialized taxonomists in the whole of Africa, he told me. 'There was one in the Ivory Coast, but I think he has retired,' he said. It takes eight to ten years to train a taxonomist, but none were coming along in Africa. 'They are the real fossils,' Maes added. He himself was soon to be let go, he told me. After seven years in Kenya, his contract was not being renewed. 'No funds,' Maes explained.

In an attempt to haul things into the modern age, in 2001 Kevin Kelly, co-founder of *Wired* magazine, launched an enterprise called the All Species Foundation with the aim of finding and recording on a database every living organism. The cost of such an exercise was estimated at anywhere from £1.3 billion to as much as £30 billion. In the event, the foundation ran out of money and became defunct in just six years. Several other initiatives – the Encyclopedia of Life, the Earth BioGenome Project, the Tree of Life Web Project, Wikispecies – have tried to take up the slack, but with modest results so far. So we remain a long way from knowing even approximately how many species of organism exist on Earth.

If, as the numbers suggest, we have perhaps 100 million types of insects yet to find, and if our rates of discovery continue at the present pace, we should have a definitive total for insects in a little over 15,000 years. The rest of the animal kingdom may take a little longer.

So why do we know as little as we do? There are nearly as many reasons as there are animals left to count, but here are a few of the principal causes:

Most living things are small and easily overlooked. In practical terms, this is not always a bad thing. You might not slumber quite so contentedly if you were aware that your mattress is home to perhaps 2 million microscopic mites, which come out in the wee hours to sup on your sebaceous oils and feast on all those lovely, crunchy flakes of skin that you shed as you doze and toss. Your pillow alone may be home to 40,000 of them. (To them your head is just one large oily bon-bon.) And don't think a clean pillowcase will make a difference. To something on the scale of bed mites, the weave of the tightest human fabric looks like ship's rigging. Indeed, if your pillow is six years old – which is apparently about the average age for a pillow – it has been estimated that one tenth of its weight will be made up of 'sloughed skin, living mites, dead mites and mite dung', to quote the man who did the measuring, Dr John Maunder of the British Medical Entomology Centre. (But at least they are *your* mites. Think of what you snuggle up with each time you climb into a hotel bed.)* These mites have been

---

* We are actually getting worse at some matters of hygiene. Dr Maunder is on the record as believing that the move towards low-temperature washing-machine detergents has encouraged bugs to proliferate. As he put it: 'If you wash lousy clothing at low temperatures, all you get is cleaner lice.'

with us since time immemorial, but they weren't discovered until 1965.

If creatures as intimately associated with us as bed mites escaped our notice until the age of colour television, it's hardly surprising that most of the rest of the small-scale world is barely known to us. Go out into the woods – any woods at all – bend down and scoop up a handful of soil, and you will be holding up to 10 billion bacteria, most of them unknown to science. Your sample will also contain perhaps a million plump yeasts, some 200,000 hairy little fungi known as moulds, perhaps 10,000 protozoans (of which the most familiar is the amoeba) and assorted rotifers, flatworms, roundworms and other microscopic creatures known collectively as cryptozoa. A large portion of these will also be unknown.

In the 1980s, a pair of Norwegian scientists, Jostein Goksøyr and Vigdis Torsvik, collected a gram of random soil from a beech forest near their lab in Bergen and carefully analysed its bacterial content. They found that this single small sample contained between 4,000 and 5,000 separate bacterial species, more than in the whole of *Bergey's Manual of Systematic Bacteriology*, the standard reference. They then travelled to a coastal location a few miles away, scooped up another gram of earth and found that it contained 4,000–5,000 *other* species. As Edward O. Wilson observed: 'If over 9,000 microbial types exist in two pinches of substrate from two localities in Norway, how many more await discovery in other, radically different habitats?'

We don't look in the right places. In *The Diversity of Life* Wilson describes how one botanist spent a few days tramping around 10 hectares of jungle in Borneo and discovered 1,000 new species of flowering plant. The plants weren't hard to find. It's just that no one had looked there before. Koen Maes of the Kenyan National Museum told me that he went to one cloud forest, as mountaintop forests are known in Kenya, and in half an hour 'of not particularly dedicated looking' found four new species of millipedes, three representing new genera, and one new species of tree. 'Big tree,' he added, and shaped his arms as if about to dance with a very large partner. Cloud forests are found on the tops of plateaux and have sometimes been isolated for millions of years. 'They provide the ideal climate for biology and they have hardly been studied,' he said.

Overall, tropical rainforests cover only about 6 per cent of Earth's surface, but they harbour more than half of its animal life and about two-thirds of its flowering plants – and most of this life remains unknown to us because too few researchers spend time in them. Not incidentally, much of this could be quite valuable. At least 99 per cent of flowering plants have never been tested for their medicinal properties. Because they can't flee from predators, plants have had to contrive elaborate chemical defences, and so are particularly rich in intriguing compounds. Even now, nearly a quarter of all prescribed medicines are derived from just forty plants, with another 16 per cent coming from animals or microbes, so there is a serious risk with every hectare of forest felled of losing medically vital possibilities. Using a method called combinatorial chemistry, chemists can generate 40,000 compounds at a time in labs, but these products are random and not uncommonly useless, whereas any natural molecule will have already passed what *The Economist* calls 'the ultimate screening programme: over three and a half billion years of evolution'.

Looking for the unknown isn't simply a matter of travelling to remote or distant places, however. In his book *Life: An Unauthorised Biography*, Richard Fortey notes how one ancient bacterium was found on the wall of a country pub 'where men had urinated for generations' – a discovery that would seem to involve rare amounts of luck *and* devotion and possibly some other quality not specified.

There aren't enough specialists. The stock of things to be found, examined and recorded very much outruns the supply of scientists available to do it. Take the hardy and little-known organisms known as bdelloid rotifers. These are microscopic animals that can survive almost anything. When conditions are tough, they curl up into a compact shape, switch off their metabolism and wait for better times. In this state, you can drop them into boiling water or freeze them almost to absolute zero – that is, the level where even atoms give up – and, when this torment has finished and they are returned to a more pleasing environment, they will uncurl and move on as if nothing has happened. So far, about 450 species have been identified, but nobody has any idea, even remotely, how many there may be altogether. For years almost all that was known about them was thanks to the work of a devoted amateur, a London clerical worker named David Bryce who studied them in his spare time. They can be found all over the world, but you could have all the bdelloid rotifer experts in the world to dinner and not have to borrow plates from the neighbours.

Even creatures as important and ubiquitous as fungi (and fungi are both) attract comparatively little notice. Fungi are everywhere and come in many forms – as mushrooms, moulds, mildews, yeasts and puffballs, to name but a sampling – and they exist in volumes that most of us little suspect. Gather together all the fungi found in a typical hectare of meadowland and you would have 2,800 kilograms of the stuff. These are not marginal organisms. Without fungi there would be no potato blights, Dutch elm disease, jock itch or athlete's foot, but also no yogurts, beers, cheeses, penicillin or many other medications. A lot of mycologists work in industry, making cheeses and yogurts and the like, so it is hard to say how many are actively involved in research, but we can safely take it that there are more species of fungi to be found than there are people to find them.

THE RICHNESS OF BEING

We are not sufficiently committed. The simple and alarming fact is that funding for nearly all areas of taxonomy is shrinking. In early 2024, to cite just one example, Duke University in North Carolina announced that it was closing down its famed herbarium, containing 825,000 specimens, because, it said, it could no longer afford to maintain the collection. The herbarium was one of the largest in the country and has played a central role for years not only in cataloguing species but also in studies of global climate, plant extinctions and the training of future botanists.

'We are a university with limited resources,' the dean of natural resources told the *New York Times*. It is perhaps worth noting that Duke University's endowment in 2024 was $11.6 billion, which rather redefines the meaning of 'limited resources'. It is perhaps also worth noting that in the same year that Duke decided it could not afford to maintain a world-class herbarium it spent $82.6 million on its sports teams, including a rather magnificent $1.3 million per player on its celebrated men's basketball team. The total cost of running the herbarium was not disclosed.

The world is a really big place. We have been lulled by the ease of air travel and other forms of communication into thinking that the world is not all that big, but at ground level, where researchers must work, it is actually enormous – enormous enough to be full of surprises. The okapi, the nearest living relative of the giraffe, is now known to exist in substantial (though worryingly shrinking) numbers in the rainforests of Central Africa, yet its existence wasn't even suspected until the twentieth century.* In 1995 a team of French and British scientists in Tibet, who were lost in a snowstorm in a remote

---

* Poaching and loss of habitat are thought to have cut the okapi's population numbers by up to half, to roughly 15,000 individuals, in twenty years. It is now listed as endangered.

valley, came across a breed of horse, called the Riwoche, that had previously been known only from prehistoric cave drawings. The valley's inhabitants were astonished to learn that the horse was considered a rarity in the wider world.

Thanks to the vastness of the world, and the inaccessibility of a good deal of it, animals that have been long feared extinct are sometimes rediscovered. The large, flightless New Zealand bird called the takahe had gone unsighted for well over 100 years and was formally declared extinct in 1898 before being found living in rugged obscurity on the country's South Island. The American Bird Conservancy, a charity, maintains a register of birds throughout the world – 144 of them as of this writing – that have not been seen for at least a decade but may still be out there. It, too, has reported several happy rediscoveries. In 2022, the black-naped pheasant pigeon, a bird the size of a chicken, was found in Papua New Guinea after having gone unrecorded since 1896. More tragically, a small relic population of yellow-eared parrots was found in the highlands of Colombia after being feared extinct for years. A conservationist named Gonzalo Cardona nursed the birds back to a healthy population of nearly 3,000, but in 2021 he was murdered by a passing criminal gang.*

Even if we groomed thousands of fieldworkers and dispatched them to the furthest corners of the world, it would not be effort enough, for wherever life can be, it is. Life's extraordinary fecundity is amazing, even gratifying, but also problematic. To survey it all, you would have to turn over every rock, sift through the litter on every forest floor, sieve unimaginable quantities of sand and dirt, climb into every forest canopy and

---

* Conservation work, it's worth noting, can be surprisingly dangerous. Two rangers and five other people were shot dead at an okapi wildlife refuge in the Democratic Republic of the Congo in 2012, and in Colombia no fewer than sixty-four wildlife workers and volunteers were killed in 2019 alone, the latest year for which records were available.

devise much more efficient ways to examine the seas. Even then you would overlook whole ecosystems. In the 1980s, amateur cave explorers entered a deep cave in Romania that had been sealed off from the outside world for a long but unknown period and found thirty-three species of insects and other small creatures – spiders, centipedes, lice – all blind, colourless and new to science. They were living off the microbes in the surface scum of pools, which in turn were feeding on hydrogen sulphide from hot springs.

Our instinct may be to see the impossibility of tracking everything down as frustrating, dispiriting, perhaps even appalling, but it can just as well be viewed as almost unbearably exciting. We live on a planet that has a more or less infinite capacity to surprise. What reasoning person could possibly want it any other way?

What is nearly always most arresting in any ramble through the scattered disciplines of modern science is realizing how many people have been willing to devote lifetimes to the most sumptuously esoteric lines of enquiry. In one of his essays, Stephen Jay Gould noted how a hero of his named Henry Edward Crampton spent fifty years, from 1906 to his death in 1956, quietly studying a genus of land snails called *Partula* in Polynesia. Over and over, year after year, Crampton measured to the tiniest degree – to eight decimal places – the whorls and arcs and gentle curves of numberless *Partula*, compiling the results into fastidiously detailed tables. A single line of text in a Crampton table could represent weeks of measurement and calculation.

Only slightly less devoted, and certainly more unexpected, was Alfred C. Kinsey, who became famous for his studies of human sexuality in the 1940s and 1950s. Before his mind became filled with sex, so to speak, Kinsey was an entomologist, and a dogged one at that. In one expedition lasting two years, he hiked 4,000 kilometres to assemble a collection of 300,000 wasps. How many stings he collected along the way is not, alas, recorded.

Something that had been puzzling me was the question of

how you assured a chain of succession in these arcane fields. Clearly there cannot be many institutions in the world that require or are prepared to support specialists in barnacles or Pacific snails. As we parted at the Natural History Museum in London, I asked Richard Fortey how science ensures that when one person goes there's someone ready to take his or her place.

He chuckled rather heartily at my naivety. 'I'm afraid it's not as if we have substitutes sitting on the bench somewhere waiting to be called in to play. When a specialist retires or, even more unfortunately, dies, that can bring a stop to things in that field, sometimes for a very long while.'

'And I suppose that's why you value someone who spends forty-two years studying a single species of plant, even if it doesn't produce anything terribly new?'

'Precisely,' he said, 'precisely.' And he really seemed to mean it.

24

CELLS

It starts with a single cell. The first cell splits to become two and the two become four and so on until you get to something over 11 trillion, by which point you are ready to spring forth as a human being.* And every one of those cells knows exactly what to do to preserve and nurture you from the moment of conception to your last breath.

You have no secrets from your cells. They know far more about you than you do. Each one carries a copy of the complete genetic code – the instruction manual for your body – so it knows how to do not only its own job but every other job in the body too. Never in your life will you have to remind a cell to keep an eye on its adenosine triphosphate levels or to find a place for the extra squirt of folic acid that's just unexpectedly

---

* The 11 trillion total does not include red blood cells, which are not true cells but essentially just containers for haemoglobin. With red blood cells included, the total number of human cells is thought to be something in the region of 37.2 trillion. That number is just an educated guess, of course. Human cells come in a variety of types, sizes and longevities, and are literally uncountable. The number will vary, in any case, depending on one's size, sex, age and other natural variations. The overall average figure of 37.2 trillion was arrived at in 2013 by a team of European scientists led by Eva Bianconi from the University of Bologna and was reported in the *Annals of Human Biology*.

turned up. It will do that for you, and millions more things besides.

Every cell in nature is a thing of wonder. Even the simplest are far beyond the limits of human ingenuity. To build the most basic yeast cell, for example, you would have to miniaturize about the same number of components as are found in a Boeing 777 jetliner and fit them into a sphere just 5 microns across; then somehow you would have to persuade that sphere to reproduce. But yeast cells are as nothing compared with human cells, which are not just more varied and complicated, but vastly more fascinating because of their complex interactions.

Your cells are a country of 11 trillion citizens, each devoted in some intensively specific way to your overall well-being. There isn't a thing they don't do for you. They let you feel pleasure and form thoughts. They enable you to stand and stretch and caper. When you eat, they extract the nutrients, distribute the energy, and carry off the wastes – all those things you learned about in school biology – but they also remember to make you hungry in the first place and reward you with a feeling of well-being afterwards so that you won't forget to eat again.

They keep your hair growing, your ears waxed, your brain quietly purring. They manage every corner of your being. They will jump to your defence the instant you are threatened. They will unhesitatingly die for you – billions of them do so daily. And not once in all your years have you thanked even one of them. So let us take a moment now to regard them with the wonder and appreciation they deserve.

We understand a little of how cells do the intricate things they do – how they lay down fat or manufacture insulin or engage in many of the other acts necessary to maintain a complicated entity like yourself – but only a little. About a fifth of your body weight is made up of proteins, comprising several hundred thousand different types (and perhaps as many as a million), and so far we understand what no more than about

2 per cent of them do. (Others put the figure at more like 50 per cent; it depends, apparently, on what you mean by 'understand'.)

Surprises at the cellular level turn up all the time. In nature, nitric oxide is a formidable toxin and a common component of air pollution. So scientists were naturally a little surprised when, in the mid-1980s, they found it being produced in a curiously devoted manner in human cells. Its purpose was at first a mystery, but then scientists began to find it all over the place – controlling the flow of blood and the energy levels of cells, attacking cancers and other pathogens, regulating the sense of smell, even assisting in penile erections. It also explained why nitroglycerine, the well-known explosive, soothes the heart pain known as angina. (It is converted into nitric oxide in the bloodstream, relaxing the muscle linings of vessels, allowing blood to flow more freely.) In barely the space of a decade this one gassy substance went from extraneous toxin to ubiquitous elixir.

You possess 'some few hundred' different types of cell, according to the Belgian biochemist Christian de Duve, but others put the number much higher; a study by the US National Institutes of Health in 2023 counted 461 distinct cell types (and more than 3,000 subtypes) just in the human brain. Our cells vary enormously in shape and come in a sumptuously wide range of sizes – nowhere more strikingly than at the moment of conception, when a single beating sperm confronts an egg 85,000 times bigger than it (which rather puts the notion of male conquest into perspective). On average, however, a human cell is about 20 microns wide – that is, about two-hundredths of a millimetre – which is too small to be seen by the naked eye but roomy enough to hold thousands of complicated structures like mitochondria, and millions upon millions of molecules.

In the most literal way, cells also vary in liveliness. Your skin cells are all dead. It's a somewhat galling notion to reflect that every centimetre of your surface is deceased. If you are an average-sized adult you are lugging around over 2 kilograms of dead skin, of which several billion tiny fragments are sloughed off each day. Run a finger along a dusty shelf and you are drawing a pattern very largely in old skin.

Most living cells seldom last more than a month or so, but there are some notable exceptions. Liver cells can survive for years, though the components within them may be renewed every few days. Brain cells last as long as you do. You are issued with 100 billion or so at birth and that is all you are ever going to get. It has been estimated that you lose 500 of them an hour, so if you have any serious thinking to do there really isn't a moment to waste. The good news is that the individual components of your brain cells are constantly renewed so that, as with the liver cells, no part of them is actually likely to be more than about a month old. Indeed, it has been suggested that there isn't a single bit of any of us – not so much as a stray molecule – that was part of us nine years ago. It may not feel like it, but at the cellular level we are all youngsters.

The first person to describe a cell was Robert Hooke, whom we last encountered squabbling with Isaac Newton over credit for the invention of the inverse square law. Hooke achieved many things in his sixty-eight years. He was a world-class theoretician and thinker, and a dab hand at making ingenious and useful instruments, but also something of a mad inventor. One of his more memorable creations (which sadly he never demonstrated) were spring-loaded boots that he said would allow the wearer to bound across the landscape with 12-foot strides. He lived for much of his life in squalid lodgings, but upon his death it was found that he had a chest containing £8,300 in cash, equivalent to more than a million pounds in today's money.

Nothing he did in life brought him greater acclaim, or had more lasting value, than his popular book *Micrographia: or Some Physiological Descriptions of Miniature Bodies Made by Magnifying*

Glasses, published in 1665. It revealed to an enchanted public a universe of the very small that was far more diverse, crowded and finely structured than anyone had ever come close to imagining.

Among the microscopic features first identified by Hooke were little chambers in plants that he called 'cells' because they reminded him of monks' cells. Hooke calculated that a one-inch square of cork would contain 1,259,712,000 of these tiny chambers – the first appearance of such a very large number anywhere in science. Microscopes by this time had been around for a generation or so, but what set Hooke's apart was their technical supremacy. They achieved magnifications of thirty times, making them the last word in seventeenth-century optical technology.

So it came as something of a shock when, just a decade later, Hooke and the other members of London's Royal Society began to receive drawings and reports from an unlettered linen draper in the Dutch city of Delft employing magnifications of up to 266 times – nearly ten times what Hooke's could achieve. The draper's name was Antoni van Leeuwenhoek. Though he had little formal education and no background in science, he was a perceptive and dedicated observer and a technical genius.

Over a period of fifty years – beginning, remarkably enough, when he was already past forty – Leeuwenhoek made almost 200 reports to the Royal Society, all written in Low Dutch, the only tongue of which he was master. He offered no interpretations, but simply the facts of what he had found, accompanied by exquisite drawings. He sent reports on almost everything that could be usefully examined – bread mould, a bee's stinger, blood cells, teeth, hair, his own saliva, excrement and semen (these last with fretful apologies for their inescapably unsavoury nature) – nearly all of which had never been seen microscopically before.

To this day it is not known how he got such magnificent magnifications from such simple handheld devices, which were little more than modest wooden dowels with a tiny bubble of glass embedded in them, far more like magnifying glasses than what most of us think of as microscopes, but really not much like either. Leeuwenhoek made a new instrument for every experiment he performed and was extremely secretive about his techniques.*

Unusually, he gave a set of twenty-six to the Royal Society in London and presented two more to Britain's Queen Mary II after she visited his workshop in Delft. Remarkably, all were subsequently lost.

The Royal Society set vanished after a Fellow of the society named Sir Everard Horne (1758–1832) borrowed them for private study at home and never returned them. Horne was the brother-in-law of the celebrated surgeon and teacher John Hunter, and became a celebrated surgeon and teacher in his own right. After Hunter's death in 1793, Horne took possession of the great man's papers with a view to cataloguing them, but became so busy producing his own papers and monographs that the project was never completed. After his death, it was discovered that all his papers were mostly literal copies of Hunter's originals, which he had claimed as his own. It turned out that he had also destroyed or discarded the twenty-six precious Leeuwenhoek microscopes. Why he would do so is an unanswerable mystery.

A little under a century later, the remaining two Leeuwenhoek originals similarly were borrowed from the royal estate by

---

* Leeuwenhoek was close friends with another Delft notable, the artist Jan Vermeer. In the mid-1600s Vermeer, who previously had been a competent but not outstanding artist, suddenly developed the mastery of light and perspective for which he has been celebrated ever since. Though it has never been proved, it has long been suspected that he used a camera obscura, a device for projecting images on to a flat surface through a lens. No such device was listed among Vermeer's personal effects after his death, but it happens that the executor of Vermeer's estate was none other than Antoni van Leeuwenhoek, the most secretive lensmaker of his day.

a wealthy lawyer and keen microscopist named Sir Frank Crisp (1843–1919)*, who essentially just absorbed them into his own massive collection. Upon his death in 1919, a London auction house called Stevens sold Crisp's entire hoard of instruments, including the two remaining Leeuwenhoeks, in several poorly catalogued lots. No one knows what became of them; the auctioneer's records were destroyed by a bomb in the Second World War.

After Leeuwenhoek reported finding 'animalcules' in a sample of pepper-water in 1676, the members of the Royal Society spent a year with the best devices English technology could produce searching for the 'little animals' before finally getting the magnification right. What Leeuwenhoek had found were protozoa. He calculated that there were 8,280,000 of these tiny beings in a single drop of water – more than the number of people in Holland. The world teemed with life in ways and numbers that no one had previously suspected. Inspired by Leeuwenhoek's fantastic findings, others began to peer into microscopes with such keenness that they sometimes found things that weren't in fact there. One respected Dutch observer, Nicolaus Hartsoecker, was convinced he saw 'tiny preformed men' in sperm cells. He called the little beings 'homunculi', and for some time many people believed that all humans – indeed, all creatures – were simply vastly inflated versions of tiny but complete precursor beings. Leeuwenhoek himself occasionally got carried away with his enthusiasms. In one of his least successful experiments he tried to study the explosive properties of gunpowder by observing a small blast at close range; he nearly blinded himself in the process.

---

* Crisp is better remembered today for his extraordinary landscaping projects at his home of Friar Park in Henley-on-Thames, which included the construction of a lake, a network of underground grottos and a 20-foot-high copy of the Matterhorn. The property later became the home of the Beatle George Harrison, who lovingly restored the grounds.

LIFE ITSELF

In 1683 Leeuwenhoek discovered bacteria – but that was about as far as progress could get for the next century and a half, because of the limitations of microscope technology. Not until 1831 would anyone first see the nucleus of a cell – it was found by the Scottish botanist Robert Brown, that frequent but always shadowy visitor to the history of science. Brown, who lived from 1773 to 1858, called it *nucleus* from the Latin *nucula*, meaning little nut or kernel. Only in 1839, however, did anyone realize that *all* living matter is cellular. It was Theodor Schwann, a German, who had this insight, and it was not only comparatively late, as scientific insights go, but not widely embraced at first. It wasn't until the 1860s, and some landmark work by Louis Pasteur in France, that it was shown conclusively that life cannot arise spontaneously but must come from pre-existing cells. The belief became known as the 'cell theory', and it is the basis of all modern biology.

The cell has been compared to many things, from 'a complex chemical refinery' (by the physicist James Trefil) to 'a vast, teeming metropolis' (the biochemist Guy Brown). A cell is both of those things and neither. It is like a refinery in that it is devoted to chemical activity on a grand scale, and like a metropolis in that it is crowded and busy and filled with interactions that seem confused and random but clearly have some system to them. But it is a much more nightmarish place than any city or factory that you have ever seen. To begin with, there is no up or down inside the cell (gravity doesn't meaningfully apply at the cellular scale), and not an atom's width of space is unused. There is activity *every*where and a ceaseless thrum of electrical energy. You may not feel terribly electrical, but you are. The food we eat and the oxygen we breathe are combined in the cells into electricity. The reason we don't give each other massive shocks or scorch the sofa when we sit down is that it is all happening on a tiny scale: a mere 0.1 volts travelling distances measured in nanometres. However, scale that up and it would translate as a jolt of 20 million volts per metre, about the same as the charge carried by the main body of a thunderstorm.

Whatever their size or shape, nearly all your cells (except red blood cells) are built to fundamentally the same plan: they have an outer casing or membrane, a nucleus wherein resides the necessary genetic information to keep you going, and a busy space between the two called the cytoplasm. The membrane is not, as most of us imagine it, a durable, rubbery casing, something that you would need a sharp pin to prick. Rather, it is made up of a type of fatty material known as a lipid, which has the approximate consistency 'of a light grade of machine oil', to quote Sherwin B. Nuland. If that seems surprisingly insubstantial, bear in mind that at the microscopic level things behave differently. To anything on a molecular scale water becomes a kind of heavy-duty gel and a lipid is like iron.

If you could visit a cell, you wouldn't like it. Blown up to a scale at which atoms were about the size of peas, a cell itself would be a sphere roughly half a mile across, and supported by a complex framework of girders called the cytoskeleton. Within it, millions upon millions of objects – some the size of basketballs, others the size of cars – would whizz about like bullets. There wouldn't be a place you could stand without being pummelled and ripped thousands of times every second from every direction. Even for its full-time occupants the inside of a cell is a hazardous place. Each strand of DNA is on average attacked or damaged once every 8.4 seconds – 10,000 times in a day – by chemicals and other agents that whack into or carelessly slice through it, and each of these wounds must be swiftly stitched up if the cell is not to perish.

The proteins are especially lively, spinning, pulsating and flying into each other up to a billion times a second. Enzymes, themselves a type of protein, dash everywhere, performing up to 1,000 tasks a second. Like greatly speeded-up worker ants, they busily build and rebuild molecules, hauling a piece off this one, adding a piece to that one. Some monitor passing proteins and mark with a chemical those that are irreparably damaged or flawed. Once so selected, the doomed proteins proceed to a structure called a proteasome, where they are stripped down and their components used to build new proteins. Some types of protein exist for less than half an hour; others survive for weeks. But all lead existences that are inconceivably frenzied. As de Duve notes, 'the molecular world must necessarily remain entirely beyond the powers of our imagination owing to the incredible speed with which things happen in it'.

But slow things down, to a speed at which the interactions can be observed, and things don't seem quite so unnerving. You can see that a cell is just millions of objects – lysosomes, endosomes, ribosomes, ligands, peroxisomes, proteins of every size and shape – bumping into millions of other objects and performing mundane tasks: extracting energy from nutrients, assembling structures, getting rid of waste, warding off intruders, sending and receiving messages, making repairs. Typically a cell will contain some 20,000 different types of protein, and of these about 2,000 types will each be represented by at least 50,000 molecules. 'This means', says Nuland, 'that even if we count only those molecules present in amounts of more than 50,000 each, the total is still a very minimum of 100 million protein molecules in each cell. Such a staggering figure gives some idea of the swarming immensity of biochemical activity within us.'

It is all an immensely demanding process. Your heart must pump 343 litres of blood an hour, over 8,000 litres every day, 3 million litres in a year – that's enough to fill an Olympic-sized swimming pool to well past overflowing – to keep all those cells freshly oxygenated. (And that's at rest. During exercise the rate can increase as much as sixfold.) The oxygen is taken up by the mitochondria. These are the cells' power stations and there are about 1,000 of them in a typical cell, though the number varies considerably depending on what a cell does and how much energy it requires.

You may recall from an earlier chapter that the mitochondria are thought to have originated as captive bacteria and that they now live essentially as lodgers in our cells, preserving their own genetic instructions, dividing to their own timetable, speaking their own language. You may also recall that we are at the mercy of their goodwill. Here's why. Virtually all the food and oxygen you take into your body are delivered, after processing, to the mitochondria, where they are converted into a molecule called adenosine triphosphate, or ATP.

You may not have heard of ATP, but it is what keeps you going. ATP molecules are essentially little battery packs that move through the cell providing energy for all the cell's processes, and you get through a *lot* of it. At any given moment, a typical cell in your body will have about 1 billion ATP molecules in it, and in two minutes every one of them will have been drained dry and another billion will have taken their place. Every day you produce and use up a volume of ATP equivalent to about half your body weight. Feel the warmth of your skin. That's your ATP at work.

When cells are no longer needed, they die with what can only be called great dignity. They take down all the struts and buttresses that hold them together and quietly devour their component parts. The process is known as apoptosis, or programmed cell death. Every day billions of your cells die for your benefit and billions of others clean up the mess. Cells can also die violently – for instance, when infected – but mostly they die because they are told to. Indeed, if not told to live – if not given some kind of active instruction from another cell – cells automatically kill themselves. Cells need a lot of reassurance.

When, as occasionally happens, a cell fails to expire in the prescribed manner, but rather begins to divide and proliferate wildly, we call the result cancer. Cancer cells are really just confused cells. Cells make this mistake fairly regularly, but the body has elaborate mechanisms for dealing with it. It is only very rarely that the process spirals out of control. On average, humans suffer one fatal malignancy for each 100 million billion cell divisions. Cancer is bad luck in every possible sense of the term.

The wonder of cells is not that things occasionally go wrong, but that they manage everything so smoothly for decades at a stretch. They do so by constantly sending and monitoring streams of messages – a cacophony of messages – from all around the body: instructions, queries, corrections, requests for assistance, updates, notices to divide or expire. Most of these signals arrive by means of couriers called hormones, chemical entities such as insulin, adrenaline, oestrogen and testosterone that convey information from remote outposts like the thyroid and endocrine glands. Still other messages arrive by telegraph from the brain or from regional centres in a process called paracrine signalling. Finally, cells communicate directly with their neighbours to make sure their actions are coordinated.

What is perhaps most remarkable is that it is all just random frantic action, a sequence of endless encounters directed by nothing more than elemental rules of attraction and repulsion. There is clearly no thinking presence behind any of the actions of the cells. It all just happens, smoothly and repeatedly and so reliably that seldom are we even conscious of it; yet somehow all this produces not just order within the cell but a perfect harmony right across the organism. In ways that we have barely begun to understand, trillions upon trillions of reflexive chemical reactions add up to a mobile, thinking, decision-making you – or, come to that, a rather less reflective but still incredibly organized dung beetle. Every living thing, never forget, is a wonder of atomic engineering.

Indeed, some organisms that we think of as primitive enjoy a level of cellular organization that makes our own look carelessly pedestrian. Disassemble the cells of a sponge (by passing them through a sieve, for instance), then dump them into a solution and they will find their way back together and build themselves into a sponge again. You can do this to them over and over and they will doggedly reassemble because, like you and me and every other living thing, they have one overwhelming impulse: to continue to be. And that's because of a curious, determined, barely understood molecule that is itself not alive and for the most part doesn't do anything at all. We call it DNA, and to begin to understand its supreme importance to science and to us we need to go back 160 years or so to Victorian England to the moment when the naturalist Charles Darwin had what has been called 'the single best idea that anyone has ever had' – and then, for reasons that take a little explaining, locked it away in a drawer for the next fifteen years.

25

DARWIN'S SINGULAR NOTION

In the late summer or early autumn of 1859, Whitwell Elwin, editor of the respected British journal the *Quarterly Review*, was sent an advance copy of a new book by the naturalist Charles Darwin. Elwin read the book with interest and agreed that it had merit, but feared that the subject matter was too narrow to attract a wide audience. He urged Darwin to write a book about pigeons instead. 'Everyone is interested in pigeons,' he observed helpfully.

Elwin's sage advice was ignored and *On the Origin of Species by Means of Natural Selection, or the Preservation of Favoured Races in the Struggle for Life* was published in late November 1859, priced at 15 shillings. The first edition of 1,250 copies sold out on the first day. It has never been out of print, and scarcely out of controversy, in all the time since – not bad going for a man whose principal other interest was earthworms and who, but for a single impetuous decision to sail around the world, would very probably have passed his life as an anonymous country parson known for – well, for an interest in earthworms.

Charles Robert Darwin was born on 12 February 1809*

---

* An auspicious date in history: on the same day in Kentucky, Abraham Lincoln was born.

in Shrewsbury, a sedate market town in the west Midlands. His father was a prosperous and well-regarded physician. His mother, who died when Charles was only eight, was the daughter of Josiah Wedgwood, of pottery fame.

Darwin enjoyed every advantage of upbringing, but continually pained his widowed father with his lacklustre academic performance. 'You care for nothing but shooting, dogs, and rat-catching, and you will be a disgrace to yourself and all your family,' wrote the elder Darwin in a line that nearly always appears just about here in any review of Charles's early life. Although his inclination was to natural history, for his father's sake he tried to study medicine at Edinburgh University, but couldn't bear the blood and suffering. The experience of witnessing an operation on an understandably distressed child – this was in the days before anaesthetics, of course – left him permanently traumatized. He tried law instead, but found that insupportably dull and finally managed, more or less by default, to acquire a degree in divinity from Cambridge.

A life in a rural vicarage seemed to await him when from out of the blue there came a more tempting offer. Darwin was invited to sail on the naval survey ship HMS *Beagle*, essentially as dinner company for the captain, Robert FitzRoy, whose rank precluded his socializing with anyone other than a gentleman. FitzRoy, who was very odd, chose Darwin in part because he liked the shape of Darwin's nose. (It betokened depth of character, he believed.) Darwin was not FitzRoy's first choice, but got the nod when FitzRoy's preferred companion dropped out. From a twenty-first-century perspective the two men's most striking shared feature was their extreme youthfulness. At the time of sailing, FitzRoy was only twenty-three, Darwin just twenty-two.

FitzRoy's formal assignment was to chart coastal waters, but his hobby – passion, really – was to seek out evidence for a literal, biblical interpretation of creation. That Darwin was trained for the ministry was central to FitzRoy's decision to have him aboard. That Darwin subsequently proved to be not only liberal of view but less than wholeheartedly devoted to Christian fundamentals became a source of lasting friction between them.

Darwin's time aboard the *Beagle*, from 1831 to 1836, was obviously the formative experience of his life, but also one of the most trying. He and his captain shared a small cabin, which can't have been easy as FitzRoy was subject to fits of fury followed by spells of simmering resentment. He and Darwin constantly engaged in quarrels, some 'bordering on insanity', as Darwin later recalled. Ocean voyages tended to become melancholy undertakings at the best of times – the previous captain of the *Beagle* had put a bullet through his brain during a moment of lonely gloom – and FitzRoy came from a family well known for a depressive instinct. His uncle, Viscount Castlereagh, had slit his throat the previous decade while serving as Chancellor of the Exchequer. (FitzRoy would himself commit suicide by the same method in 1865.) Even in his calmer moods, FitzRoy proved strangely unknowable. Darwin was astounded to learn upon the conclusion of their voyage that almost at once FitzRoy married a young woman to whom he had long been betrothed. In five years in Darwin's company, he had not once hinted at an attachment or even mentioned her name.

In every other respect, however, the *Beagle* voyage was a triumph. Darwin experienced adventure enough to last a lifetime and accumulated a hoard of specimens sufficient to make his reputation and keep him occupied for years. He found a magnificent trove of giant ancient fossils, including the finest megatherium known to date; survived a lethal earthquake in Chile; discovered a new species of dolphin (which he dutifully named *Delphinus fitzroyi*); conducted diligent and useful geological investigations throughout the Andes; and developed a new and much-admired theory for the formation of coral atolls, which suggested, not incidentally, that atolls could not form in less than a million years – the first hint of his long-standing attachment to the extreme antiquity of earthly processes. In 1836, aged twenty-seven, he returned home, having been away for five years and two days. He never left England again.

One thing Darwin didn't do on the voyage was propound the theory (or even a theory) of evolution. For a start, evolution as a concept was already decades old by the 1830s. Darwin's own grandfather, Erasmus, had paid tribute to evolutionary principles in a poem of inspired mediocrity called 'The Temple of Nature' years before Charles was even born. It wasn't until the younger Darwin was back in England and read Thomas Malthus's *Essay on the Principle of Population* (which proposed that increases in food supply could never keep up with population growth for mathematical reasons) that the idea began to percolate through his mind that life is a perpetual struggle and that natural selection was the means by which some species prospered while others failed. Specifically, what Darwin saw was that all organisms compete for resources, and those that had some innate advantage would prosper and pass on that advantage to their offspring. By such means would species continuously improve.

It seems an awfully simple idea – it *is* an awfully simple idea – but it explained a great deal, and Darwin was prepared to devote his life to it. 'How stupid of me not to have thought of it!' T. H. Huxley cried upon reading *On the Origin of Species*. It is a view that has been echoed ever since.

Interestingly, Darwin didn't use the phrase 'survival of the fittest' in any of his work (though he did express his admiration for it). The expression was coined, in 1864, five years after the publication of *On the Origin of Species*, by Herbert Spencer in *Principles of Biology*. Nor did he employ the word 'evolution' in print until the sixth edition of *Origin* (by which time its use had become too widespread to resist), preferring instead 'descent with modification'. Nor, above all, were his conclusions in any way inspired by his noticing, during his time in the Galápagos

Islands, an interesting diversity in the beaks of finches. The story as conventionally told (or, at least, as frequently remembered by many of us) is that Darwin, while travelling from island to island, noticed that on each island the finches' beaks were marvellously adapted for exploiting local resources – that on one island beaks were sturdy and short and good for cracking nuts, while on the next island beaks were perhaps long and thin and well suited for winkling food out of crevices – and it was this that set him to thinking that perhaps the birds had not been created this way, but had in a sense created themselves.

In fact, the birds *had* created themselves, but it wasn't Darwin who noticed it. At the time of the *Beagle* voyage, Darwin was fresh out of university and not yet an accomplished naturalist, and so failed to see that the Galápagos birds were all of a type. It was his friend the ornithologist John Gould who realized that what Darwin had found was lots of finches with different talents. Unfortunately, in his inexperience Darwin had not noted which birds came from which islands. (He had made a similar error with tortoises.) It took years to sort the muddles out.

Because of these various oversights, and the need to sort through crates and crates of other *Beagle* specimens, it wasn't until 1842, five years after his return to England, that Darwin finally began to sketch out the rudiments of his new theory. These he expanded into a 230-page 'sketch' two years later. And then he did an extraordinary thing: he put his notes away and for the next decade and a half busied himself with other matters. He fathered ten children, devoted nearly eight years to writing an exhaustive opus on barnacles ('I hate a barnacle as no man ever did before,' he sighed, understandably, upon the work's conclusion) and fell prey to strange disorders that left him chronically listless, faint, and 'flurried', as he put it. The symptoms nearly always included a terrible nausea and generally also incorporated palpitations, migraines, exhaustion, trembling, spots before the eyes, shortness of breath, 'swimming of the head' and, not surprisingly, depression.

The cause of the illness has never been established. The most romantic and perhaps likely of the many suggested possibilities is that he suffered from Chagas's disease, a lingering tropical malady that he could have acquired from the bite of a Benchuga bug in South America. A more prosaic explanation is that his condition was psychosomatic. In either case, the misery was not. Often he could work for no more than twenty minutes at a stretch, sometimes not even that.

Much of the rest of his time was devoted to a series of increasingly desperate treatments – icy plunge baths, dousings in vinegar, draping himself with 'electric chains' that subjected him to small jolts of current. He became something of a hermit, seldom leaving his home in Kent, Down House. One of his first acts upon moving to the house was to erect a mirror outside his study window so that he could identify, and if necessary avoid, callers.

Darwin kept his theory to himself because he well knew the storm it would cause. In 1844, the year he locked his notes away, a book called *Vestiges of the Natural History of Creation* roused much of the thinking world to fury by suggesting that humans might have evolved from lesser primates without the assistance of a divine creator. Anticipating the outcry, the author had taken careful steps to conceal his identity, which he kept a secret from even his closest friends for the next forty years. Some wondered if Darwin himself might be the author. Others suspected Prince Albert. In fact, the author was a successful and generally unassuming Scottish publisher named Robert Chambers whose reluctance to reveal himself had a practical dimension as well as a personal one: his firm was a leading publisher of Bibles.* *Vestiges* was warmly blasted from

---

* Darwin was one of the few to guess correctly. He happened to be visiting Chambers one day when an advance copy of the sixth edition of *Vestiges* was delivered. The keenness with which Chambers checked the revisions was something of a giveaway, though it appears the two men did not discuss it.

pulpits throughout Britain and far beyond, but also attracted a good deal of more scholarly ire. The *Edinburgh Review* devoted nearly an entire issue – eighty-five pages – to pulling it to pieces. Even T. H. Huxley, a believer in evolution, attacked the book with some venom, unaware that the author was a friend.

Darwin's own manuscript might have remained locked away till his death but for an alarming blow that arrived from the Far East in the early summer of 1858 in the form of a packet containing a friendly letter from a young naturalist named Alfred Russel Wallace and the draft of a paper, 'On the Tendency of Varieties to Depart Indefinitely from the Original Type', outlining a theory of natural selection that was uncannily similar to Darwin's secret jottings. Even some of the phrasing echoed Darwin's own. 'I never saw a more striking coincidence,' Darwin reflected in dismay. 'If Wallace had my manuscript sketch written out in 1842, he could not have made a better short abstract.'

Wallace didn't drop into Darwin's life quite as unexpectedly as is sometimes suggested. The two were already corresponding, and Wallace had more than once generously sent Darwin specimens that he thought might be of interest. In the process of these exchanges Darwin had discreetly warned Wallace that he regarded the subject of species creation as his exclusive territory. 'This summer will mark the 20th year (!) since I opened my first note-book, on the question of how & in what way do species & varieties differ from each other,' he had written to Wallace some time earlier. 'I am now preparing my work for publication,' he added, even though he wasn't really. Wallace failed to grasp what Darwin was trying to tell him – and in any case, of course, he could have had no idea that his own theory was so nearly identical to one that Darwin had been evolving, as it were, for two decades.

Darwin was placed in an agonizing quandary. If he rushed into print to preserve his priority, he would be taking advantage of an innocent tip-off from a distant admirer. But if he stepped aside, as gentlemanly conduct arguably required, he would lose credit for a theory that he had independently propounded. Wallace's theory was, by Wallace's own admission, the result of a flash of insight; Darwin's was the product of years of careful, plodding, methodical thought. It was all crushingly unfair.

To compound his misery, Darwin's youngest son, also named Charles, had contracted scarlet fever and was critically ill. At the height of the crisis, on 28 June, the child died. Despite the distraction of his son's illness, Darwin found time to dash off letters to his friends Charles Lyell and Joseph Hooker, offering to step aside but noting that to do so would mean that all his work, 'whatever it may amount to, will be smashed'. Lyell and Hooker came up with the compromise solution of presenting a summary of Darwin's and Wallace's ideas together. The venue they settled on was a meeting of the Linnaean Society, which at the time was struggling to find its way back into fashion as a seat of scientific eminence. On 1 July 1858, Darwin's and Wallace's theory was unveiled to the world. Darwin himself was not present. On the day of the meeting, he and his wife were burying their son.

The Darwin–Wallace presentation was one of seven that evening – one of the others was on the flora of Angola – and if the thirty or so people in the audience had any idea that they were witnessing the scientific highlight of the century, they showed no sign of it. No discussion followed. Nor did the event attract much notice elsewhere. Darwin cheerfully noted later that only one person, a Professor Haughton of Dublin, mentioned the two papers in print and his conclusion was that 'all that was new in them was false, and what was true was old.'

Wallace, still in the distant East, learned of these manoeuvrings long after the event, but was remarkably equable, and seemed pleased to have been included at all. He even referred to the theory for ever after as 'Darwinism'.

Much less amenable to Darwin's claim of priority was a Scottish gardener named Patrick Matthew who had, rather remarkably, also come up with the principles of natural selection more than twenty years earlier – in fact, in the very year that Darwin had set sail in the *Beagle*. Unfortunately, Matthew had published these views in a book called *Naval Timber and Arboriculture*, which had been missed not just by Darwin but by the entire world. Matthew kicked up in a lively manner, with a letter to *Gardener's Chronicle*, when he saw Darwin gaining credit everywhere for an idea that really was his. Darwin apologized without hesitation, though he did note for the record: 'I think that no one will feel surprised that neither I, nor apparently any other naturalist, has heard of Mr Matthew's views, considering how briefly they are given, and they appeared in the Appendix to a work on Naval Timber and Arboriculture.'

Wallace continued for another fifty years as a naturalist and thinker, occasionally a very good one, but increasingly fell from scientific favour by taking up dubious interests such as spiritualism and the possibility of life existing elsewhere in the universe. So the theory became, essentially by default, Darwin's alone.

Darwin never ceased being tormented by his ideas. He referred to himself as 'the Devil's Chaplain' and said that revealing the theory felt 'like confessing a murder'. Apart from all else, he knew it deeply pained his beloved and pious wife. Even so, he set to work at once expanding his manuscript into a book-length work. Provisionally he called it *An Abstract of an Essay on the Origin of Species and Varieties through Natural Selection* – a title so tepid and tentative that his publisher, John Murray, decided to issue just 500 copies. But once presented with the manuscript, and a slightly more arresting title, Murray reconsidered and increased the initial print run to 1,250.

On the Origin of Species was an immediate commercial success, but rather less of a critical one. Darwin's theory presented two intractable difficulties. It needed far more time than Lord Kelvin was willing to concede, and it was scarcely supported by fossil evidence. Where, asked Darwin's more thoughtful critics, were the transitional forms that his theory so clearly called for? If new species were continuously evolving, then there ought to be lots of intermediate forms scattered across the fossil record, but there were not.* In fact, the record as it existed then (and for a long time afterwards) showed no life at all right up to the moment of the famous Cambrian explosion.

But now here was Darwin, without any evidence, insisting that the earlier seas must have had abundant life and that we just hadn't found it yet because, for whatever reason, it hadn't been preserved. It simply could not be otherwise, Darwin maintained. 'The case at present must remain inexplicable; and may be truly urged as a valid argument against the views here entertained,' he allowed most candidly, but he refused to entertain an alternative possibility. By way of explanation he speculated – inventively but incorrectly – that perhaps the Precambrian seas had been too clear to lay down sediments and thus had preserved no fossils.

Even Darwin's closest friends were troubled by the blitheness of some of his assertions. Adam Sedgwick, who had taught Darwin at Cambridge and taken him on a geological tour of Wales in 1831, said the book gave him 'more pain than pleasure'. Louis Agassiz, the celebrated Swiss palaeontologist, dismissed it as poor conjecture. Even Lyell concluded gloomily: 'Darwin goes too far.'

T. H. Huxley disliked Darwin's insistence on huge amounts of geological time because Huxley was a saltationist, which is to say a believer in the idea that evolutionary changes happen not gradually but suddenly. Saltationists (the word comes from

---

* By coincidence, in 1861, at the height of the controversy, just such evidence turned up when workers in Bavaria found the bones of an ancient archaeopteryx, a creature halfway between a bird and a dinosaur. (It had feathers, but it also had teeth.) It was an impressive and helpful find, and its significance much debated, but a single discovery could hardly be considered conclusive.

the Latin for 'leap') couldn't accept that complicated organs could ever emerge in slow stages. What good, after all, is one-tenth of a wing or half an eye? Such organs, they thought, made sense only if they appeared in a finished state.

The belief was a little surprising in as radical a spirit as Huxley because it closely recalled a very conservative religious notion first put forward by the English theologian William Paley in 1802 and known as argument from design. Paley contended that if you found a pocket-watch on the ground, even if you had never seen such a thing before, you would instantly perceive that it had been made by an intelligent entity. So it was, he believed, with nature: its complexity was proof of its design. The notion was a powerful one in the nineteenth century, and it gave Darwin trouble too. 'The eye to this day gives me a cold shudder,' he acknowledged in a letter to a friend. In the *Origin* he conceded that it 'seems, I freely confess, absurd in the highest possible degree' that natural selection could produce such an instrument in gradual steps.

Even so, and to the unending exasperation of his supporters, Darwin not only insisted that all change was gradual, but in nearly every edition of *Origin* stepped up the amount of time he supposed necessary to allow evolution to progress, which pushed his ideas increasingly out of favour. 'Eventually,' according to the scientist and historian Jeffrey Schwartz, 'Darwin lost virtually all the support that still remained among the ranks of fellow natural historians and geologists.'

Ironically, considering that Darwin called his book *On the Origin of Species*, the one thing he couldn't explain was how species originated. Darwin's theory suggested a mechanism for how a species might become stronger or better or faster – in a word, fitter – but gave no indication of how it might throw up a new species. A Scottish engineer, Fleeming Jenkin, considered the problem and noted an important flaw in Darwin's argument. Darwin believed that any beneficial trait that arose in one generation would be passed on to subsequent generations, thus strengthening the species. Jenkin pointed out that a favourable trait in one parent wouldn't become dominant in succeeding generations, but in fact would be diluted through blending. If you pour whisky into a tumbler of water, you don't make the whisky stronger, you make it weaker. And if you pour that dilute solution into another glass of water, it becomes weaker still. In the same way, any favourable trait introduced by one parent would be successively watered down by subsequent matings until it ceased to be apparent at all. Thus Darwin's theory was a recipe not for change, but for constancy. Lucky flukes might arise from time to time, but they would soon vanish under the general impulse to bring everything back to a stable mediocrity. If natural selection were to work, some alternative, unconsidered mechanism was required.

Unknown to Darwin and everyone else, 1,200 kilometres away in a tranquil corner of Middle Europe a retiring monk named Gregor Mendel was coming up with the solution.

Mendel was born in 1822 to a humble farming family in a backwater of the Austrian Empire in what is now the Czech Republic. Schoolbooks once portrayed him as a simple but observant provincial monk whose discoveries were largely serendipitous — the result of noticing some interesting traits of inheritance while pottering about with pea plants in the monastery's kitchen garden. In fact, Mendel was a trained scientist — he had studied physics and mathematics at the Olmütz Philosophical Institute and University of Vienna — and he brought scientific discipline to all he did. Moreover, the monastery at Brno where he lived from 1843 was known as a learned institution. It had a library of 20,000 books and a tradition of careful scientific investigation.

Before embarking on his experiments, Mendel spent two years preparing his control specimens, seven varieties of pea, to make sure they bred true. Then, helped by two full-time assistants, he repeatedly bred and cross-bred hybrids from 30,000 pea plants. It was delicate work, requiring the three men to take the most exacting pains to avoid accidental cross-fertilization and to note every slight variation in the growth and appearance of seeds, pods, leaves, stems and flowers. Mendel knew what he was doing.

He never used the word 'gene' – it wasn't coined until 1913, in an English medical dictionary – though he did invent the terms 'dominant' and 'recessive' with respect to inheritance. What he established was that every seed contained two 'factors' or *Elemente*, as he called them – a dominant one and a recessive one – and these factors, when combined, produced predictable patterns of inheritance.

The results he converted into precise mathematical formulae. Altogether Mendel spent eight years on the experiments, then confirmed his results with similar experiments on flowers, corn and other plants. If anything, Mendel was *too* scientific in his approach, for when he presented his findings at the February and March meetings of the Natural History Society of Brno in 1865, the audience of about forty listened politely but was conspicuously unmoved, even though the breeding of plants was a matter of great practical interest to many of the members.

When Mendel's report was published, he eagerly sent a copy to the great Swiss botanist Karl-Wilhelm von Nägeli, whose support was more or less vital for the theory's prospects. Unfortunately, Nägeli failed to perceive the importance of what Mendel had found. He suggested that Mendel try breeding hawkweed. Mendel dutifully obeyed, but quickly realized that hawkweed had none of the requisite features for studying heritability. It was evident that Nägeli had not read the paper closely, or possibly at all. Frustrated, Mendel retired from investigating heritability and spent the rest of his life growing outstanding vegetables and studying bees, mice and sunspots, among much else. Eventually he was made abbot.

Mendel's findings weren't quite as widely ignored as is

sometimes suggested. His study received a glowing entry in the *Encyclopaedia Britannica* – then a more leading record of scientific thought than now – and was cited repeatedly in an important paper by the German Wilhelm Olbers Focke. Indeed, it was because Mendel's ideas never entirely sank below the waterline of scientific thought that they were so easily recovered when the world was ready for them.

Together, without realizing it, Darwin and Mendel laid the groundwork for all of life sciences in the twentieth century. Darwin saw that all living things are connected, that ultimately they 'trace their ancestry to a single, common source'; Mendel's work provided the mechanism to explain how that could happen. The two men could easily have helped each other. Mendel owned a German edition of the *Origin of Species*, which he is known to have read, so he must have realized the applicability of his work to Darwin's, yet he appears to have made no effort to get in touch. And Darwin, for his part, is known to have studied Focke's influential paper with its repeated references to Mendel's work, but didn't connect them to his own studies.

The one thing everyone thinks featured in Darwin's argument, that humans are descended from apes, didn't feature at all except as one passing allusion. Even so, it took no great leap of imagination to see the implications for human development in Darwin's theories, and it became an immediate talking point.

The showdown came on Saturday, 30 June 1860, at a meeting of the British Association for the Advancement of Science in Oxford. Huxley had been urged to attend by Robert Chambers, author of *Vestiges of the Natural History of Creation*, though he was still unaware of Chambers's connection to that contentious tome. Darwin, as ever, was absent. The meeting was held at the Oxford Zoological Museum. More than 1,000 people crowded into the chamber; hundreds more were turned away. People knew that something big was going to happen, though they had first to wait while a slumber-inducing speaker named John William Draper of New York University bravely slogged his way through two hours of introductory remarks on 'The Intellectual Development of Europe Considered with Reference to the Views of Mr Darwin'.

Finally, the Bishop of Oxford, Samuel Wilberforce, rose to speak. Wilberforce had been briefed (or so it is generally assumed) by the ardent anti-Darwinian Richard Owen, who had been a guest in his home the night before. As nearly always with events that end in uproar, accounts of what exactly transpired vary widely. In the most popular version, Wilberforce, when properly in flow, turned to Huxley with a dry smile and demanded of him whether he claimed attachment to the apes by way of his grandmother or grandfather. The remark was doubtless intended as a quip, but it came across as an icy challenge. According to his own account, Huxley turned to his neighbour and whispered, 'The Lord hath delivered him into my hands,' then rose with a certain relish.

Others, however, recalled a Huxley trembling with fury and indignation. At all events, Huxley declared that he would rather claim kinship to an ape than to someone who used his eminence to propound uninformed twaddle in what was supposed to be a serious scientific forum. Such a riposte was a scandalous impertinence, as well as an insult to Wilberforce's office, and the proceedings instantly collapsed into tumult. A Lady Brewster fainted. Robert FitzRoy, Darwin's companion on the *Beagle* twenty-five years before, wandered through the hall with a Bible held aloft, shouting, 'The Book, the Book!' (He was at the conference to present a paper on storms in his capacity as head of the newly created Meteorological Department.) Interestingly, each side afterwards claimed to have routed the other.

Darwin did eventually make his belief in our kinship with the apes explicit in *The Descent of Man* in 1871. The conclusion was a bold one, since nothing in the fossil record supported such a notion. The only known early human remains of that time were the famous Neanderthal bones from Germany and a few uncertain fragments of jawbones, and many respected authorities refused to believe even in their antiquity. *The Descent of Man* was altogether a more controversial book than the *Origin*, but by the time of its appearance the world had grown less excitable and its arguments caused much less of a stir.

For the most part, however, Darwin passed his twilight years on other projects, most of which touched only tangentially on questions of natural selection. He spent amazingly long periods picking through bird droppings, scrutinizing the contents in an attempt to understand how seeds spread between continents, and spent years more studying the behaviour of worms. One of his experiments was to play the piano to them – not to amuse them but to study the effects on them of sound and vibration. He was the first to realize how vitally important worms are to soil fertility. 'It may be doubted whether there are many other animals which have played so important a part in the history of the world,' he wrote in his masterwork on the subject, *The Formation of Vegetable Mould Through the Action of Worms* (1881), which was actually more popular than *On the Origin of Species* had ever been. Among his other books were *On the Various Contrivances by which British and Foreign Orchids Are Fertilised by Insects* (1862), *Expressions of the Emotions in Man and Animals* (1872), which sold almost 5,300 copies on its first day, *The Effects of Cross and Self Fertilization in the Vegetable Kingdom* (1876) – a subject that came improbably close to Mendel's own work, without attaining anything like the same insights – and *The Power of Movement in Plants* (1880).

Finally, but not least, he devoted much effort to studying the consequences of inbreeding – a matter of private interest to him. Having married his own cousin, Darwin glumly suspected that certain physical and mental frailties among his children arose from a lack of diversity in his family tree.

Darwin was often honoured in his lifetime, but never for

On the Origin of Species or *The Descent of Man*. When the Royal Society bestowed on him the prestigious Copley Medal it was for his geology, zoology and botany, not evolutionary theories, and the Linnaean Society was similarly pleased to honour Darwin without embracing his radical notions. He was never knighted, though he was buried in Westminster Abbey – next to Newton. He died at home in Kent in April 1882. Mendel died two years later.

Darwin's theory didn't really gain widespread acceptance until the 1930s and 1940s, with the advance of a refined theory called, with a certain hauteur, the Modern Synthesis, combining Darwin's ideas with those of Mendel and others. For Mendel, appreciation was also posthumous, though it came somewhat sooner. In 1900, three scientists working separately in Europe rediscovered Mendel's work more or less simultaneously. It was only because one of them, a Dutchman named Hugo de Vries, seemed set to claim Mendel's insights as his own that a rival made it noisily clear that the credit really lay with the forgotten monk.

The world was almost – but not quite – ready to begin to understand how we got here: how we made each other. It is fairly amazing to reflect that at the beginning of the twentieth century, and for some years beyond, the best scientific minds in the world couldn't actually tell you, in any meaningful way, where babies came from.

And these, you may recall, were men who thought science was nearly at an end.

26

THE STUFF OF LIFE

If your two parents hadn't bonded just when they did – possibly to the second, possibly to the nanosecond – you wouldn't be here. And if their parents hadn't bonded in a precisely timely manner, you wouldn't be here either. And if their parents hadn't done likewise, and their parents before them, and so on, obviously and indefinitely, you wouldn't be here.

Push backwards through time and these ancestral debts begin to add up. Go back just eight generations to about the time that Charles Darwin and Abraham Lincoln were born, and already there are over 250 people on whose timely couplings your existence depends. Continue further, to the time of Shakespeare and the Mayflower pilgrims, and you have no fewer than 16,384 ancestors earnestly exchanging genetic material in a way that would, eventually and miraculously, result in you.

At twenty generations ago, the number of people procreating on your behalf has risen to 1,048,576. Five generations before that, and there are no fewer than 33,554,432 men and women on whose devoted couplings your existence depends. By thirty generations ago, your total number of forebears – remember, these aren't cousins and aunts and other incidental relatives, but only parents and parents of parents in a line leading ineluctably to you – is over 1 billion (1,073,741,824, to be precise). If you go back sixty-four generations, to the time of the Romans, the number of people on whose cooperative efforts your eventual existence depends has risen to approximately 18 quintillion, which is of course vastly more than the number of people who have ever lived.

Clearly something has gone wrong with our maths here. The answer, it may interest you to learn, is that your line is not pure. You couldn't be here without a little incest – actually quite a lot of incest – albeit at a genetically discreet remove. With so many millions of ancestors in your background, there will have been many occasions when a relative from your mother's side of the family procreated with some distant cousin from your father's side of the ledger. In fact, if you are in a partnership now with someone from your own race and country, the chances are excellent that you are at some level related. Indeed, if you look around you on a bus or in a park or café or any crowded place, *most* of the people you see are very probably relatives. In point of fact, if you go back just 7,000 years, you reach a point of convergence where all our genealogical lines intersect – that is to say, where everyone alive back then with an unbroken line of descendants is directly related to every single person on Earth today, amazingly enough. So when someone boasts to you that he is descended from Shakespeare or William the Conqueror, you should answer at once: 'Me, too!' In the most literal and fundamental sense we are all family.

We are also uncannily alike. Compare your genes with any other human being's and on average they will be about 99.9 per cent the same. That is what makes us a species. The tiny differences in that remaining 0.1 per cent – 'roughly one nucleotide base in every thousand', to quote the British geneticist and Nobel laureate John Sulston – are what endow us with our individuality. Much has been made in recent years of the piecing together of the human genome. In fact, there is no such thing as 'the' human genome. Every human genome is different. Otherwise we would all be identical. It is the endless recombinations of our genomes – each nearly identical to all the others, but not quite – that make us what we are, both as individuals and as a species.

But what exactly is this thing we call the genome? And what, come to that, are genes? Well, start with a cell again. Inside the cell is a nucleus and inside each nucleus are the chromosomes – forty-six little bundles of complexity, of which twenty-three come from your mother and twenty-three from your father. With a few important exceptions, every cell in your body holds the same complement of chromosomes. (The exceptions are red blood cells, some immune system cells and egg and sperm cells, which for various organizational reasons don't carry the full genetic package.) Chromosomes constitute the complete set of instructions necessary to make and maintain you and are made of long strands of the little wonder chemical called deoxyribonucleic acid or DNA – 'the most extraordinary molecule on Earth', as it has been called.

DNA exists for just one reason – to create more DNA – and you have a lot of it inside you: nearly 2 metres squeezed into each of several trillion cells. You have so much, in fact, that if you took all the DNA within you and wove it into a single fine strand, it would stretch 10 billion miles, to well beyond Pluto. Think of it: there is enough of you to reach the outer solar system. Each length of DNA comprises some 3.2 billion letters of coding, enough to provide $10^{3,480,000,000}$ possible combinations, 'guaranteed to be unique against all conceivable odds', in the words of Christian de Duve. That's a lot of possibility.

Your body, in short, loves to make DNA, and without it you couldn't live. Yet DNA is not itself alive. No molecule is, but DNA is, as it were, especially unalive. It is 'among the most nonreactive, chemically inert molecules in the living world', in the words of the geneticist Richard Lewontin. That is why it can be recovered from patches of long-dried blood or semen in murder investigations and coaxed from the bones of ancient

Neanderthals. It also explains why it took scientists so long to work out how a substance so mystifyingly low-key – so, in a word, lifeless – could be at the very heart of life itself.

As a known entity, DNA has been around longer than you might think. It was discovered as far back as 1869 by Johann Friedrich Miescher, a Swiss scientist working at the University of Tübingen in Germany. While delving microscopically through the pus in surgical bandages, Miescher found a substance he didn't recognize and called it nuclein (because it resided in the nuclei of cells). At the time, Miescher did little more than note its existence, but nuclein clearly remained on his mind, for twenty-three years later, in a letter to his uncle, he raised the possibility that such molecules could be the agents behind heredity. This was an extraordinary insight, but one so far in advance of the day's scientific requirements that it attracted no attention at all.

For most of the next half-century the common assumption was that DNA had at most a subsidiary role in matters of heredity. It was too simple. It had just four basic components, called nucleotides, which was like having an alphabet of just four letters. How could you possibly write the story of life with such a rudimentary alphabet? (The answer is that you do it in much the way that you create complex messages with the simple dots and dashes of Morse code – by combining them.) DNA didn't do anything at all, as far as anyone could tell. It just sat there in the nucleus, possibly binding the chromosome in some way or adding a splash of acidity on command or fulfilling some other trivial task that no one had yet thought of. The necessary complexity, it was thought, had to exist in proteins in the nucleus.

There were, however, two problems with dismissing DNA. First, there was just so much of it, so clearly the cells esteemed it in some important way. On top of this, it kept turning up, like the suspect in a murder mystery, in experiments. In two studies in particular, one involving the *Pneumonococcus* bacterium and another involving bacteriophages (viruses that infect bacteria),

THE STUFF OF LIFE

DNA betrayed an importance that could be explained only if its role were more central than prevailing thought allowed. The evidence suggested that DNA was somehow involved in the making of proteins, a process vital to life, yet it was also clear that proteins were being made *outside* the nucleus, well away from the DNA that was supposedly directing their assembly.

No one could understand how DNA could possibly be getting messages to the proteins. The answer, we now know, was RNA, or ribonucleic acid, which acts as an interpreter between the two. It is a notable oddity of biology that DNA and proteins don't speak the same language. For almost 4 billion years they have been the living world's great double act, and yet they answer to mutually incompatible codes, as if one spoke Spanish and the other Hindi. To communicate they need a mediator in the form of RNA. Working with a kind of chemical clerk called a ribosome, RNA translates information from a cell's DNA into terms proteins can understand and act upon.

However, by the early 1900s, where we resume our story, we were still a very long way from understanding that, or indeed almost anything else to do with the confused business of heredity.

Clearly there was a need for some inspired and clever experimentation, and happily the age produced a young person with the diligence and aptitude to undertake it. His name was Thomas Hunt Morgan and in 1904, just four years after the timely rediscovery of Mendel's experiments with pea plants and still almost a decade before *gene* would even become a word, he began to do remarkably dedicated things with chromosomes.

Chromosomes had been discovered by chance in 1888 and were so called because they readily absorbed dye and thus were easy to see under the microscope. By the turn of the century it was strongly suspected that they were involved in the passing on of traits, but no one knew how, or even really whether, they did this.

Morgan chose as his subject of study a tiny, delicate fly formally called *Drosophila melanogaster*, but more commonly known as the fruit fly (or vinegar fly, banana fly or garbage fly). *Drosophila* is familiar to most of us as that frail, colourless insect that seems to have a compulsive urge to drown in our drinks. As laboratory specimens fruit flies had certain very attractive advantages: they cost almost nothing to house and feed, could be bred by the millions in milk bottles, went from egg to productive parenthood in ten days or less and had just four chromosomes, which kept things conveniently simple.

Working out of a small lab (which became known, inevitably, as the Fly Room) in Schermerhorn Hall at Columbia University in New York, Morgan and his team embarked on a programme of meticulous breeding and cross-breeding involving millions of flies (one biographer says billions, though that is probably an exaggeration), each of which had to be captured with tweezers and examined under a jeweller's glass for any tiny variations in inheritance. For six years they tried to produce mutations by any means they could think of – zapping the flies with radiation and X-rays, rearing them in bright light and darkness, baking them gently in ovens, spinning them crazily in centrifuges – but nothing worked. Morgan was on the brink of giving up when there occurred a sudden and repeatable mutation – a fly that had white eyes rather than the usual red ones. With this breakthrough, Morgan and his assistants were able to generate useful deformities, allowing them to track a trait through successive generations. By such means they could work out the correlations between particular characteristics and individual chromosomes, eventually proving to more or less everyone's satisfaction that chromosomes were at the heart of inheritance.

The problem, however, remained the next level of biological intricacy: the enigmatic genes and the DNA that composed them. These were much trickier to isolate and understand. As late as 1933, when Morgan was awarded a Nobel Prize for his work, many researchers still weren't convinced that genes even existed. As Morgan noted at the time, there was no consensus 'as to what the genes are — whether they are real or purely fictitious'. It may seem surprising that scientists could struggle to accept the physical reality of something so fundamental to cellular activity, but, as Wallace, King and Sanders point out in *Biology: The Science of Life* (that rarest thing: a readable textbook), we are in much the same position today in respect of mental processes such as thought and memory. We know that we have them, of course, but we don't know what, if any, physical form they take. So it was for a very long time with genes. The idea that you could pluck one from your body and take it away for study was as absurd to many of Morgan's peers as the idea that scientists today might capture a stray thought and examine it under a microscope.

What was certainly true was that *something* associated with chromosomes was directing cell replication. Finally, in 1944, after fifteen years of effort, a team at the Rockefeller Institute in Manhattan, led by a brilliant but diffident Canadian named Oswald Avery, succeeded with an exceedingly tricky experiment in which an innocuous strain of bacteria was made permanently infectious by crossing it with alien DNA, proving that DNA was far more than a passive molecule and almost certainly was the active agent in heredity. The Austrian-born biochemist Erwin Chargaff later quite seriously suggested that Avery's discovery was worth two Nobel Prizes.

Unfortunately, Avery was opposed by one of his own colleagues at the institute, a strong-willed and disagreeable protein enthusiast named Alfred Mirsky, who did everything in his power to discredit Avery's work — including, it has been said, lobbying the authorities at the Karolinska Institute in Stockholm not to give Avery a Nobel Prize. Avery by this time was sixty-six years old and tired. Unable to deal with the stress and controversy, he resigned his position and never went near a lab again. But other experiments elsewhere overwhelmingly supported his conclusions, and soon the race was on to find the structure of DNA.

Had you been a betting person in the early 1950s, your money would almost certainly have been on Linus Pauling of Caltech, America's leading chemist, to crack the structure of DNA. Pauling was unrivalled in determining the architecture of molecules and had been a pioneer in the field of X-ray crystallography, a technique that would prove crucial to peering into the heart of DNA. In an exceedingly distinguished career he would win two Nobel Prizes (in Chemistry in 1954 and Peace in 1962), but with DNA he became convinced that the structure was a triple helix, not a double one, and never quite got on the right track. Instead, victory fell to an unlikely quartet of scientists in England who didn't work as a team, often weren't on speaking terms and were for the most part novices in the field.

Of the four, the nearest to a conventional boffin was Maurice Wilkins, who had spent much of the Second World War helping to design the atomic bomb. Two of the others, Rosalind Franklin and Francis Crick, had passed their war years working for the British government on mines – Crick on the type that blow up, Franklin on the type that produce coal. The most unconventional of the foursome was James Watson, an American prodigy who as a boy had distinguished himself as a member of a highly popular radio programme called *The Quiz Kids* (and thus could claim to be at least part of the inspiration for some of the members of the Glass family in *Franny and Zooey* and other works by J. D. Salinger) and who had entered the University of Chicago aged just fifteen. He had earned his PhD by the age of twenty-two and was now attached to the famous Cavendish Laboratory in Cambridge. In 1951, he was a gawky twenty-three-year-old with a strikingly lively head of hair that appears in photographs to be straining to attach itself to some powerful magnet just out of frame.

Crick, twelve years older and still without a doctorate, was less memorably hirsute and slightly more tweedy. In Watson's account he is presented as blustery, nosy, cheerfully argumentative, impatient with anyone slow to share a notion, and constantly in danger of being asked to go elsewhere. Neither was formally trained in biochemistry. They assumed – correctly, as it turned out – that if you could determine the shape of a DNA molecule you would be able to see how it did what it did. They hoped to achieve this, it would appear, by doing as little work as possible beyond thinking, and no more of that than was absolutely necessary. As Watson cheerfully (if a touch disingenuously) remarked in his autobiographical book *The Double Helix*, 'it was my hope that the gene might be solved without my learning any chemistry'. They weren't actually assigned to work on DNA, and at one point were ordered to stop doing it. Watson was ostensibly mastering the art of crystallography; Crick was supposed to be completing a thesis on the X-ray diffraction of large molecules.

Although Crick and Watson enjoy nearly all the credit in popular accounts for solving the mystery of DNA, their breakthrough was crucially dependent on experimental work done by their competitors, the results of which were obtained 'fortuitously', in the tactful words of the historian Lisa Jardine. Far ahead of them, at least at the beginning, were two academics at King's College in London, Wilkins and Franklin.

The New Zealand-born Wilkins was a retiring figure, almost to the point of invisibility. A 1998 PBS documentary on the discovery of the structure of DNA – a feat for which he shared the 1962 Nobel Prize with Crick and Watson – managed to overlook him entirely.

Franklin was the most enigmatic character of them all. In a severely unflattering portrait, Watson in *The Double Helix* depicted Franklin as a woman who was unreasonable, secretive, chronically uncooperative and – this seemed especially to irritate him – almost wilfully unsexy. He allowed that she 'was not unattractive and might have been quite stunning had she taken even a mild interest in clothes', but in this she disappointed all expectations. She didn't even use lipstick, he noted in wonder, while her dress sense 'showed all the imagination of English blue-stocking adolescents'.*

However, she did have the best images in existence of the possible structure of DNA, achieved by means of X-ray crystallography, the technique perfected by Linus Pauling. Crystallography had been used successfully to map atoms in crystals (hence 'crystallography'), but DNA molecules were a much more finicky proposition. Only Franklin was managing to get good results from the process, but to Wilkins's perennial exasperation she refused to share her findings.

If Franklin was not warmly forthcoming with her findings, she cannot be altogether blamed. Female academics at King's in the 1950s were treated with a formalized disdain that dazzles modern sensibilities (actually, any sensibilities). However senior or accomplished, they were not allowed into the college's senior common room but instead had to take their meals in a more utilitarian chamber that even Watson conceded was 'dingily pokey'. On top of this she was being constantly pressed – at times actively harassed – to share her results with a trio of men whose desperation to get a peek at them was seldom matched by more engaging qualities, like respect. 'I'm afraid we always used to adopt – let's say a patronizing attitude towards her,' Crick later recalled. Two of these men were from a competing institution and the third was more or less openly siding with them. It should hardly come as a surprise that she kept her results locked away.

That Wilkins and Franklin did not get along was a fact that

---

* In 1968, Harvard University Press cancelled publication of *The Double Helix* after Crick and Wilkins complained about its characterizations, which Lisa Jardine has described as 'gratuitously hurtful'. The descriptions quoted above are as worded after Watson had softened his comments.

Watson and Crick seem to have exploited to their benefit. Although the two of them were trespassing rather unashamedly on Wilkins's territory, it was with them that he increasingly sided – not altogether surprisingly, since Franklin herself was beginning to act in a decidedly queer way. Although her results showed that DNA definitely was helical in shape, she insisted to all that it was not. To Wilkins's presumed dismay and embarrassment, in the summer of 1952 she posted a mock notice around the King's physics department that said: 'It is with great regret that we have to announce the death, on Friday 18th July 1952 of D.N.A. helix . . . It is hoped that Dr. M. H. F. Wilkins will speak in memory of the late helix.'

The outcome of all this was that in January 1953 Wilkins showed Watson Franklin's images, 'apparently without her knowledge or consent'. It would be an understatement to call it a significant help to him. Years later, Watson conceded that it 'was the key event . . . it mobilized us'. Armed with the knowledge of the DNA molecule's basic shape and some important elements of its dimensions, Watson and Crick redoubled their efforts. Everything now seemed to go their way. At one point Pauling was en route to a conference in England at which he would in all likelihood have met Wilkins and learned enough to correct the misconceptions that had put him on the wrong line of enquiry; but this was the McCarthy era and Pauling found himself detained at Idlewild Airport in New York, his passport confiscated, on the grounds that he was too liberal of temperament to be allowed to travel abroad. Crick and Watson also had the no less convenient good fortune that Pauling's son was working at the Cavendish and innocently kept them abreast of any news of developments and setbacks at home.

Still facing the possibility of being trumped at any moment, Watson and Crick applied themselves feverishly to the problem. It was known that DNA had four chemical components – called adenine, guanine, cytosine and thymine – and that these paired up in particular ways. By playing with pieces of cardboard cut into the shapes of molecules, Watson and Crick were able to work out how the pieces fitted together. From this they made a Meccano-like model — perhaps the most famous in modern science — consisting of metal plates bolted together in a spiral, and invited Wilkins, Franklin and the rest of the world to have a look. Any informed person could see at once that they had solved the problem. It was without question a brilliant piece of detective work, with or without the boost of Franklin's picture.

The 25 April 1953 edition of *Nature* carried a 900-word article by Watson and Crick titled 'A Structure for Deoxyribose Nucleic Acid'. Accompanying it were separate articles by Wilkins and Franklin. It was an eventful time in the world — Edmund Hillary was just about to clamber to the top of Everest, while Elizabeth II was shortly to be crowned Queen — so the discovery of the secret of life was mostly overlooked. It received a small mention in the *News Chronicle* and was ignored elsewhere.

Rosalind Franklin did not share in the Nobel Prize. She died of ovarian cancer at the age of just thirty-seven in 1958, four years before the award was granted. Nobel Prizes are not awarded posthumously. The cancer almost certainly arose as a result of chronic overexposure to X-rays through her work and could have been avoided. Her biographer Brenda Maddox noted that Franklin rarely wore a lead apron and often stepped carelessly in front of a beam. Oswald Avery never won a Nobel Prize either and was also largely overlooked by posterity, though he did at least have the satisfaction of living just long enough to see his findings vindicated. He died in 1955.

Watson and Crick's discovery wasn't actually confirmed until the 1980s. As Crick said in one of his books: 'It took over twenty-five years for our model of DNA to go from being only rather plausible, to being very plausible . . . and from there to being virtually certainly correct.'

Even so, with the structure of DNA understood, progress

in genetics was swift, and by 1968 the journal *Science* could run an article entitled 'That Was the Molecular Biology That Was', suggesting – it hardly seems possible, but it is so – that the work of genetics was nearly at an end. In fact, of course, it was only just beginning. Even now there is a great deal about DNA that we scarcely understand, not least why so much of it doesn't actually seem to *do* anything. A good 98 per cent of your DNA consists of nothing but long stretches of seemingly meaningless garble – 'non-coding DNA' as biochemists put it. Only here and there along each strand do you find sections that control and organize vital functions. These are the curious and long-elusive genes.

Genes are nothing more (nor less) than instructions to make proteins. This they do with a certain dull fidelity. In this sense, they are rather like the keys of a piano, each playing a single note and nothing else, which is obviously a trifle monotonous. But combine the genes, as you would combine piano keys, and you can create chords and melodies of infinite variety. Put all these genes together and you have (to continue the metaphor) the great symphony of existence known as the human genome.

An alternative and more common way to regard the genome is as a kind of instruction manual for the body. Viewed this way, the chromosomes can be imagined as the book's chapters and the genes as individual instructions for making proteins. The words in which the instructions are written are called codons and the letters are known as bases. The bases – the letters of the genetic alphabet – consist of the four nucleotides mentioned a page or two back: adenine, thymine, guanine and cytosine. Despite the importance of what they do, these substances are not made of anything exotic. Guanine, for instance, is the same stuff that abounds in, and takes its name from, guano.

The shape of a DNA molecule, as everyone knows, is rather like a spiral staircase or twisted rope ladder: the famous double helix. The uprights of this structure are made of a type of sugar called deoxyribose and the whole of the helix is a nucleic acid – hence the name 'deoxyribonucleic acid'. The rungs (or steps) are formed by two bases joining across the space between, and they can combine in only two ways: guanine is always paired with cytosine and thymine always with adenine. The order in which these letters appear as you move up or down the ladder constitutes the DNA code; logging all of that was the challenge of the Human Genome Project.

Now, the particular brilliance of DNA lies in its manner of replication. When it is time to produce a new DNA molecule, the two strands part down the middle, like the zip on a jacket, and each half goes off to form a new partnership. Because each nucleotide along a strand pairs up with a specific other nucleotide, each strand serves as a template for the creation of a new matching strand. If you possessed just one strand of your own DNA, you could easily enough reconstruct the matching side by working out the necessary partnerships: if the topmost rung on one strand was made of guanine, then you would know that the topmost rung on the matching strand must be cytosine.

Work your way down the ladder through all the nucleotide pairings and eventually you would have the code for a new molecule. That is just what happens in nature, except that nature does it really quickly – in only a matter of seconds, which is quite a feat. Most of the time our DNA replicates with dutiful accuracy, but just occasionally – about one time in a million – a letter gets into the wrong place. This is known as a single nucleotide polymorphism, or SNP, familiarly known to biochemists as a 'Snip'. Generally these SNPs are buried in stretches of non-coding DNA and have no detectable consequence for the body. But occasionally they make a difference. They might leave you predisposed to some disease, but equally they might confer some slight advantage – more protective pigmentation. Over time, these slight modifications accumulate both in individuals and in populations, contributing to the distinctiveness of both.

The balance between accuracy and errors in replication is a

fine one. Too many errors and the organism can't function, but too few and it sacrifices adaptability. A similar balance must exist between stability and innovation in an organism. An increase in red blood cells can help a person or group living at high elevations to move and breathe more easily, because more red cells can carry more oxygen. But additional red cells also thicken the blood. Add too many 'and it's like pumping oil', in the words of the anthropologist Charles Weitz. That's hard on the heart. Thus, those designed to live at high altitude get increased breathing efficiency, but pay for it with higher-risk hearts. By such means does Darwinian natural selection look after us. It also helps to explain why we are all so similar. Evolution simply won't let you become too different – not without becoming a new species, anyway.

The 0.1 per cent difference between your genes and mine is accounted for by our SNPs. Now, if you compared your DNA with a third person's, there would also be 99.9 per cent correspondence, but the SNPs would, for the most part, be in different places. Add more people to the comparison and you will get yet more SNPs in yet more places. For every one of your 3.2 billion bases, somewhere on the planet there will be a person, or group of persons, with different coding in that position. So not only is it wrong to refer to 'the' human genome, but in a sense we don't even have 'a' human genome. We have 7 billion of them. We are all 99.9 per cent the same, but equally, in the words of the biochemist David Cox, 'you could say all humans share nothing, and that would be correct, too.'

But we have still to explain why so little of that DNA has any readily discernible purpose. In recent years, it has become evident that some non-coding DNA plays an important role in what is called gene expression – that is, regulating the timing and quantities of protein output, since these must be pumped out in just the right amounts at the right times if the cell is to function smoothly. And it is altogether possible, even probable, that other useful functions will be revealed as scientists get to understand these very complex entities better.* But even so it seems that a very large proportion of our DNA consists of long stretches of letters that, in Matt Ridley's words, 'exist for the pure and simple reason that they are good at getting themselves duplicated'. Much of your DNA, in other words, is devoted not to you but to itself. It starts to get a little unnerving, but it really can seem that the purpose of life is to perpetuate DNA. You are a machine for the benefit of it, not it for you.

All organisms are in some sense slaves to their genes. That's why salmon and spiders and other types of creature more or less beyond counting are prepared to die in the process of mating. The desire to breed, to disperse one's genes, is the most powerful impulse in nature. As Sherwin B. Nuland has put it: 'Empires fall, ids explode, great symphonies are written, and behind all of it is a single instinct that demands satisfaction.' From an evolutionary point of view, sex is really just a reward mechanism to encourage us to pass on our genetic material.

Scientists had only barely absorbed the surprising news that a lot of our DNA doesn't seem to do anything terribly worthwhile when even more unexpected findings began to turn up. First in Germany and then in Switzerland, researchers performed some rather bizarre experiments that produced curiously unbizarre outcomes. In one, they took the gene that controlled the development of a mouse's eye and inserted it into the larva of a fruit fly. The thought was that it might produce something

---

* Non-coding DNA is employed in DNA fingerprinting. Its practicality for this purpose was discovered accidentally by Alec Jeffreys, a scientist at the University of Leicester. In 1986 Jeffreys was studying DNA sequences for genetic markers associated with heritable diseases when he was approached by the police and asked if he could help connect a suspect to two murders. He realized his technique ought to work perfectly for solving criminal cases – and so it proved. A young baker with the improbable name of Colin Pitchfork was sentenced to two life terms in prison for the murders.

interestingly grotesque. In fact, the mouse-eye gene not only made a viable eye in the fruit fly, it made a *fly's* eye. Here were two creatures that hadn't shared a common ancestor for 500 million years, yet could swap genetic material as if they were sisters.

The story was the same wherever researchers looked. They found that they could insert human DNA into certain cells of flies and the flies would accept it as if it were their own. Over 60 per cent of human genes, it turns out, are fundamentally the same as those found in fruit flies. At least 90 per cent correlate at some level with those found in mice. (We even have the same genes for making a tail, if only they would switch on.) In field after field, researchers found that whatever organism they were working on – whether nematode worms or human beings – they were often studying essentially the same genes. Life, it appeared, was drawn up from a single set of blueprints.

Further probings revealed the existence of a clutch of master control genes, each directing the development of a section of body, which were dubbed homeotic (from a Greek word meaning 'similar') or hox genes. Hox genes answered the long-bewildering question of how billions of embryonic cells, all arising from a single fertilized egg and carrying identical DNA, know where to go and what to do – that this one should become a liver cell, this one a stretchy neuron, this one a bubble of blood, this one part of the shimmer on a beating wing. It is the hox genes that instruct them, and they do it for all organisms in much the same way. Interestingly, the amount of genetic material and how it is organized doesn't necessarily, or even generally, reflect the level of sophistication of the creature that contains it. We have forty-six chromosomes, but some ferns have more than 600. The lungfish, one of the least evolved of all complex animals, has forty times as much DNA as we have. Even the common newt is more genetically splendorous than we are, by a factor of five.

Clearly it is not the number of genes you have that matters, so much as what you do with them. Initially it was thought that humans had at least 100,000 genes, possibly a good many more, but that number has been drastically reduced now to fewer than 20,000. It won't have escaped your attention that genes have been commonly implicated in any number of human frailties. Exultant scientists have at various times declared themselves to have found the genes responsible for obesity, schizophrenia, homosexuality, criminality, violence, alcoholism, even shoplifting and homelessness. Perhaps the apogee (or nadir) of this faith in biodeterminism was a study published in the journal *Science* in 1980 contending that women are genetically inferior at mathematics. In fact, we now know, almost nothing about you is so accommodatingly simple.

This is clearly a pity in one important sense, for if you had individual genes that determined height or propensity to diabetes or to baldness or any other distinguishing trait, then it would be easy – comparatively easy, anyway – to isolate and tinker with them. Unfortunately, 20,000-odd genes functioning independently is not nearly enough to produce the kind of physical complexity that makes a satisfactory human being. Genes clearly, therefore, must cooperate. A few disorders – haemophilia, Parkinson's disease, Huntington's disease and cystic fibrosis, for example – are caused by lone dysfunctional genes, but as a rule disruptive genes are weeded out by natural selection long before they can become permanently troublesome to a species or population. For the most part our fate and comfort – and even our eye colour – are determined not by individual genes but by complexes of genes working in alliance, often in intricate ways. To take one example, retinitis pigmentosa, a common cause of blindness, has been associated with 3,000 rogue mutations involving seventy different genes. That's why it is so hard to work out how it all fits together and why we won't be producing designer babies any time soon.

In fact, the more we have learned in recent years, the more complicated matters have tended to become. Even thinking, it turns out, affects the ways genes work. How fast a man's beard grows, for instance, is partly a function of how much he thinks about sex (because thinking about sex produces a testosterone surge). In the early 1990s, scientists made an even more profound discovery when they found they could knock out supposedly vital genes from embryonic mice, and still see the mice often not only born healthy, but sometimes actually fitter than their brothers and sisters who had not been tampered with. When certain important genes were destroyed, it turned out, others were stepping in to fill the breach. This was excellent news for us as organisms, but not so good for our understanding of how cells work, since it introduced an extra layer of complexity to something that we had barely begun to understand anyway.

It is largely because of these complicating factors that cracking the human genome came to be seen almost at once as only a beginning. The genome, as Eric Lander of MIT has put it, is like a parts list for the human body: it tells us what we are made of, but says nothing about how we work. What's needed now is the operating manual – instructions for how to make it go. We are not close to that point yet.

So now the quest is to crack the human proteome – the library of information that creates proteins. This is, however, a much bigger challenge. Proteins, you will remember, are the workhorses of all living systems; as many as 100 million of them may be busy in any cell at any moment. That's a lot of activity to try to figure out. Worse, proteins' behaviour and functions are based not simply on their chemistry, as with genes, but also on their shapes. To function, a protein not only must have the necessary chemical components, properly assembled, but then must also be folded into an extremely specific shape. 'Folding' is the term that's used, but it's a misleading one as it suggests a geometrical tidiness that doesn't in fact apply. Proteins loop and coil and crinkle into shapes that are at once extravagant and complex. They are more like furiously mangled coat hangers than folded towels.

Moreover, proteins are (if I may be permitted to use a handy archaism) the swingers of the biological world. Depending on mood and metabolic circumstance, they will allow themselves to be phosphorylated, glycosylated, acetylated, ubiquitinated, farneysylated, sulphated and linked to glycophosphatidylinositol anchors, among rather a lot else. Often it takes relatively little to get them going, it appears. Drink a glass of wine, as *Scientific American* notes, and you materially alter the number and types of proteins at large in your system. This is pleasant for drinkers, but not nearly so helpful for geneticists who are trying to understand what is going on.

It can all begin to seem impossibly complicated, and in some ways it *is* impossibly complicated. But there is also an underlying simplicity in all this, too, owing to an equally elemental unity in the way life works. All the tiny, deft chemical processes that animate cells – the cooperative efforts of nucleotides, the transcription of DNA into RNA – evolved just once and have stayed pretty well fixed ever since across the whole of nature. As the great French geneticist Jacques Monod put it, only half in jest: 'Anything that is true of *E. coli* must be true of elephants, except more so.'

Every living thing is an elaboration on a single original plan. As humans we are mere increments – each of us a musty archive of adjustments, adaptations, modifications and providential tinkerings stretching back 3.8 billion years. Remarkably, we are even quite closely related to fruit and vegetables. About half the chemical functions that take place in a banana are fundamentally the same as the chemical functions that take place in you.

It cannot be said too often: all life is one. That is, and I suspect will for ever prove to be, the most profound true statement there is.

VI

THE ROAD TO US

27

ICE TIME

I had a dream, which was not all a dream.
The bright sun was extinguish'd, and the stars
Did wander...

Byron, 'Darkness'

In 1815, on the island of Sumbawa in Indonesia, a handsome and long quiescent mountain named Tambora exploded spectacularly, killing 100,000 people with its blast and associated tsunamis. No one living now has ever seen such fury. Tambora was the biggest volcanic explosion in 10,000 years – 150 times the size of the Mount St Helens eruption in Washington State in 1980, equivalent to 60,000 Hiroshima-sized atom bombs.

News didn't travel terribly fast in those days. In London, *The Times* ran a small story – actually a letter from a merchant – seven months after the event. But by this time Tambora's effects were already being felt. Two hundred and forty cubic kilometres of smoky ash, dust and grit had diffused through the atmosphere, obscuring the Sun's rays and causing the Earth to cool. Sunsets were unusually but blearily colourful, an effect memorably captured by the artist J. M. W. Turner, who could not have been happier, but mostly the world existed under an oppressive, dusky pall. It was this deathly dimness that inspired Byron to write the lines quoted above.

Spring never came and summer never warmed: 1816 became known as the year without summer. Crops everywhere failed to grow. In Ireland a famine and associated typhoid epidemic killed 65,000 people. In New England, the year became popularly known as Eighteen Hundred and Froze to Death. Morning frosts continued until June and almost no planted seed would grow. Short of fodder, livestock died or had to be prematurely slaughtered. In every way it was a dreadful year – almost certainly the worst for farmers in modern times. Yet globally the temperature fell by less than 1 degree Celsius. The Earth's natural thermostat, as scientists would learn, is an exceedingly delicate instrument.

The nineteenth century was already a chilly time. For 200 years Europe and North America had been experiencing a Little Ice Age, as it has become known, which permitted all kinds of wintry events – frost fairs on the Thames, ice skating races along Dutch canals – that are mostly impossible now. It was a period, in other words, when frigidity was much on people's minds. So we may perhaps excuse nineteenth-century geologists for being slow to realize that the world they lived in was in fact balmy compared with former epochs, and that much of the land around them had been shaped by crushing glaciers and cold that would wreck even a frost fair.

They knew there was something odd about the past. The European landscape was littered with inexplicable anomalies – the bones of Arctic reindeer in the warm south of France, huge rocks stranded in improbable places – and they often came up with inventive but not terribly plausible explanations. One French naturalist named de Luc, trying to explain how granite boulders had come to rest high up on the limestone flanks of the Jura Mountains, suggested that perhaps they had been shot there by compressed air in caverns, like corks out of a popgun. The term for a displaced boulder is an erratic, but in the nineteenth century the expression seemed to apply more often to the theories than to the rocks.

If James Hutton, the eighteenth-century father of geology, had visited Switzerland, he would very probably have seen at once the significance of the carved valleys, the polished striations, the telltale strand lines where rocks had been dumped, and the other abundant clues that point to passing ice sheets. Unfortunately, Hutton was not a traveller. But even with nothing better at his disposal than second-hand accounts, Hutton rejected the idea that huge boulders had been carried 1,000 metres up mountainsides by floods – all the water in the world won't make a boulder float, he pointed out – and became one of the first to argue for widespread glaciation. Unfortunately his ideas escaped notice, and for another half-century most naturalists continued to insist that the gouges on rocks could be attributed to passing carts or even the scrape of hobnailed boots.

Local peasants, uncontaminated by scientific orthodoxy, knew better, however. The naturalist Jean de Charpentier told the story of how in 1834 he was walking along a country lane with a Swiss woodcutter when they got to talking about the rocks along the roadside. The woodcutter matter-of-factly told him that the boulders had come from the Grimsel, a zone of granite some distance away. 'When I asked him how he thought that these stones had reached their location, he answered without hesitation: "The Grimsel glacier transported them on both sides of the valley, because that glacier extended in the past as far as the town of Bern."'

Charpentier was delighted, for he had come to such a view himself; but when he raised the notion at scientific gatherings, it was dismissed. One of Charpentier's closest friends was another Swiss naturalist, Louis Agassiz, who after some initial scepticism came to embrace, and eventually all but appropriate, the theory.

Agassiz had studied under Cuvier in Paris and now held the post of Professor of Natural History at the College of

Neuchâtel in Switzerland. Another friend of Agassiz's, a botanist named Karl Schimper, was actually the first to coin the term 'ice age' (in German, *Eiszeit*), in 1837, and to propose that there was good evidence to show that ice had once lain heavily not just across the Swiss Alps but over much of Europe, Asia and North America. It was a radical notion. He lent Agassiz his notes – then came very much to regret it as Agassiz increasingly got the credit for what Schimper felt, with some legitimacy, was his theory.

Charpentier likewise ended up a bitter enemy of his old friend. Alexander von Humboldt, yet another friend, may have had Agassiz at least partly in mind when he observed that there are three stages in scientific discovery: first, people deny that it is true; then they deny that it is important; finally they credit the wrong person.

At all events, Agassiz made the field his own. In his quest to understand the dynamics of glaciation, he went everywhere – deep into dangerous crevasses and up to the summits of the craggiest Alpine peaks, often apparently unaware that he and his team were the first to climb them. Nearly everywhere Agassiz encountered an unyielding reluctance to accept his theories. Humboldt urged him to return to his area of real expertise, fossil fish, and give up this mad obsession with ice, but Agassiz was a man possessed by an idea.

Agassiz's theory found even less support in Britain, where most naturalists had never seen a glacier and often couldn't grasp the crushing forces that ice in bulk exerts. 'Could scratches and polish just be due to *ice*?' asked Roderick Murchison in a mocking tone at one meeting, evidently imagining the rocks as covered in a kind of light and glassy rime. To his dying day, he expressed the frankest incredulity at those 'ice-mad' geologists who believed that glaciers could account for so much. William Hopkins, a Cambridge professor and leading member of the Geological Society, endorsed this view, arguing that the notion that ice could transport boulders presented 'such obvious mechanical absurdities' as to make it unworthy of the society's attention. Undaunted, Agassiz travelled tirelessly to promote his theory. In 1840 he read a paper to a meeting of the British Association for the Advancement of Science in Glasgow, at which he was openly criticized by the great Charles Lyell. The following year the Geological Society of Edinburgh passed a resolution conceding that there might be some general merit in the theory but that certainly none of it applied to Scotland.

Lyell did eventually come round. His moment of epiphany came when he realized that a moraine, or line of rocks, near his family estate in Scotland, which he had passed hundreds of times, could be understood only if one accepted that a glacier had dropped it there. But, having become converted, Lyell then lost his nerve and backed off public support of the ice age idea. It was a frustrating time for Agassiz. His marriage was breaking up, Schimper was hotly accusing him of the theft of his ideas, Charpentier wouldn't speak to him and the greatest living geologist offered support of only the most tepid and vacillating kind.

In 1846 Agassiz travelled to America to give a series of lectures, and there at last found the esteem he craved. Harvard gave him a professorship and built him a first-rate museum, the Museum of Comparative Zoology. Doubtless it helped that he had settled in New England, where the long winters encouraged a certain sympathy for the idea of interminable periods of cold. It also helped that six years after his arrival the first scientific expedition to Greenland reported that nearly the whole of that semi-continent was covered in an ice sheet just like the ancient one imagined in Agassiz's theory. At long last, his ideas began to find a real following. The one central defect of Agassiz's theory was that his ice ages had no cause. But assistance was about to come from an unlikely quarter.

In the 1860s, journals and other learned publications in Britain began to receive papers on hydrostatics, electricity and other scientific subjects from a James Croll of Anderson's University in Glasgow. One of the papers, on how variations in the

Earth's orbit might have precipitated ice ages, was published in the *Philosophical Magazine* in 1864 and was recognized at once as a work of the highest standard. So there was some surprise, and perhaps just a touch of embarrassment, when it turned out that Croll was not an academic at the university, but a janitor.

Born in 1821, Croll grew up poor and his formal education lasted only to the age of thirteen. He worked at a variety of jobs – as a carpenter, insurance salesman, keeper of a temperance hotel – before taking a position as a janitor at Anderson's (now the University of Strathclyde) in Glasgow. By somehow inducing his brother to do much of his work, he was able to pass many quiet evenings in the university library teaching himself physics, mechanics, astronomy, hydrostatics and the other fashionable sciences of the day, and gradually began to produce a string of papers with a particular emphasis on the motions of the Earth and their effect on climate.

Croll was the first to suggest that cyclical changes in the shape of Earth's orbit, from elliptical (which is to say, slightly oval) to nearly circular to elliptical again, might explain the onset and retreat of ice ages. No one had ever thought before to consider an astronomical explanation for variations in Earth's weather. Thanks almost entirely to Croll's persuasive theory, people in Britain began to become more responsive to the notion that at some former time parts of the Earth had been in the grip of ice. When his ingenuity and aptitude were recognized, Croll was given a job at the Geological Survey of Scotland and widely honoured: he was made a Fellow of the Royal Society in London and of the New York Academy of Science, and awarded an honorary degree from the University of St Andrews, among much else.

Unfortunately, just as Agassiz's theory was at last beginning to find converts in Europe, he was busy taking it into ever more exotic territory in America. He began to find evidence for glaciers practically everywhere he looked, including near the equator. Eventually he became convinced that ice had once covered the whole Earth, extinguishing all life, which God had then recreated. None of the evidence Agassiz cited supported such a view. Nonetheless, in his adopted country his stature grew and grew until he was regarded as only slightly below a deity. When he died in 1873 Harvard felt it necessary to appoint three professors to take his place.

Yet, as sometimes happens, his theories fell swiftly out of fashion. Less than a decade after his death his successor in the chair of geology at Harvard wrote that the 'so-called glacial epoch . . . so popular a few years ago among glacial geologists may now be rejected without hesitation'.

Part of the problem was that Croll's computations suggested that the most recent ice age occurred 80,000 years ago, whereas the geological evidence increasingly indicated that the Earth had undergone some sort of dramatic perturbation much more recently than that. Without a plausible explanation for what might have provoked an ice age, the whole theory fell into abeyance. There it might have remained for some time had it not been for a Serbian academic named Milutin Milanković, who had no background in celestial motions at all – he was a mechanical engineer by training – but who in the early 1900s developed an unexpected interest in the matter. Milanković realized that the problem with Croll's theory was not that it was incorrect but that it was too simple.

As the Earth moves through space, it is subject not just to variations in the length and shape of its orbit, but also to rhythmic shifts in its angle of orientation to the Sun – its tilt and pitch and wobble – all affecting the duration and intensity of sunlight falling on any patch of land. In particular it is subject to three changes in position, known formally as its obliquity, precession and eccentricity, over long periods of time. Milanković wondered if there might be a relationship between these complex cycles and the comings and goings of ice ages. The difficulty was that the cycles were of widely different lengths – of approximately 20,000, 40,000 and 100,000 years respectively, but varying in each case by up to a few thousand years – which meant that determining their points of intersection over long spans of time involved a nearly endless amount of exceedingly devoted computation. Essentially, Milanković had to work out the angle and duration of incoming solar radiation at every latitude on Earth, in every season, for a million years, adjusted for three ever-changing variables.

Happily, this was precisely the sort of repetitive toil that suited Milanković's temperament. For the next twenty years, even while on holiday, he worked ceaselessly with pencil and slide rule computing the tables of his cycles – work that now could be completed in hours with a computer. The calculations all had to be made in his spare time, but in 1914 Milanković suddenly got a great deal of that when the First World War broke out and he was arrested owing to his position as a reservist in the Serbian Army. He spent most of the next four years under loose house arrest in Budapest, required only to report to the police once a week. The rest of his time was spent working in the library of the Hungarian Academy of Sciences. He was possibly the happiest prisoner of war in history.

The eventual outcome of his diligent scribblings was the 1930 book *Mathematical Climatology and the Astronomical Theory of Climatic Changes*. Milanković was right that there was a relationship between ice ages and planetary wobble, though like most people he assumed that it was a gradual increase in harsh winters that led to these long spells of coldness. It was a Russian-German meteorologist, Wladimir Köppen – father-in-law of our tectonic friend Alfred Wegener – who saw that the process was more subtle, and rather more unnerving, than that.

The cause of ice ages, Köppen decided, is to be found in cool summers, not brutal winters. If summers are too cool to melt all the snow that falls on a given area, more incoming sunlight is bounced back by the reflective surface, exacerbating the cooling effect and encouraging yet more snow to fall.

The consequence would tend to be self-perpetuating. As snow accumulated into an ice sheet the region would grow cooler, prompting more ice to accumulate. As the glaciologist Gwen Schultz has noted: 'It is not necessarily the *amount* of snow that causes ice sheets but the fact that snow, however little, lasts.' It is thought that an ice age could start from a single unseasonal summer. The leftover snow reflects heat and exacerbates the chilling effect. 'The process is self-enlarging, unstoppable, and once the ice is really growing it moves,' says McPhee. You have advancing glaciers and an ice age.

In the 1950s, because of imperfect dating technology, scientists were unable to correlate Milanković's carefully worked-out cycles with the supposed dates of ice ages as then perceived, and so Milanković and his calculations increasingly fell out of favour. He died in 1958, unable to prove that his cycles were correct. By this time, in the words of one history of the period, 'you would have been hard pressed to find a geologist or meteorologist who regarded the model as being anything more than an historical curiosity'. Not until the 1970s and the refinement of a potassium-argon method for dating ancient sea-floor sediments were his theories finally vindicated.

The Milanković cycles alone are not enough to explain cycles of ice ages. Many other factors are involved – not least the disposition of the continents, in particular the presence of land masses over the poles – but the specifics of these are imperfectly understood. It has been suggested, however, that if you hauled North America, Eurasia and Greenland just 500 kilometres north we would have permanent and inescapable ice ages. We are very lucky, it appears, to get any good weather at all. Even less well understood are the cycles of comparative balminess within ice ages, known as interglacials. It is mildly disconcerting to reflect that the whole of meaningful human history – the development of farming, the creation of towns, the rise of mathematics and writing and science and all the rest – has taken place within an atypical patch of fair weather.

Previous interglacials have lasted as little as 8,000 years. Our own has already passed its ten-thousandth anniversary.

The fact is, we are still very much in an ice age; it's just a somewhat shrunken one – though less shrunken than many people realize. At the height of the last period of glaciation, around 20,000 years ago, about 30 per cent of the Earth's land surface was under ice. Ten per cent still is. (And a further 14 per cent is in a state of permafrost.) Three-quarters of all the fresh water on Earth is locked up in ice even now, and we have ice caps at both poles – a situation that may be unique in the Earth's history outside of actual ice ages. That there are snowy winters through much of the world and permanent glaciers even in temperate places such as New Zealand may seem quite natural, but in fact it is a most unusual situation for the planet.

For most of its history until fairly recent times the general pattern for the Earth was to be hot, with no permanent ice anywhere. The current ice age – ice epoch, really – started about 40 million years ago, and has ranged from murderously bad to not bad at all. We live in one of the few spells of the latter. Ice ages tend to wipe out evidence of earlier ice ages, so the further back you go the more sketchy the picture grows, but it appears that we have had at least seventeen severe glacial episodes in the last 2.5 million years or so. Two commonly cited culprits for the present epoch are the rise of the Himalayas and the formation of the Isthmus of Panama, the first disrupting air flows, the second ocean currents. India, once an island, has pushed 2,000 kilometres into the Asian land mass over the past 45 million years, raising not only the Himalayas, but also the vast Tibetan Plateau behind it. The hypothesis is that the higher landscape was not only cooler, but diverted winds in a way that made them flow north and towards North America, making it more susceptible to long-term chills. Then, beginning about 5 million years ago, Panama rose from the sea, closing the gap between North and South America, disrupting the flows of warming currents between the Pacific and the

Atlantic and changing patterns of precipitation across at least half the world. One consequence was a drying out of Africa, which caused apes to climb down out of trees and go looking for a new way of living on the emerging savannahs.

At all events, with the oceans and continents arranged as they are now, it appears that ice will be a long-term part of our future – unless, of course, we carelessly melt it all away with global warming, with consequences that we can't comfortably predict.

Before 50 million years ago the Earth had no regular ice ages, but when we did have them they tended to be colossal. A massive freezing occurred about 2.2 billion years ago, followed by a billion years or so of warmth. Then there was another ice age even larger than the first – so large that scientists refer to the period in which it occurred as the Cryogenian, or super ice age. The condition is more popularly known as Snowball Earth.

Snowball, however, barely captures the murderousness of conditions. The theory is that because of a fall in solar radiation of about 6 per cent and a drop-off in the production (or retention) of greenhouse gases, the Earth essentially lost its ability to hold on to its heat. It became a kind of all-over Antarctica. Temperatures plunged by as much as 45 degrees centigrade. The entire surface of the planet may have frozen solid, with ocean ice up to 800 metres thick at higher latitudes and tens of metres thick even in the tropics.

There is a serious problem in all this in that the geological evidence indicates ice everywhere, including around the equator, while the biological evidence suggests just as firmly that there must have been open water somewhere. For one thing, cyanobacteria survived the experience and they photosynthesize. For that they needed sunlight, but as you will know if you have ever tried to peer through it, ice very quickly becomes opaque and after only a few yards would pass on no light at all. Two possibilities have been suggested. One is that a little ocean water did remain exposed (perhaps because of some kind of localized warming at a hot spot); the other is that maybe the ice formed in such a way that it remained translucent – a condition that does sometimes happen in nature.

The long freezing of Earth raises the interesting and difficult question of how it ever got warm again. An icy planet should reflect so much heat that it would stay frozen for ever. It appears that rescue may have come from our molten interior. Once again we may be indebted to tectonics for allowing us to be here. The idea is that we were saved by volcanoes, which pushed through the buried surface, pumping out lots of heat and gases that melted the snows and re-formed the atmosphere. But it is thought that this was not a peaceful process. As Earth warmed, it probably had the wildest weather it has ever experienced, with hurricanes powerful enough to raise waves to the heights of skyscrapers and rainfalls of indescribable intensity. Interestingly, this spell of climatic extremity coincided with the Ediacaran period, when complex visible life was just bursting forth. It may be that the extreme conditions acted as a spur to evolution – testing the new fauna's mettle, as it were – or it may be that the new life forms were protected at the bottom of shallow seas. What is certain is that for some 10 million years life thrived as it never had before – some specimens grew up to a metre high – before all the Ediacaran biota disappeared totally and altogether mysteriously during the Cambrian period.

Compared with a Cryogenian outburst, the ice ages of more recent times seem pretty small-scale, but of course they were immensely grand by the standards of anything to be found on Earth today. The Wisconsin ice sheet, which covered much of Europe and North America, was over 3 kilometres thick in places and marched forward at a rate of about 120 metres a year. What a thing it must have been to behold. Even at their leading edge, the ice sheets could be nearly 800 metres thick. Imagine standing at the base of a wall of ice that high. Behind this edge, over an area measuring in the millions of square kilometres, would be nothing but more ice, with only a few of the tallest mountain summits poking through here and there. Whole continents sagged under the weight of so much ice and even now, 12,000 years after the glaciers' withdrawal, are still rising back into place. The ice sheets didn't just dribble out boulders and long lines of gravelly moraines, but dumped entire land masses – Long Island and Cape Cod and Nantucket, among others – as they slowly swept along. It's little wonder that geologists before Agassiz had trouble grasping their monumental capacity to rework landscapes.

For a long time it was thought that we moved into and out of ice ages gradually, over hundreds or thousands of years, but we now know that that has not always been the case. Thanks to ice cores from Greenland we have a detailed record of climate for something over 100,000 years, and what is found there is not comforting. It shows that for most of its recent history the Earth has been nothing like the stable and tranquil place that civilization has known, but rather has lurched violently between periods of warmth and brutal chill.

Towards the end of the last big glaciation, some 12,000 years ago, Earth began to warm, and quite rapidly, but then abruptly plunged back into bitter cold for 1,000 years or so in an event known to science as the Younger Dryas. (The name comes from the Arctic plant the dryas, which is one of the first to recolonize land after an ice sheet withdraws. There was also an Older Dryas period, but it wasn't so sharp.) At the end of this 1,000-year onslaught average temperatures leaped again, by as much as 4 degrees centigrade in twenty years, which may not sound terribly dramatic but is equivalent to exchanging the climate of Scandinavia for that of the Mediterranean in just two decades. Locally, changes have been even more dramatic. Greenland ice cores show the temperatures there changing by as much as 8 degrees centigrade in ten years, drastically altering rainfall patterns and growing conditions. This must have been unsettling enough on a thinly populated planet. Today the consequences would be pretty well unimaginable.

What is most alarming is that we have no idea – none – what natural phenomena could so swiftly rattle the Earth's thermometer. As Elizabeth Kolbert, writing in the *New Yorker*, has observed: 'No known external force, or even any that has been hypothesized, seems capable of yanking the temperature back and forth as violently, and as often, as these cores have shown to be the case.' There seems to be, she adds, 'some vast and terrible feedback loop', probably involving the oceans and disruptions of the normal patterns of ocean circulation, but all this is a long way from being understood.

One theory is that the heavy inflow of meltwater to the seas at the beginning of the Younger Dryas reduced the saltiness (and thus density) of northern oceans, causing the Gulf Stream to swerve to the south, like a driver trying to avoid a collision. Deprived of the Gulf Stream's warmth, the northern latitudes returned to chilly conditions. But this doesn't begin to explain why 1,000 years later, when the Earth warmed once again, the Gulf Stream didn't veer as before. Instead, we were given the period of unusual tranquillity known as the Holocene, the time in which we live now.

There is no reason to suppose that this stretch of climatic stability should last much longer. In fact, some authorities believe that we are in for even worse. It is natural to suppose that global warming would act as a useful counterweight to the Earth's tendency to plunge back into glacial conditions. However, as Kolbert has pointed out, when you are confronted with a fluctuating and unpredictable climate, 'the last thing you'd want to do is conduct a vast unsupervised experiment on it'. It has even been suggested, with more plausibility than would at first seem evident, that an ice age might actually be induced by a rise in temperatures. The idea is that a slight warming would enhance evaporation rates and increase cloud cover, leading in the higher latitudes to more persistent accumulations of snow. In fact, global warming could plausibly, if paradoxically, lead to powerful localized cooling in North America and northern Europe.

Climate is the product of so many variables – rising and falling carbon dioxide levels, the shifts of continents, solar activity, the stately wobbles of the Milankovitch cycles – that it is as difficult to comprehend the events of the past as it is to predict those of the future. Much is simply beyond us. Take Antarctica. For at least 20 million years after it settled over the South Pole Antarctica remained covered in plants and free of ice. That simply shouldn't have been possible.

No less intriguing are the known ranges of some late dinosaurs. The British geologist Stephen Drury has noted that forests within 10 degrees' latitude of the North Pole were home to great beasts, including *Tyrannosaurus rex*. 'That is bizarre,' he writes, 'for such a high latitude is continually dark for three months of the year.' Moreover, there is now evidence that these high latitudes suffered severe winters. Oxygen isotope studies suggest that the climate around Fairbanks, Alaska, was about the same in the late Cretaceous period as it is now. So what was tyrannosaurus doing there? Either it migrated seasonally over enormous distances or it spent much of the year in snowdrifts in the dark. How dinosaurs managed to survive in such conditions can only be guessed.

In the long run – the very long run – ice ages are by no means altogether bad news for the planet. They grind up rocks, leaving behind new soils of sumptuous richness, and gouge out freshwater lakes that provide abundant nutritive possibilities for hundreds of species of being. They act as a spur to migration and keep the planet dynamic. As Tim Flannery has remarked: 'There is only one question you need ask of a continent in order to determine the fate of its people: "Did you have a good ice age?"' And with that in mind, it's time at last to look at a species of ape that truly did.

28

THE MYSTERIOUS BIPED

Just before Christmas 1887, a young Dutch doctor with an un-Dutch name, Marie Eugène François Thomas Dubois,* arrived in Sumatra, in the Dutch East Indies, with the intention of finding the earliest human remains on Earth.

Several things were extraordinary about this. To begin with, to go looking for ancient human bones was a rare activity. Almost everything that had been found to this point had been found accidentally, and nothing in Dubois' background suggested that he was the ideal candidate to make the process intentional. He was an anatomist by training with no background in palaeontology. Nor was there any special reason to suppose that the East Indies would hold early human remains. Logic dictated that if ancient people were to be found at all, it would be on a large and long-populated land mass, not in the comparative fastness of an archipelago. Dubois was driven to the East Indies on nothing stronger than a hunch, the availability of employment and the knowledge that Sumatra was full of caves, the environment in which most of the important hominid fossils had so far been found. What is most extraordinary

---

* Though Dutch, Dubois was from Eijsden, a town bordering the French-speaking part of Belgium.

in all this – nearly miraculous, really – is that he found what he was looking for.

At the time Dubois conceived his plan to search for a missing link, the human fossil record consisted of very little: five incomplete Neanderthal skeletons, one partial jawbone of uncertain provenance and half a dozen ice-age humans recently found by railway workers in a cave at a cliff called Cro-Magnon near Les Eyzies, France. Of the Neanderthal specimens, the best preserved was sitting unremarked on a shelf in London. It had been found by workers blasting rock from a quarry in Gibraltar in 1848, so its preservation was a wonder, but unfortunately no one yet appreciated just how special it was.

After being briefly described at a meeting of the Gibraltar Scientific Society, the fossil was sent to London where it was passed to Hugh Falconer, an eminent natural historian. Falconer quickly recognized that it appeared to be from an unknown species of human, but unfortunately, and indeed rather amazingly, took the matter no further, though he may perhaps be excused on grounds of ill health: he was chronically unwell and would be dead in just over a year. At all events, the skull ended up in the Hunterian Museum in London where it remained undisturbed, apart from an occasional light dusting, for half a century. Much later, it was realized that remains that had been recovered from a cave in Belgium in 1829 were also Neanderthal, but they weren't recognized as such for more than 100 years.

So instead the name and credit for the discovery of the first early humans went to the Neander Valley in Germany – not altogether unfittingly, as it happens, for by uncanny coincidence *Neander* in Greek means 'new man'. There, in 1856, eight years after the unappreciated Gibraltar discovery and twenty-seven after the similarly unappreciated find in Belgium, workmen at another quarry, in a cliff face overlooking the Düssel River, found some curious-looking bones which they passed to a local schoolteacher, knowing he had an interest in all things natural. To his great credit the teacher, Johann Karl Fuhlrott, saw that he had some new type of human, but lacked the influence and contacts to get the discovery taken seriously. Such was the lack of interest that the naming of this new species – the first-ever addition to the genus *Homo* – fell to an obscure university professor in the west of Ireland who had never seen the original Neanderthal skull, had no specialized knowledge in palaeontology, and lost interest in the matter almost immediately afterwards.

His name was William King and, though just a footnote to history, he was an extraordinary figure. Raised in poverty in Sunderland, in the northeast of England, as a youth he showed a remarkable aptitude for finding and interpreting fossils, to such an extent that he was put in charge of the Museum of Natural History in Newcastle-upon-Tyne while still a young man. It was soon noticed, however, that the museum's collections were visibly shrinking under his care. It turned out that King was quietly selling items to other institutions and collectors and pocketing the proceeds.

Encouraged to find new employment elsewhere, King moved to Queen's College at Galway, on the western edge of Ireland, where he became a professor of mineralogy and natural sciences despite a lack of academic qualifications. There, King somehow acquired a plaster cast of the German fossil, then known as the Feldhofer fossil, after the cave in which it was found. Using nothing more than this plaster cast and a few published drawings – and with absolutely no expertise in the area – he confidently concluded that this was a new type of human, which he called *Homo neanderthalensis*. He was also the first in what would become a very long line to suggest that Neanderthals were brutish and dull. King introduced the name and his findings at a meeting of the geology section of the British Association for the Advancement of Science in 1863 and followed up with a paper in 1864. And then he appears to have lost all interest in the matter. Neanderthals are never mentioned again in any of his writings.

Many people refused to accept that the Neanderthal bones were ancient at all. August Mayer, a professor at the University of Bonn and a man of influence, insisted that the bones were merely those of a Mongolian Cossack soldier who had been wounded while fighting in Germany in 1814 and had crawled into the cave to die. Hearing of this, T. H. Huxley in England drily observed how remarkable it was that the soldier, though mortally wounded, had climbed 60 feet up a cliff, divested himself of his clothing and personal effects, sealed the cave opening and buried himself under 2 feet of soil. Another anthropologist, puzzling over the Neanderthal's heavy brow ridge, suggested that it was the result of long-term frowning arising from a poorly healed forearm fracture. In their eagerness to reject the idea of earlier humans, authorities were often willing to embrace the most singular possibilities. At about the time that Eugène Dubois was setting out for Sumatra, a skeleton found in Périgueux was confidently declared to be that of an Eskimo. Quite what an ancient Eskimo was doing in southwest France was never comfortably explained. It was actually an early Cro-Magnon.

It was against this background that Dubois began his search for ancient human bones. He did no digging himself, but instead used fifty convicts lent by the Dutch authorities. For a year they worked on Sumatra, then transferred to Java. And there, in 1891, Dubois – or rather his team, for Dubois himself seldom visited the sites – found a section of ancient human cranium now known as the Trinil skullcap. Though only part of a skull, it showed that the owner had had distinctly non-human features but a much larger brain than any ape. Dubois called it *Anthropithecus erectus* (later changed for technical reasons to *Pithecanthropus erectus*) and declared it the missing link between apes and humans. It quickly became popularized as 'Java Man'. Today we know it as *Homo erectus*.

The next year Dubois' workers found a virtually complete thighbone that looked surprisingly modern. In fact, many anthropologists think it *is* modern and has nothing to do with Java Man. If it is an *erectus* bone, it is unlike any other found since. Nonetheless Dubois used the thighbone to deduce – correctly, as it turned out – that *Pithecanthropus* walked upright. He also produced, with nothing but a scrap of cranium and one tooth, a model of the complete skull, which also proved uncannily accurate.

In 1895 Dubois returned to Europe, expecting a triumphal reception. In fact, he met nearly the opposite reaction. Most scientists disliked both his conclusions and the arrogant manner in which he presented them. The skullcap, they said, was that of an ape, probably a gibbon, and not of any early human. Hoping to bolster his case, in 1897 Dubois allowed a respected anatomist from the University of Strasbourg, Gustav Schwalbe, to make a cast of the skullcap. To Dubois' dismay, Schwalbe thereupon produced a monograph that received far more sympathetic attention than anything Dubois himself had written, and followed it with a lecture tour in which he was celebrated nearly as warmly as if he had dug up the skull himself. Appalled and embittered, Dubois withdrew into an undistinguished position as a professor of geology at the University of Amsterdam and for the next two decades refused to let anyone examine his precious fossils again. He died in 1940, an unhappy man.

As for the first Neanderthals, it would take years for them to be widely accepted into the genus *Homo*, and decades more before any authority allowed that they might have been more than knuckle-dragging brutes. In the early years, several other names were proposed for the species, of which perhaps the most telling, if not most generous, was *Homo stupidus*. The first formal description of Neanderthals wasn't written until 1907, by a geologist named William Sollas, who had only a passing competency in anatomy.

So, as the twentieth century got underway, we were finally – and in many cases reluctantly – aware that we *Homo sapiens* had some ancient cousins. No one, however, had any idea just how

THE MYSTERIOUS BIPED

vastly more complicated, and endlessly fractious, the field of human origins would become in the years ahead.

In 1924, in South Africa, Raymond Dart, head of anatomy at the University of the Witwatersrand, was sent the skull of a child from a place called Taung, on the edge of the Kalahari Desert. He realized that it was unlike anything ever seen before. It came from a being that had a mix of ape-like and human characteristics – exactly the sort of transitional figure that palaeoanthropology needed to explain the evolution of modern peoples. Dart dubbed it *Australopithecus africanus*, or 'southern ape man of Africa', and suggested the need for an entirely new family, *Homo simiadae* ('the man-apes'), to accommodate the find.*

Dart's conclusions were almost universally dismissed. For one thing, everyone knew that humans arose in Asia, not Africa. In addition, South Africa was considered something of an academic backwater and Dart an underqualified interloper in what was essentially a European field. Dart spent five years working up a monograph, but could find no one to publish it. Eventually, following a nervous breakdown, he gave up the quest to publish altogether. For the next thirty years the received wisdom was that the Taung child was an ape and nothing more. Most textbooks didn't even mention it. For years,

---

* Humans are put in the family Hominidae. Its members, called hominids, traditionally included any creatures (including extinct ones) that are more closely related to us than to any surviving chimpanzees. The apes, meanwhile, are lumped together in a family called Pongidae. Many authorities believe that chimps, gorillas and orangutans should also be included in the family Hominidae, with humans and chimps in a subfamily called Homininae. The upshot is that the creatures historically called hominids become, under this arrangement, hominins. Hominoidea is the name of the ape superfamily, which includes us. Put more simply, *hominid* describes all apes and humans, living and extinct, while *hominin* applies only to members of the human line (i.e. those designated *Homo*).

the skull — today recognized as one of the supreme treasures of anthropology — sat as a paperweight on a colleague's desk.*

So the search for the earliest humans remained focused on Asia, particularly after a gifted Canadian named Davidson Black began to poke around at a place called Dragon Bone Hill, 25 miles from Beijing (then known in the West as Peking) which was locally famous as a hunting ground for old bones. Unfortunately, rather than preserving the bones for study, the Chinese ground them up to make medicines. We can only guess how many priceless ancient bones ended up as a sort of Chinese equivalent of Beecham's powder. The site had been much denuded by the time Black arrived, but he found a single fossilized molar and on the basis of that alone quite brilliantly announced the discovery of *Sinanthropus pekinensis*, which quickly became known as Peking Man.

At Black's urging, more determined excavations were undertaken and many other bones found. Unfortunately all were lost the day after the Japanese attack on Pearl Harbor in 1941, when a contingent of US marines, trying to spirit the bones (and themselves) out of the country, was intercepted by the Japanese and imprisoned. Seeing that their crates held nothing but bones, the Japanese soldiers left them at the roadside. It was the last that was ever seen of them.

Meantime, back in Java, a team led by Ralph von Koenigswald had found another group of early humans which became known as the Solo People, from the site of their discovery on the Solo River at Ngandong. Koenigswald's discoveries might have been more impressive still but for a tactical error that was realized too late. He had offered locals 10 cents for every piece

---

* Rather remarkably, both Dubois and Dart almost managed to lose their precious treasures through carelessness. Dubois in 1895 absent-mindedly left the remains of Java Man in a Parisian café where he had just lunched, while Dart's wife in 1931 left the skull of the Taung child in a London taxi. Luckily, in both cases the missing possessions were recovered.

of hominid bone they could come up with, then discovered to his horror that they had been enthusiastically smashing large pieces into small ones to maximize their income.

In the following years, as more bones were found and identified, there came a flood of new names – *Homo aurignacensis, Australopithecus transvaalensis, Paranthropus crassidens, Zinjanthropus boisei* and scores of others, nearly all involving a new genus type as well as a new species. By the 1950s, the number of named hominid types had risen to comfortably over 100. To add to the confusion, individual forms often went by a succession of different names as palaeoanthropologists refined, reworked and squabbled over classifications. The Solo People were known variously as *Homo soloensis, Homo primigenius asiaticus, Homo neanderthalensis soloensis, Homo sapiens soloensis, Homo erectus erectus* and, finally, plain *Homo erectus*.

In an attempt to introduce some order, in 1960 F. Clark Howell of the University of Chicago, following the suggestions of Ernst Mayr and others the previous decade, proposed cutting the number of genera to just two – *Australopithecus* and *Homo* – and rationalizing many of the species. The Java and Peking men both became *Homo erectus*. For a time order prevailed in the world of the hominids. It didn't last.

After about a decade of comparative calm, palaeoanthropology embarked on another period of swift and numerous discovery. The 1960s produced *Homo habilis*, thought by some to be the missing link between apes and humans, but thought by others not to be a separate species at all. Then came (among others) *Homo ergaster, Homo louisleakeyi, Homo rudolfensis, Homo microcranus* and *Homo antecessor*, as well as a raft of australopithecines: *A. afarensis, A. praegens, A. walkeri, A. anamensis* and still others.

A big part of the problem, paradoxically, is a shortage of evidence. Since the dawn of time, it's estimated that 110 billion human (or human-like) beings have lived, each contributing a little genetic variability to the total human stock. Out of this vast number, the whole of our understanding of human prehistory is based on the remains of just 6,000 or so individuals (according to the Smithsonian Institution), and much of what they left behind is often extremely fragmentary – a tooth, a sliver of shinbone, a ragged fragment of skull.

'You could fit it all into the back of a pickup truck if you didn't mind how much you jumbled everything up,' Ian Tattersall, then curator (now retired) of anthropology at the American Museum of Natural History in New York, told me when I asked him in 2002 the size of the total world archive of hominid and early human bones. More bones have been found since then, of course, so you might now need, say, a delivery van, but the general point still holds – the number of fossil hominid bones in the world is modest indeed.

The shortage wouldn't be so bad if the bones were distributed evenly through time and space, but of course they are not. They appear randomly, often in the most tantalizing fashion. *Homo erectus* walked the Earth for well over a million years and inhabited territory from the Atlantic edge of Europe to the Pacific side of China, yet if you brought back to life every *Homo erectus* individual whose existence we can vouch for, they would barely fill a couple of school buses.

It is the patchiness of the record that makes each new find look so sudden and distinct from all the others. If we had tens of thousands of skeletons distributed at regular intervals through the historical record, there would be appreciably more degrees of shading. Whole new species don't emerge instantaneously, as the fossil record implies, but gradually out of other, existing species. The closer you go back to a point of divergence, the closer the similarities are, so that it becomes exceedingly difficult, and sometimes impossible, to distinguish a late *Australopithecus* from an early *Homo*, since it is likely to be both and neither.

So, bearing in mind that there is little you can say about human prehistory that won't be disputed by someone somewhere, other than that we most certainly had one, what we think we know about who we are and where we come from is roughly this.

For the first 99.99999 per cent of our history as organisms, we were in the same ancestral line as chimpanzees. Virtually nothing is known about the prehistory of chimpanzees, but whatever they were, we were. Then, about 7 million years ago, something major happened. A group of new beings emerged from the tropical forests of Africa into more open woodland.

These would evolve into the australopithecines – the people that Raymond Dart had discovered with the Taung child – and for the next 2 million years they would be the world's dominant hominid group. (*Austral* is from the Latin for 'southern' and has no connection in this context with Australia.) Australopithecines came in several varieties, some slender and gracile, like Dart's Taung child, others more sturdy and robust, but all were capable of walking upright. Some of these species existed for well over a million years, others for a more modest few hundred thousand, but it is worth bearing in mind that even the least successful had histories longer than we have yet achieved.

The most famous hominid remains are those of a 3.18-million-year-old australopithecine found at Hadar in Ethiopia in 1974 by a team led by the American Donald Johanson. Formally known as A. L. (for 'Afar Locality') 288-1, the skeleton became more familiarly known as Lucy, after the Beatles song 'Lucy In The Sky With Diamonds'. Johanson has never doubted her importance. 'She is our earliest ancestor, the missing link between ape and human,' he has said.

Lucy was small – just 3.5 feet tall. She could walk, though how well is a matter of some contention. She was evidently a good climber too. Much else is unknown. Her skull was almost entirely missing, so little could be said with confidence about her brain size, though skull fragments suggested it was small. Most books describe Lucy's skeleton as being 40 per cent complete, though some put it closer to half and one produced by the American Museum of Natural History describes Lucy as two-thirds complete. The BBC television series *Ape Man* actually called it 'a complete skeleton', even while showing that it was anything but.

A human body has 206 bones, but many of these are repeated. If you have the left femur from a specimen, you don't need the right to know its dimensions. Strip out all the redundant bones and the total you are left with is 120 – what is called a half-skeleton. Even by this fairly accommodating standard, and even counting the slightest fragment as a full bone, Lucy constituted only 28 per cent of a half skeleton (and only about 20 per cent of a full one).

In *The Wisdom of the Bones*, Alan Walker recounts how he once asked Johanson how he had come up with a figure of 40 per cent. Johanson breezily replied that he had discounted the 106 bones of the hands and feet – more than half the body's total, and a fairly important half, too, one would have thought, since Lucy's principal defining attribute was the use of those hands and feet to deal with a changing world. At all events, rather less is known about Lucy than is generally supposed. It isn't even actually certain that she was a female. Her sex is merely presumed from her diminutive size.

But it is certain from her anatomy that she walked. Bipedalism is a demanding and risky strategy. It means refashioning the pelvis into a full load-bearing instrument. To preserve the required strength, the birth canal in the female must be comparatively compressed. This has two very significant immediate consequences and one longer-term one. First, it means a lot of pain for any birthing mother and a greatly increased danger of fatality to mother and baby both. Moreover, to get the baby's head through such a tight space it must be born while its brain is still small – and while the baby, therefore, is still helpless. This means long-term infant care, which in turn implies solid male–female bonding. All this is problematic enough when you are the intellectual master of the planet, but when you are a small, vulnerable australopithecine, with a brain about the size of an orange, the risk must have been enormous.

So why did Lucy and her kind come down from the trees and out of the dense forests? Probably they had no choice. The slow rise of the Isthmus of Panama had cut the flow of waters from the Pacific into the Atlantic, diverting warming currents away from the Arctic and leading to the onset of an exceedingly sharp ice age in northern latitudes. In Africa, this would have produced seasonal drying and cooling, gradually turning jungle into savannah. 'It was not so much that Lucy and her like left the forests,' John Gribbin has written, 'but that the forests left them.'

But stepping out on to the open savannah also clearly left the early hominids much more exposed. An upright hominid could see better, but could also be seen better. Even now, as a species we are almost preposterously vulnerable in the wild. Nearly every large animal you care to name is stronger, faster and toothier than us. Faced with attack, modern humans have only two advantages. We have a good brain, with which we can devise strategies; and we have hands, with which we can fling or brandish hurtful objects. We are the only creature that can harm at a distance. We can thus afford to be physically vulnerable.

All the elements would appear to have been in place for the rapid evolution of a potent brain, and yet that seems not to have happened. For some 2 million years, Lucy and her fellow australopithecines scarcely changed at all.

Thanks to Lucy and one or two other significant discoveries, by the beginning of the present century some twenty types of hominid were recognized (though almost no two experts recognized exactly the same twenty) and it looked as if our understanding of the long path to becoming us was heading towards some kind of completion.

We couldn't have been more wrong. A run of startling fossil

discoveries, plus the development of an entirely new science, palaeogenomics, revealed that early humans had lived, bred and moved about in ways that no palaeontologist would have dreamed possible – and sometimes in ways that none can plausibly explain. Nothing, it turns out, was nearly as linear or straightforward as once thought. We now know that there were once far more human species than anyone had ever suspected, that they spread to the far corners of the Earth much earlier than supposed, that they sometimes ended up in places so difficult to get to that no one can explain how they got there, and that enough of them bred with us sufficiently heartily or often that we still carry a good deal of their DNA within us. Populations of archaic humans of various types arose, moved about, procreated with other archaic humans, and moved about some more. Gradually, all but one of these wandering tribes died out or were absorbed by us, *Homo sapiens*. All of this happened over immense spans of time, with genomes mixed and remixed through encounters that may have happened millennia apart.

Palaeogenomics* – the ability to extract and compare DNA from ancient fossils – means that scientists can now begin to tease apart these ancient bumpings-together. The technique was more or less single-handedly invented by the Swedish geneticist Svante Pääbo, who won a Nobel Prize in 2022 for his work.

Pääbo and his new science first attracted widespread notice in 2010 following a small, largely unremarked discovery two years earlier at a cave in the Altai Mountains of south-central Russia not far from the border with Kazakhstan. The discovery was a tiny fragment of finger bone, stained a deep brown by age and no bigger than a shirt button. The researchers put it to one side, unaware that this was one of the great finds in palaeontological history. Eventually the bone ended up in Pääbo's lab at the Max Planck Institute for Evolutionary

---

* Sometimes called palaeogenetics or evolutionary genetics.

Anthropology in Leipzig, Germany. When Pääbo and his team tested the bone for Neanderthal DNA they were astounded to discover that it was from a child who was neither Neanderthal nor *Homo sapien*, but a type of person not previously encountered. These new people were dubbed Denisovans, after an eighteenth-century hermit named Denis who had lived in the cave.

The Denisovans, it turned out, had interbred (or engaged in 'introgression', as anthropologists sometimes delicately put it) with both Neanderthals and *Homo sapiens* – a fact that is preserved in our DNA to this day. Studies of the cave-floor sediments showed that the comings and goings of the three groups over thousands of years was anything but straightforward. Denisovans, it appears, had the cave to themselves from 250,000 to 170,000 years ago; Neanderthals then joined them for about 40,000 years; then the Denisovans vanished, leaving the cave to Neanderthals alone for a further 30,000 years; Denisovans returned to the cave about 100,000 years ago, but this time it was a different population group; finally, modern humans moved in about 45,000 years ago, just about the time that all the other early humans on the planet vanished for ever. It is curious that Denisovans and Neanderthals shared the cave for thousands of years, and incontestably had sex and bore offspring, yet remained genetically distinct and didn't blend into a single species. This complicated relationship between Neanderthals and Denisovans is in large part why the Denisovans have not been designated as a new species. (Initially for a time the group was known as *Homo altaiensis*, after the surrounding mountains.)

In 2019 another big surprise came when an ancient jawbone from the highlands of Tibet, 2,250 kilometres from the Denisova cave, was identified as Denisovan. It had been found at an elevation of nearly 11,000 feet in an area so inhospitable that it was thought that no early humans could possibly have lived there. It is hard to say why – never mind how – these people elected to pass their lives at such a bleak and forbidding height when so much of the rest of the planet was sparsely occupied. In the spring and summer of 2025, the Denisovans' range was extended further still when a jawbone dredged from the sea floor off the coast of Taiwan and a skull found years earlier in Harbin, China, were both identified as Denisovan. So it is quite obvious that Denisovans thrived in multiple environments for a very long time and covered a lot of ground. Their DNA has been found in modern populations thousands of miles away in Australia and the western Pacific.

It is clearly extraordinary that an entire race of previously unimagined people could be conjured up from a tiny fragment of bone, but at the same time genetics can only take us so far. It tells us massive amounts about Denisovans at a granular level – what diseases they were susceptible to, how well they could breathe at altitude, what colour hair they had – but virtually nothing about how they lived, whether they were capable of speech, love, art, compassion or friendship, or nearly anything else that made them human.

Which makes them tantalizingly enigmatic – though nothing like as enigmatic as the next group we must consider: the tiny, wholly mysterious beings discovered in a cave on the Indonesian island of Flores in 2003 and formally known as *Homo floresiensis*, but more familiarly referred to as 'Hobbits'.

Virtually everything about them is unexpected. They were, for a start, tiny – just 3 feet 5 inches tall, or about the size of a modern toddler. They had small brains, even allowing for their compact body size, but left evidence that they crafted tools for hunting and butchering, and utilized fire. They had short legs and long feet, which would have given them a high-stepped gait, rather like someone walking in swimming flippers. Despite this, they hunted the island's swift and ferocious Komodo dragons and a species of dwarf but now extinct elephants called stegodons – and when I say 'dwarf' that is only when compared to conventional elephants. The Flores elephants weighed

800 pounds – a formidable challenge for any modern human, never mind a group of people who were no bigger (or brainier) than modern preschool children.

As Ian Tattersall has put it: 'No hominid fossil find . . . had ever come from farther out in left field than this strange phenomenon. The skeleton was unlike anything anyone had ever seen before.'

The discoverers suggested that the Flores people were dwarfed *Homo erectus* – that is, that they shrank (as sometimes happens with island species) after reaching Flores more than a million years ago. But others have pointed out that nearly all their most salient features – the size of their brain, the proportions of their limbs, the thickness of their bones, their ape-like toes – have no affinities with *Homo erectus* or any other *Homo* species, but instead recall the earlier, more ape-like (and already small) australopithecines. As Tattersall has pointed out, 'any known dwarfing process would have left the Flores form with a much larger brain than it actually has. Moreover, it is hard to see how reducing your body size could have been at all advantageous in an island crawling with Komodo dragons.'

All this raises the possibility that some primitive, small-brained hominids left Africa well over a million years ago, made their way across some 10,000 kilometres of highly variable terrain without leaving any evidence of their passing along the way, before finally reaching tropical Asia, where they crossed at least two formidable stretches of open water to get to a remote island filled with dangerous animals, and then decided to go no further ever again. And if that doesn't give us pause for thought, it is hard to imagine what would.

Although reaching Flores may have been the single most extraordinary human achievement in the whole of prehistory, paleoanthropologists have been curiously vague about how that was done. Here is a typical, breezily brisk summation from one study: 'Regardless of the route taken, transit is thought to have been accidental.' With respect, that is hardly a satisfactory answer. The most commonly offered hypothesis is that a male and female of breeding age were washed ashore after perhaps clinging to a fallen tree in a tsunami, but this overlooks the obvious fact that it would take more than one couple to provide enough genetic diversity to establish a colony. A study by Corey J. A. Bradshaw of Flinders University in Australia and others in 2019 concluded that it would require between 1,300 and 1,550 individuals to establish a lasting colony in a new land. Other studies have suggested that the numbers could be much smaller than that – perhaps as few as twenty or so people of breeding age – but however you calculate it, it is not easy to come up with plausible circumstances that involve sufficient numbers of people alighting on distant beaches twice, by accident, a million years or so before the presumed invention of ocean-going craft.

In 2019 the puzzle grew even more complicated when another small-bodied archaic species – again, more like australopithecines than anything in the *Homo* line – was found on the island of Luzon in the Philippines, several hundred miles to the north, and dubbed *Homo luzonensis*. Researchers found thirteen bones, about 130,000 years old, making them contemporaneous with – and possibly just a variant of – *Homo floresiensis*. As with the Hobbits of Flores, how they got there and why they came are questions that cannot be answered. It's interesting to consider how different our palaeontological perceptions might be today if Eugène Dubois had found either of the dwarf species in 1891 instead of Java Man.

So far no one has been able to extract DNA from any of the Flores or Luzon bones. 'It's a real challenge,' Matt Tocheri, a professor of anthropology at Lakehead University in Ontario and a member of the Human Origins programme at the Smithsonian Institution in Washington, told me. 'You can't just keep grinding up ancient bones in the hope that eventually you'll get some usable DNA, but without the DNA we will probably never know quite who these people were.'

The palaeontological astonishments did not end with the

Flores and Luzon people, however. Halfway around the world in Africa, in 2013, our understanding of early human development took another sideways lurch with the discovery of a large cache of bones in caves in Gauteng province in South Africa. The bones came from at least fifteen individuals of a species never before seen and subsequently dubbed *Homo naledi*.

Homo naledi were only slightly smaller than modern humans – they were about 5 feet tall and weighed about 100 pounds – but their brains were less than half the size of ours. What was particularly notable, and puzzling, was that their bones were found deep within a tortuous cave system that is exceptionally difficult to negotiate even with the benefit of flashlights and modern caving equipment. The chamber where the bones were found is more than 250 feet from the cave entrance, reachable only through a maze of very tight squeezes (some no more than 10 inches wide), culminating in a sheer drop of almost 40 feet. How primitive beings with no control of fire managed to reach the chamber is a mystery – and why they took the trouble is a greater mystery still.

One theory – not universally accepted – is that it was a burial chamber and that the 1,500 bones found there were placed there on purpose by those who survived them. Those who dispute the burial hypothesis point out that the bones were not reverently arranged in a funereal fashion but scattered about chaotically, which rather argues against the idea of ceremonial interment. Whatever was going on there, so far this is the only site where *Homo naledi* fossils have been found – a reminder that entire peoples may have come and gone in our long history and left only the most tantalizing hint of their existence, or none at all.

All these disparate, scattered species have one very obvious feature in common: they are no longer here. It may come as a surprise to know that we *Homo sapiens* may once almost have gone extinct ourselves. By extrapolating backwards through genomes, researchers at the Chinese Academy of Sciences in

Shanghai in 2023 concluded that nearly 99 per cent of early humans vanished during a period of intense climatic instability some 930,000 years ago. By their calculations, the number of adults on Earth capable of breeding fell to as low as 1,280, and that number didn't rebound from those imperilled depths for 100,000 years or so.

Until quite recently, it was a commonplace to consider our success as a species as more or less preordained – think of the famous illustration called the 'March of Progress', which shows a parade of bipeds advancing through evolutionary stages from knuckle-dragging ape on the left to tall, muscular, confidently striding modern man on the right. But as we now know, our rise to eminence wasn't straightforward at all and almost certainly depended as much on luck as on our superior brains and strength of reasoning. The most humbling certainty of our supremacy, as we are about to see, is that it wasn't inevitable at all.

29

THE RESTLESS APE

Some time about a million and a half years ago, some forgotten genius of the hominid world did an unexpected thing. He (or very possibly she) took one stone and carefully used it to shape another. The result was a simple teardrop-shaped hand-axe, but it was the world's first piece of advanced technology. It was so superior to existing tools that soon others were following the inventor's lead and making hand-axes of their own. Eventually whole societies existed that seemed to do little else.

Stunning evidence of this survives in a little-visited site just beyond the blue Ngong Hills in Kenya, to the south and west of Nairobi. Drive out of the city on the main highway to Uganda and there comes a moment of startling glory when the ground abruptly falls away and you are presented with a hang-glider's view of boundless, pale-green African plain. This is the Great Rift Valley, which arcs across 3,000 miles of East Africa, marking the tectonic rupture that is setting Africa adrift from Asia. Here, along the baking valley floor, is an ancient site called Olorgesailie, which once stood beside a large and pleasant lake. In 1919, long after the lake had dried up, a geologist named J. W. Gregory was scouting the area for mineral prospects when he came across a stretch of open ground littered – in places positively heaped – with anomalous dark stones that had clearly been shaped by human hand. He had found one of the great and original sites of Acheulean tool manufacture.*

After its discovery by Gregory, Olorgesailie lay undisturbed for years before the famed husband-and-wife team of Louis and Mary Leakey began an excavation. What the Leakeys found was a site stretching to 10 acres or so, where tools were made in incalculable numbers for roughly a million years, from about 1.2 million years ago to 200,000 years ago. Today the main tool beds are sheltered from the worst of the elements beneath large tin lean-tos and fenced off with chicken wire to discourage opportunistic scavenging by visitors, but otherwise the tools are left just where their creators dropped them and where the Leakeys found them.

Jillani Ngalli, a keen young man from the Kenyan National Museum who had been dispatched to act as guide, told me when I visited in 2002 that the quartz and obsidian rocks from which the axes were made were never found on the valley floor. 'They had to carry the stones from there,' he said, nodding at a pair of mountains in the hazy middle distance, in opposite directions from the site: Olorgesailie and Ol Esakut. Each was about 10 kilometres away – a long way to carry an armload of stone.

Why the early Olorgesailie people went to such trouble we can only guess, of course. Not only did they lug hefty stones considerable distances to the lakeside, but they then organized the worksite. The Leakeys' excavations revealed that there were areas where axes were fashioned and others where blunt axes were brought to be resharpened. Olorgesailie was, in short, a kind of factory; one that stayed in business for a million years.

Various replications have shown that the axes were not easy to make – even in skilled hands, an axe could take hours to fashion – and yet they were not particularly good for cutting or chopping or scraping or any of the other tasks to which they

---

* After St Acheul, a suburb of Amiens in northern France, where the first examples were found in the nineteenth century.

were presumably put. So we are left with the position that for a million years – far, far longer than our own species has even been in existence, much less engaged in continuous cooperative efforts – early people came in considerable numbers to this particular site to make extravagantly large volumes of tools that appear to have been rather curiously pointless.

And who were these people? We have no idea, actually. We assume they were *Homo erectus* because there are no other known candidates, which means that at their peak – their *peak* – the Olorgesailie workers would have had the brains of a modern infant. But there is no physical evidence on which to base a conclusion. Despite over eighty years of looking, just three fragments of bone from a single human have ever been found in or around the vicinity of Olorgesailie. However much time they spent there shaping rocks, it appears they went elsewhere to die.

'It's all a mystery,' Jillani Ngalli told me, beaming happily.

The Olorgesailie people disappeared from the scene about 200,000 years ago when the lake dried out and the Rift Valley started to become the hot and challenging place it is today. But by this time their days as a species were numbered. They would be succeeded – here and everywhere else across the world – by a lither and ultimately smarter species of newcomers, the ancestors of every one of us alive today.

These first *Homo sapiens* are surprisingly shadowy. We know less about ourselves, curiously enough, than about almost any other line of hominids. It is odd indeed, as Ian Tattersall has noted, 'that the most recent major event in human evolution – the emergence of our own species – is perhaps the most obscure of all'. Nobody can even quite agree where truly modern humans first appear in the fossil record.

Until as recently as the 1980s, it was widely believed that *Homo sapiens* arose as a distinct species only about 40,000 years ago. That estimate was gradually pushed back to about 200,000 years as more evidence came to light. But in 2017, our collective venerability got a substantial boost when researchers at the Max Planck Institute for Evolutionary Anthropology in Germany announced that fossil bones of *Homo sapiens* found at Jebel Irhoud in Morocco were about 300,000 years old. This not only made *Homo sapiens* a good deal more ancient than previously thought, but put them in a corner of Africa where no one expected them to be. Interestingly, nothing in the artefacts those early people left behind indicated that they were any more sophisticated than other hominins of the time. They would need another 200,000-odd years of unhurried evolution to become the clever, ambitious, kindly, vain, gabby, emotional, art-loving beings that we consider properly human. Some authorities, it should be noted, think the Jebel Irhoud fossils were not from *Homo sapiens* at all but from a near cousin.

So it is not easy to say *when* we became a dominant species, though we can make a pretty good guess as to *why* it happened, and the answer is a bit surprising. We became good at running. A gracile build and a few anatomical modifications (such as developing something called the nuchal ligament, at the back of your neck, which exists solely to hold your head steady when running) allowed us to trot along more efficiently and for far greater distances than any human species had before. It was this elemental skill – serious, dogged, long-distance running – that very probably launched us on our way to planetary pre-eminence.

Most large mammals – the ones you would want to eat – are built to sprint, but are not terribly good at covering distances. A typical antelope or wildebeest will become exhausted and collapse after about 15 kilometres if it is not allowed to rest. So early *Homo sapiens* learned the trick of jogging relentlessly after their prey, keeping them constantly harried until they finally dropped and could be easily finished off, cut into hunks and carried home. A diet rich in meat provided the protein necessary to support a larger brain, and a larger brain gave us the cognitive wherewithal to develop speech, refine our cooperative endeavours, and master new skills like cooking over fire, which made meat easier to chew and digest, helping our brains to grow bigger still.

Then, some 60,000–70,000 years ago, probably in response to climatic changes, our direct ancestors began to leave Africa in comparatively large numbers and spread into Eurasia. Maps and diagrams showing human dispersal across the globe nearly always involve big arrows depicting sweeping movements across continents, as if describing the advances of an army, but of course it was nothing like as calculated as that. Wherever they went, according to traditional theory, these new *Homo sapiens* displaced their duller, less adept predecessors. In Europe, these dominant new people became known as Cro-Magnons after the rock shelter in southwestern France where bones of these early arrivals were first uncovered in 1868.

No signs of slaughter have ever been found, so the default explanation became that the newcomers must have used their larger brains and superior wiles to outwit and outcompete the older ones. Here is a typical comment by a science reporter from the *New York Times* in a book published at the very end of the twentieth century: 'Modern humans neutralized this advantage [the Neanderthal's more robust physique] with better clothing, better fires and better shelter; meanwhile the Neanderthals were stuck with an oversize body that required more food to sustain.'

In other words, the very factors that had allowed them to survive for hundreds of thousands of years suddenly became an insuperable handicap. As even a moment's reflection should have reminded the author, Neanderthals obviously knew how to dress for an ice age. They knew the topography and weather patterns – the best times and places to shelter, where and when to hunt – far more intimately than any recent arrivals possibly could. For tens of thousands of years they lived through conditions that no modern human outside a few polar scientists and explorers has experienced. During the worst of the ice ages, blizzards with hurricane-force winds were common. Temperatures routinely fell to minus 45 degrees centigrade. Polar bears padded across the snowy vales of southern England. Neanderthals naturally retreated from the worst of it, but even so they will have experienced weather that was at least as bad as a modern Siberian winter. They suffered, to be sure – a Neanderthal who lived much past thirty was lucky indeed – but as a species they were magnificently resilient and practically indestructible. They survived for at least 400,000 years over an area stretching from the Straits of Gibraltar to Siberia, which is a pretty successful run for any species of being.

Above all, a point that is seldom much dwelt upon is that Neanderthals had brains that were significantly larger than those of modern people – 1.8 litres for Neanderthals versus 1.4 for modern people, according to one calculation. This is more than the difference between modern *Homo sapiens* and late *Homo erectus*, a species we are happy to regard as barely human. It is true that absolute brain size doesn't tell you everything – men have bigger brains than women, but clearly aren't any smarter – but the conclusion that Neanderthals' brains were dimmer is moot, to put it mildly, and wholly unscientific.

These days few would argue that Neanderthals lost the ability to survive simply because we had better clothing, better shelter and 'better fires' (whatever in the world that might mean). Nonetheless it is still largely assumed that the Cro-Magnons, in some undefined way, using some ineffable superiority, drove the Neanderthals into an ever-shrinking corner of southwest Europe where ultimately they had no choice but to fall in the sea or go extinct.

There's a good chance, however, that that is completely wrong.

In the late summer of 2024 I flew to Gibraltar, the snug and charming little city-state (formally a British Overseas Territory) at the foot of the Rock of Gibraltar. It is a delightfully disorienting place – a provincial English market town magically transported to the shores of the Mediterranean. I was there to see Clive Finlayson, the long-serving director of the Gibraltar Museum, who had offered to take me to the caves where the last known Neanderthals resided. A keen and genial Oxford-educated biologist and historical polymath, Clive knows more – and certainly cares more – about Neanderthals than perhaps anyone alive. 'I'm sometimes referred to as the defence lawyer for the Neanderthals,' he joked when he picked me up at my hotel early on a hot August morning.

We had met once before – almost thirty years earlier when I was in Gibraltar on assignment for the *National Geographic* magazine. But Clive has the engaging manner, which seems to be more or less universal among Gilbraltarians, of making visitors feel like old friends.

'We've learned a few things since you were last here,' he said, as if it were a matter of months rather than decades since we last met. 'You've some catching up to do!'

In the driver's seat was Clive's son Stewart, who is an evolutionary biologist in his own right and also is employed by the museum – as indeed is Clive's wife and Stewart's mother, Geraldine, a distinguished biologist and undersea archaeologist.

'So I grew up with all this,' Stewart said with a broad smile, vaguely indicating the world around us. 'Caves and Neanderthals and archaeology and natural history and everything else that the museum is interested in, which is practically everything. It was a pretty wonderful way to grow up.'

We drove along a coast road that climbed from Gibraltar town to the area known as the Back of the Rock. The caves we were headed for, called Vanguard and Gorham's, are at sea level, but access is from high above. We parked outside a military checkpoint – the Rock of Gibraltar is still literally a fortress – passed through a locked gate, donned obligatory hard hats, and began a long descent down 347 very steep, twisting, occasionally precarious steps. Even early in the morning the heat off the rockface was formidable and the iron railing was hot to the touch. Because of the difficulty of access, the caves cannot be opened to the public.

'When I first came to these caves, I had nothing to do with Neanderthals,' Clive said as we picked our way down. 'I was here to ring birds as part of my work as a young ecologist. This was the early 1970s and there wasn't much interest in Neanderthals then. No archaeologist had done any work in the caves since the 1950s. But in about 1980 some researchers from the Natural History Museum in London asked to have a look around, and I volunteered to take them down since I knew the way. That was where my interest started. Everything I've learned since then has only deepened my fascination with the Neanderthals. I think it is safe to say that no early humans have been more misunderstood for longer than they have been.'

We were sweating heavily by the time we arrived at the foot of the Rock, but Vanguard Cave was wonderfully cool. It had a cathedral-like spaciousness, with a ceiling perhaps 50 feet high and a sloping interior that ended in shadows. An arched opening framed a view of rocky shoreline and broad sun-spangled sea beyond. If you were going to live in a cave, this is the one you would choose. The excavating season had just ended, and the cave floor was covered in sandbags to protect the recently exposed surfaces. This, and Gorham's Cave around the corner, were the last known refuges of Neanderthals.

'It would have been very different when the Neanderthals were here,' Clive said. 'The Mediterranean was much lower, so there would have been about four kilometres of open plain between here and the water's edge, which would have been dotted with watering holes and vegetation and lots of game for hunting.'

The caves have yielded an amazingly sumptuous range of foodstuffs. 'They hunted deer, ibex, rabbits, birds and basking seals, and from the seas they took mussels, fish of every type and even tuna and two species of dolphin. They lived lives of abundance and comfort. It was a kind of paradise.'

'How did they catch tuna?' I asked.

'No idea,' Clive answered cheerfully. 'Catching tuna is not easy. Even now it's not easy. They must somehow have lured them into the shallows, but more than that it is hard to say.'

'And they smoked what they caught,' Stewart added. He showed me a recently excavated area which they believe served as a kind of smokehouse. 'They must have been preserving the food,' Clive added. 'It was too much to eat without going off. We think they may have been smoking it with seal blubber. These were not unsophisticated people. They were knowledgeable and enterprising and intimately acquainted with the natural world.'

They also collected fruit, nuts and mushrooms, and caught and roasted birds in abundance – partridge, pigeons, ducks and almost anything else that flew past, which was a large proportion of all the bird types in Europe. Because of its proximity to Africa – so close that you can see individual farmsteads in Morocco from the Rock – Gibraltar has always been a major migratory route for birds. 'We have identified 161 species of bird in these caves,' Stewart noted.

In short, the Gibraltar Neanderthals may have been the most comfortable and well-fed people in prehistory. They were certainly among the most successful. They lived in these caves for about 100,000 years – a span of continuity that is, in modern human terms, almost inconceivable.

'We know they were here 130,000 years ago,' Clive told me. 'They may have been here earlier, but everything was washed away by a sea-level rise 130,000 years ago. We believe they were last here around 32,000 years ago. Most people say they went extinct around 40,000 years ago, so that's a bit controversial.'

But then Clive is used to controversy. 'From the very beginning, we were treated as heretics because we were challenging the orthodoxy about Neanderthals and their place in prehistory.'

I asked him if there was resentment because he wasn't an archaeologist.

'Absolutely! I tried to publish in one journal early on and one of the referees was very dismissive. The objections were not about the paper itself – that was fine apparently – but about my qualifications. I was the wrong kind of scientist! It's better now, but in the beginning it was sometimes very challenging.'

Being an outsider has certain advantages, however. Clive has been a passionate birder all his life, as well as a trained naturalist, which clearly is helpful in understanding the behaviours and strategies of people whose lives were so closely intertwined with nature.

'You know, it's not easy to catch 161 different species of bird,' Clive said. 'You need a whole armoury of skills, and in this respect the Neanderthals were quite exceptional. They were catching and making use of nearly everything that came past, and not just for food. We can tell that they were selecting golden eagles for their feathers – in particular the black feathers. We can't say with any certainty why they were doing that, but they were definitely selecting, possibly for ceremonial purposes. We can also tell from cut marks that they were cutting the bones of vultures and eagles in a way that would not have yielded meat, but would have made a kind of poncho, which they could have worn, presumably also for ceremonial purposes.'

One thing is certain: the Neanderthals who lived in these caves were not driven here by invading Cro-Magnons. They were here long before *Homo sapiens* entered Europe. 'They didn't overlap with *Homo sapiens* at all,' Clive said. There's no evidence that they ever even met.

'The whole argument that modern humans outcompeted Neanderthals', Clive went on, 'is based on the idea that wherever you do excavations you find a dividing line with Neanderthal remains underneath and a more recent level above where there are only modern humans. From this it is assumed that the modern humans wiped out the Neanderthals. But you could also interpret it that the modern humans only managed to move in *after* the Neanderthals had departed, and you cannot tell the difference from the stratigraphy.'

Clive powerfully believes that there is absolutely no reason to suppose that Neanderthals were in any way our inferiors. By all the evidence, their lives were as comfortable and secure, and far more settled, than those of any *Homo sapiens* of the period.

So what then became of them? Clive thinks it's largely to do with the environments in which the two competing species arose. 'Neanderthals evolved in Eurasia in woodlands where they could get close to their prey by ambush hunting,' he says. They were stocky and muscular, so would have been particularly proficient at thrusting with spears from close range. But about 60,000 years ago, in the build-up to the last ice age, a lot of Eurasia became steppe tundra and turned into treeless plains, making it much harder for them to sneak up on prey.

Modern humans, by contrast, had come into their own on African savannahs like the Serengeti. When they spread into Eurasia, it was into a fundamentally familiar environment. Instead of wildebeest they now encountered huge herds of reindeer, but they could hunt them using the same methods they had in Africa. 'This doubtless gave them an advantage over the Neanderthals, who couldn't easily change their hunting methods because they were not built to run,' Clive suggests.

'It was just luck that favoured the modern humans,' he added. 'Had the climate gone the other way, and the forests expanded rather than shrunk, it could have been the Neanderthals who thrived. We could all be Neanderthals now.'

And how different would that make us? I asked

'Very possibly not very different at all,' Clive said. 'After all, we weren't so different that we couldn't interbreed.'

Indeed, to some extent Neanderthals live on with us today through our mingled DNA. Depending on where you come from in the world (as long as it is outside Africa), you will have somewhere between 1 and 4 per cent Neanderthal DNA within you. Although individually we carry just a small amount, collectively something like 20 per cent of Neanderthal DNA, and some authorities think as much as 50 per cent, is preserved in modern humans.

Although the barren, windswept steppes eventually covered nearly all of Europe, including most of what is now Spain, they never quite reached the coast.

'It never got that cold at Gibraltar,' Clive told me. 'Even at the height of the ice age, you still had olive trees growing here. The last Neanderthals here had the same environment as the first ones.'

No one can say what became of Gibraltar's Neanderthals. One possibility is that a prolonged drought dried up their watering holes. At all events, at some point they packed up and moved on, presumably unaware that there was nowhere better to go. They can of course have had no idea how singular they were, but they may well have been the last surviving archaic humans. All the others had disappeared in a few tens of thousands of years just as *Homo sapiens* emerged from the mists, strongly suggesting that our rise and their extinctions were more than coincidental. As Svante Pääbo has put it: 'We were their bad luck.'

Precisely how we managed to absorb or exterminate every last non-modern human on the planet is a mystery that will probably never be solved. These were beings who had survived for hundreds of thousands, even millions of years, sometimes in the most challenging of conditions, and yet they yielded to us with what seems to have been scarcely a murmur.

However it happened, Planet Earth was now ours. As we are about to see, it's not always clear that that was the best possible outcome.

30

GOODBYE

In the early 1680s, at just about the time that Edmond Halley and his friends Christopher Wren and Robert Hooke were settling down in a London coffee house and embarking on the casual wager that would result eventually in Isaac Newton's *Principia*, Henry Cavendish's weighing of the Earth and many of the other inspired and commendable undertakings that have occupied us for the past 400 or so pages, a rather less desirable milestone was being passed on the island of Mauritius, far out in the Indian Ocean some 1,300 kilometres off the east coast of Madagascar.

There, some forgotten sailor or sailor's pet was harrying to death the last of the dodos, the famously flightless bird whose dim but trusting nature and lack of leggy zip made it a rather irresistible target for bored young tars on shore leave. Millions of years of peaceful isolation had not prepared it for the erratic and deeply unnerving behaviour of human beings.

We don't know precisely the circumstances, or even the year, attending the last moments of the last dodo, so we don't know which arrived first, a world that contained a *Principia* or one that had no dodos, but we do know that they happened at more or less the same time. You would be hard pressed, I would submit, to find a better pairing of occurrences to illustrate the simultaneously divine and felonious nature of the human being – a species of organism that is capable of unravelling the deepest secrets of the heavens while at the same time pounding into extinction, for no purpose at all, a creature that never did us any harm and wasn't even remotely capable of understanding what we were doing to it as we did it. Indeed, dodos were so spectacularly short on insight, it is reported, that if you wished to find all the dodos in a vicinity you had only to catch one and set it to squawking, and all the others would waddle along to see what was up.

The indignities to the poor dodo didn't end quite there. In 1755, some seventy years after the last dodo's death, the director of the Ashmolean Museum in Oxford decided that the institution's stuffed dodo was becoming unpleasantly musty and ordered it tossed on a bonfire. This was a surprising decision as it was by this time the only dodo in existence, stuffed or otherwise. A passing employee, aghast, tried to rescue the bird but could save only its head and part of one limb.

As a result of this and other departures from common sense, we are not now entirely sure what a living dodo was like. We possess much less information than most people suppose – a handful of crude descriptions by 'unscientific voyagers, three or four oil paintings, and a few scattered osseous fragments', in the somewhat aggrieved words of the nineteenth-century naturalist H. E. Strickland.

As Strickland wistfully observed, we have more physical evidence of some ancient sea monsters and lumbering sauropods than we do of a bird that lived into modern times and required nothing of us to survive except our absence.

So what is known of the dodo is this: it lived on Mauritius, was plump but not tasty, and was the biggest-ever member of the pigeon family, though by quite what margin is unknown as its weight was never accurately recorded. Extrapolations from Strickland's 'osseous fragments' and the Ashmolean's modest remains show that it was a little over 2.5 feet tall and about the same distance from beak-tip to backside. Being flightless, it nested on the ground, leaving its eggs and chicks tragically easy prey for pigs, dogs and monkeys brought to the island by outsiders. It was probably extinct by 1683 and was most certainly gone by 1693. Beyond that we know almost nothing except, of course, that we will not see its like again. We know nothing of its reproductive habits and diet, where it ranged, what sounds it made in tranquillity or alarm. We don't possess a single dodo egg.

From beginning to end, our acquaintance with animate dodos lasted just seventy years. That is a breathtakingly scanty period – though it must be said that by this point in our history we did have thousands of years of practice behind us in the matter of irreversible eliminations. Nobody knows quite how destructive human beings are, but it is a fact that over the last 50,000 years or so, wherever we have gone animals have tended to vanish, often in astonishingly large numbers.

In America, thirty genera of large animals – some very large indeed – disappeared practically at a stroke after the arrival of modern humans on the continent between 10,000 and 20,000 years ago. Altogether North and South America between them lost about three-quarters of their big animals once human hunters arrived with their flint-headed spears and keen organizational capabilities. Europe and Asia, where the animals had had longer to evolve a useful wariness of humans, lost between a third and a half of their big creatures. Australia, for exactly the opposite reasons, lost no less than 95 per cent.

Because the early hunter populations were comparatively small and the animal populations truly monumental – as many as 10 million mammoth carcasses are thought to lie frozen in the tundra of northern Siberia alone – some authorities think there must be other explanations, possibly involving climate change or some kind of pandemic. As Ross MacPhee of the American Museum of Natural History put it: 'There's no material benefit to hunting dangerous animals more often than you need to – there are only so many mammoth steaks you can eat.' Others believe it may have been almost criminally easy to catch and clobber prey. 'In Australia and the Americas,' says Tim Flannery, 'the animals probably didn't know enough to run away.'

Some of the creatures that were lost were singularly spectacular and would take a little managing if they were still around. Imagine ground sloths that could look into an upstairs window, tortoises nearly the size of a small Fiat, monitor lizards 6 metres long basking beside desert highways in Western Australia. Alas, they are gone, and we live on a much-diminished planet. Today, across the whole world, only four types of really hefty (a tonne or more) land animals survive: elephants, rhinos, hippos and giraffes. Not for tens of millions of years has life on Earth been so diminutive and tame.

The question that arises is whether the disappearances of the stone age and disappearances of more recent times are in effect part of a single extinction event – whether, in short, humans are inherently bad news for other living things. The sad likelihood is that we may well be. According to the University of Chicago palaeontologist David Raup, the background rate of extinction on Earth throughout biological history has been one species lost every four years on average. According to Richard Leakey and Roger Lewin in *The Sixth Extinction*, human-caused extinction now may be running at as much as 120,000 times that level.

In the mid-1990s, the Australian naturalist Tim Flannery, who was then head of the South Australian Museum in Adelaide, became struck by how little we seemed to know about many extinctions, including relatively recent ones. 'Wherever you looked, there seemed to be gaps in the records – pieces missing, as with the dodo, or not recorded at all,' he told me in Melbourne in early 2002.

Flannery recruited his friend Peter Schouten, an artist and fellow Australian, and together they embarked on a slightly obsessive quest to scour the world's major collections to find out what was lost, what was left and what had never been known at all. They spent four years picking through old skins, musty specimens, old drawings and written descriptions – whatever was available. Schouten made life-sized paintings of every animal they could reasonably recreate and Flannery wrote the words. The result was an extraordinary book called *A Gap in Nature*, constituting the most complete – and, it must be said, moving – catalogue of animal extinctions from the last 300 years.

For some animals, records were good, but nobody had done anything much with them, sometimes for years, sometimes for ever. Steller's sea cow, a walrus-like creature related to the dugong, was one of the last really big animals to go extinct. It was truly enormous – an adult could reach lengths of nearly 9 metres and weigh 10 tonnes – but we are acquainted with it only because in 1741 a Russian expedition happened to be shipwrecked on the sole place where the creatures still survived in any numbers, the remote and foggy Commander Islands in the Bering Sea.

Happily, the expedition had a naturalist, Georg Steller, who was fascinated by the animal. 'He took the most copious notes,' says Flannery. 'He even measured the diameter of its whiskers. The only thing he wouldn't describe was the male genitals – though, for some reason, he was happy enough to do the female's. He even saved a piece of skin, so we had a good idea of its texture. We weren't always so lucky.'

The one thing Steller couldn't do was save the sea cow itself. Already hunted to the brink of extinction, it would be gone altogether within twenty-seven years of Steller's discovery of it. Many other animals, however, couldn't be included because too little is known about them. The Darling Downs hopping mouse, Chatham Islands swan, Ascension Island flightless crake, at least five types of large turtle and many others are for ever lost to us except as names.

A great deal of extinction, Flannery and Schouten discovered, hasn't been cruel or wanton, but just kind of majestically foolish. In 1894, when a lighthouse was built on a lonely rock called Stephens Island in the tempestuous strait between the North and South Islands of New Zealand, the lighthouse keeper's cat kept bringing him strange little birds that it had caught. The keeper dutifully sent some specimens to the museum in Wellington. There a curator grew very excited because the bird was a relic species of flightless wrens – the only example of a flightless perching bird ever found anywhere. He set off at once for the island, but by the time he got there the cat had killed them all. Twelve stuffed museum specimens of the Stephens Island flightless wren are all that now exist.

At least we have those. All too often, it turns out, we are not much better at looking after species after they have gone than we were before they went. Take the case of the lovely Carolina parakeet. Emerald-green, with a golden head, it was arguably the most striking and beautiful bird ever to live in North America – parrots don't usually venture so far north, as you may have noticed – and at its peak it existed in vast numbers, exceeded only by the passenger pigeon. But the Carolina parakeet was also considered a pest by farmers and easily hunted because it flocked tightly and had a peculiar habit of flying up at the sound of gunfire (as you would expect), but then returning almost at once to the same spot, as if reluctant to leave its fallen loved ones.

In his classic *American Ornithology*, written in the early nineteenth century, Charles Willson Peale describes an occasion in which he repeatedly empties a shotgun into a tree in which they roost:

> At each successive discharge, though showers of them fell, yet the affection of the survivors seemed rather to increase; for, after a few circuits around the place, they again alighted near me, looking down on their slaughtered companions with such manifest symptoms of sympathy and concern, as entirely disarmed me.

GOODBYE

By the second decade of the twentieth century, the birds had been so relentlessly hunted that only a few remained alive in captivity. The last one, named Inca, died in Cincinnati Zoo in 1918 (not quite four years after the last passenger pigeon died in the same zoo) and was reverently stuffed. And where would you go to see poor Inca now? Nobody knows. The zoo lost it.

What is both most intriguing and puzzling about the story above is that Peale was a lover of birds, and yet did not hesitate to kill them in large numbers for no better reason than that it interested him to do so. It is a truly astounding fact that for a very long time the people who were most intensely fascinated by the world's living things were the ones most likely to extinguish them.

No one represented this position on a larger scale (in every sense) than Lionel Walter Rothschild, the second Baron Rothschild. Scion of the great banking family, Rothschild was a strange and reclusive fellow. He lived his entire life, from 1868 to 1937, in the nursery wing of his home at Tring, in Buckinghamshire, using the furniture of his childhood – even sleeping in his childhood bed, though eventually he weighed 300 pounds (or about 135 kilos). His passion was natural history and he became a devoted accumulator of objects. He sent hordes of trained men – as many as 400 at a time – to every quarter of the globe to clamber over mountains and hack their way through jungles in the pursuit of new specimens – particularly things that flew. These were crated or boxed up and sent back to Rothschild's estate at Tring, where he and a battalion of assistants exhaustively logged and analysed everything that came before them, producing a constant stream of books, papers and monographs – some 1,200 in all. Altogether, Rothschild's natural history factory processed well over 2 million specimens and added 5,000 species of creature to the scientific archive.

Remarkably, Rothschild's collecting efforts were neither the most extensive nor the most generously funded of the nineteenth century. That title almost certainly belongs to a slightly earlier but also very wealthy British collector named Hugh Cuming, who became so preoccupied with accumulating objects that he built a large ocean-going ship and employed a crew to sail the world full-time picking up whatever they could find – birds, plants, animals of all types, and especially shells. It was his unrivalled collection of barnacles that passed to Darwin and served as the basis for his seminal study.

However, Rothschild was easily the most scientific collector of his age, though also the most regrettably lethal, for in the 1890s he became interested in Hawaii, perhaps the most temptingly vulnerable environment Earth has yet produced. Millions of years of isolation had allowed Hawaii to evolve 8,800 unique species of animals and plants. Of particular interest to Rothschild were the islands' colourful and distinctive birds, often consisting of very small populations inhabiting extremely specific ranges.

The tragedy for many Hawaiian birds was that they were not only distinctive, desirable and rare – a dangerous combination in the best of circumstances – but also often heartbreakingly easy to take. The greater koa finch, an innocuous member of the honeycreeper family, lurked shyly in the canopies of koa trees, but if someone imitated its song it would abandon its cover at once and fly down in a show of welcome. The last of the species vanished in 1896, killed by Rothschild's ace collector Harry Palmer, five years after the disappearance of its cousin the lesser koa finch, a bird so sublimely rare that only one has ever been seen: the one shot for Rothschild's collection. Altogether, during the decade or so of Rothschild's most intensive collecting, at least nine species of Hawaiian birds vanished, but it may have been more.

Rothschild was by no means alone in his zeal to capture birds at more or less any cost. Others, in fact, were more ruthless. In 1907, when a well-known collector named Alanson Bryan realized that he had shot the last three specimens of black mamos, a species of forest bird that had only been discovered the previous decade, he noted that the news filled him with 'joy'.

It was, in short, a difficult age to fathom – a time when almost any animal was persecuted if it was deemed the least bit intrusive. In 1890, New York State paid out over 100 bounties for eastern mountain lions, even though it was clear that the much-harassed creatures were on the brink of extermination. Right up until the 1940s many states continued to pay bounties for almost any kind of predatory creature. West Virginia gave out an annual college scholarship to whoever brought in the greatest number of dead pests – and 'pests' was liberally interpreted to mean almost anything that wasn't grown on farms or kept as pets.

Perhaps nothing speaks more vividly for the strangeness of the times than the fate of the lovely little Bachman's warbler. A native of the southern United States, the warbler was famous for its unusually thrilling song, but its population numbers, never robust, gradually dwindled until by the 1930s the warbler vanished altogether and went unseen for many years. Then, in 1939, by happy coincidence two separate birding enthusiasts, in widely separated locations, came across lone survivors just two days apart. They both shot the birds.

The impulse to exterminate was by no means exclusively American. In Australia, bounties were paid on the Tasmanian tiger (properly the thylacine), a dog-like creature with distinctive 'tiger' stripes across its back, until shortly before the last one died, forlorn and nameless, in a private Hobart zoo in 1936. Go to the Tasmanian Museum and Art Gallery today and ask to see the last of this species – the only large carnivorous marsupial to live into modern times – and all they can show you are photographs and sixty-one seconds of old film footage. Upon its death, the last surviving thylacine was thrown out with the weekly trash.

I mention all this to make the point that if you were designing an organism to look after life in our lonely cosmos, to monitor where it is going and keep a record of where it has been, you wouldn't choose human beings for the job.

But here's an extremely salient point: we have been chosen, by fate or providence or whatever you wish to call it. As far as we can tell, we are the best there is. We may be all there is. It's an unnerving thought that we may be the living universe's supreme achievement and its worst nightmare at the same time.

Because we are so remarkably careless about looking after things, both when they are alive and when they are not, we have very little idea how many have died off permanently, or may do so soon, or will with luck squeak through. Most estimates suggest that we lose at least 1,000 species of living thing every year, though some put the number much higher. The United Nations Convention on Biological Diversity in 2007 stated that the world was losing up to 150 species every day, though most authorities think that estimate way too high. More plausibly, the World Wildlife Fund has put the number of critically endangered species – which is to say, those unquestionably on the brink – at about 6,800 as of 2020. But as many as a million species of living things could fall into the looser but still worrisome category of 'threatened with extinction'. Scarcely less alarming is the global loss of abundance. According to a report for Britain's Royal Society, between 1970 and 2016 the number of vertebrates on the planet shrank by 68 per cent.

But even the best such numbers are estimates and many are little more than guesses. The fact is, we just don't know. What we do know is that there is only one known planet with life, and only one species of being capable of making a considered difference to the health and abundance of that life. Edward

O. Wilson expressed it with unimprovable brevity in *The Diversity of Life*: 'One planet, one experiment.'

If this book has a lesson, it is that we are awfully lucky to be here – and by 'we' I mean every living thing. To attain any kind of life at all in this universe of ours appears to be quite an achievement. As humans we are doubly lucky, of course. We enjoy not only the privilege of existence, but also the singular ability to appreciate it and even, in a multitude of ways, to make it better. It is a trick we have only just begun to grasp. We have arrived at this position of eminence in a stunningly short time. Behaviourally modern humans have been around for less than 0.01 per cent of Earth's history – almost nothing, really – but even existing for that little while has required a nearly endless string of good fortune.

We really are at the beginning of it all. The trick, of course, is to make sure we never find the end. And that, almost certainly, will require a lot more than lucky breaks.

ACKNOWLEDGEMENTS

I cannot begin to thank adequately those who helped me in the preparation of this book, both in 2000–2003 and in 2024–2025 for this updated edition.

I wish to give special thanks to my saintly editor Steph Duncan for her endless assistance and encouragement, and to Tom Whipple, science editor of *The Times*, and to Chris Stringer of the Natural History Museum, London, and Andy Buckley of Glasgow University for reading all or parts of the revised manuscript and offering many corrections and improvements (though any errors that remain are of course mine).

In addition, the following were uniformly generous and kindly and showed the most heroic reserves of patience in answering one simple, endlessly repeated question: 'I'm sorry, but can you explain that again?'

In England: Daniel Davis, Richard Craster, Cathal Hosty and David Caplin of Imperial College London; Carlos Frenk of Durham University; Sir Paul Nurse, chief executive of the Francis Crick Institute; Martin Rees (Baron Rees of Ludlow), former Astronomer Royal; Ross Anderson and George Wedlake of Oxford University; Richard Fortey, Len Ellis, Jo Wilbraham, and Kathy Way of the Natural History Museum; Martin Raff of University College London; Dr Laurence

ACKNOWLEDGEMENTS

Smaje, formerly of the Wellcome Institute; and Keith Blackmore, formerly of *The Times*.

In the United States: Ray Anderson and Brian Witzke of the Iowa Department of Natural Resources, Iowa City; Ian Tattersall of the American Museum of Natural History in New York; John Thorstensen, Mary K. Hudson and David Blanchflower of Dartmouth College in Hanover, New Hampshire; Dr William Abdu and Dr Bryan Marsh of Dartmouth–Hitchcock Medical Center in Lebanon, New Hampshire; Mike Voorhies of the University of Nebraska and Ashfall Fossil Beds State Park near Orchard, Nebraska; Chuck Offenburger of Buena Vista University, Storm Lake, Iowa; Ken Rancourt, director of research, Mount Washington Observatory, Gorham, New Hampshire; Paul Doss, former geologist of Yellowstone National Park, and his wife, Heidi; Frank Asaro of the University of California at Berkeley; Oliver Payne and Lynn Addison of the National Geographic Society; James O. Farlow, Indiana–Purdue University; and Roger L. Larson, professor of marine geophysics, University of Rhode Island.

In Australia: Tim Flannery, author and former director of the South Australian Museum in Adelaide; the Reverend Robert Evans of Hazelbrook, New South Wales; Dr Jill Cainey, Australian Bureau of Meteorology; Anne Milne, formerly of the *Sydney Morning Herald*; Ian Nowak, formerly of the Geological Society of Western Australia; Thomas H. Rich of Museum Victoria; Natalie Papworth and Alan MacFadyen of the Royal Tasmanian Botanical Gardens, Hobart; and the very helpful staff of the State Library of New South Wales in Sydney.

And elsewhere: Clive, Geraldine and Stewart Finlayson of the Gibraltar Museum; Matt Tocheri of Lakehead University, Thunder Bay, Ontario; Sue Superville, information centre manager at the Museum of New Zealand in Wellington; Dr Emma Mbua, Dr Koen Maes and Jillani Ngalla of the Kenya National Museum in Nairobi; and Kanishk Sharma of Class 10 at the Air Force Bal Bharati School in New Delhi for correct-

ACKNOWLEDGEMENTS

ing an etymological error that sat in this book for more than twenty years apparently unnoticed by anyone.

Finally, I am deeply and variously indebted to Susanna Wadeson, Gillian Blake, Amy Black, Larry Finlay, Patrick Janson-Smith, Gerald Howard, Marianne Velmans, Katrina Whone, Elizabeth Dobson, Antony Topping, David Bryson, Sam Bryson, Felicity Gould, Catherine Williams, Dan McLean, Patrick Gallagher, and the very helpful staff of the London Library.

Above all, and as always, my profoundest thanks to my dear, patient, incomparable wife, Cynthia.

BIBLIOGRAPHY

Aczel, Amir D., *God's Equation: Einstein, Relativity, and the Expanding Universe*. London: Piatkus Books, 2002.

Alberts, Bruce, Dennis Bray, Alexander Johnson, Julian Lewis, Martin Raff, Keith Roberts and Peter Walter, *Essential Cell Biology: An Introduction to the Molecular Biology of the Cell*. New York and London: Garland Publishing, 1998.

Allen, Oliver E., *Atmosphere*. Alexandria, Va.: Time-Life Books, 1983.

Alvarez, Walter, *T. Rex and the Crater of Doom*. Princeton, NJ: Princeton University Press, 1997.

Annan, Noel, *The Dons: Mentors, Eccentrics and Geniuses*. London: HarperCollins, 2000.

Ashcroft, Frances, *Life at the Extremes: The Science of Survival*. London: HarperCollins, 2000.

Asimov, Isaac, *The History of Physics*. New York: Walker & Co., 1966.

——*Exploring the Earth and the Cosmos: The Growth and Future of Human Knowledge*. London: Penguin Books, 1984.

——*Atom: Journey Across the Subatomic Cosmos*. New York: Truman Talley/Dutton, 1991.

Atkins, P. W., *The Second Law*. New York: *Scientific American*, 1984.

——*Molecules*. New York: *Scientific American*, 1987.

——*The Periodic Kingdom*. London: Weidenfeld & Nicolson, 1995.

BIBLIOGRAPHY

Attenborough, David, *Life on Earth: A Natural History*. London: Collins, 1979.

—— *The Living Planet: A Portrait of the Earth*. London: Collins, 1984.

—— *The Private Life of Plants: A Natural History of Plant Behaviour*. London: BBC Books, 1984.

Baeyer, Hans Christian von, *Taming the Atom: The Emergence of the Visible Microworld*. London: Viking, 1993.

Bakker, Robert T., *The Dinosaur Heresies: New Theories Unlocking the Mystery of the Dinosaurs and their Extinction*. New York: William Morrow, 1986.

Ball, Philip, *H_2O: A Biography of Water*. London: Phoenix/Orion, 1999.

Ballard, Robert D., *The Eternal Darkness: A Personal History of Deep-Sea Exploration*. Princeton, NJ: Princeton University Press, 2000.

Barber, Lynn, *The Heyday of Natural History: 1820–1870*. London: Jonathan Cape, 1980.

Barry, Roger G., and Richard J. Chorley, *Atmosphere, Weather and Climate*, 7th edn. London: Routledge, 1998.

Biddle, Wayne, *A Field Guide to the Invisible*. New York: Henry Holt, 1998.

Bodanis, David, *The Body Book*. London: Little, Brown, 1984.

—— *The Secret House: Twenty-Four Hours in the Strange and Unexpected World in Which We Spend our Nights and Days*. New York: Simon & Schuster, 1984.

—— *The Secret Family: Twenty-Four Hours Inside the Mysterious World of Our Minds and Bodies*. New York: Simon & Schuster, 1997.

—— *$E = mc^2$: A Biography of the World's Most Famous Equation*. London: Macmillan, 2000.

Bolles, Edmund Blair, *The Ice Finders: How a Poet, a Professor and a Politician Discovered the Ice Age*. Washington DC: Counterpoint/Perseus, 1999.

Boorse, Henry A., Lloyd Motz and Jefferson Hane Weaver, *The Atomic Scientists: A Biographical History*. New York: John Wiley & Sons, 1989.

Boorstin, Daniel J., *The Discoverers*. London: Penguin Books, 1986.

—— *Cleopatra's Nose: Essays on the Unexpected*. New York: Random House, 1994.

Bracegirdle, Brian, *A History of Microtechnique: The Evolution of the Microtome and the Development of Tissue Preparation*. London: Heinemann, 1978.

Breen, Michael, *The Koreans: Who They Are, What They Want, Where Their Future Lies*. London: Texere, 1998.

BIBLIOGRAPHY

Broad, William J., *The Universe Below: Discovering the Secrets of the Deep Sea*. New York: Simon & Schuster, 1997.

Brock, William H., *The Norton History of Chemistry*. London: W. W. Norton, 1993.

Brockman, John, and Katinka Matson (eds), *How Things Are: A Science Tool-Kit for the Mind*. London: Weidenfeld & Nicolson, 1995.

Brookes, Martin, *Fly: The Unsung Hero of Twentieth-Century Science*. London: Phoenix, 2002.

Brown, Guy, *The Energy of Life*. London: Flamingo/HarperCollins, 2000.

Browne, Janet, *Charles Darwin: A Biography*, vol. 1. London: Jonathan Cape, 1995.

Brusatte, Steve, *The Rise and Reign of the Mammals*, London: Picador, 2022.

Burenhult, Göran (ed.), *The First Americans: Human Origins and History to 10,000 BC*. London: HarperCollins, 1993.

Cadbury, Deborah, *Terrible Lizard: The First Dinosaur Hunters and the Birth of a New Science*. New York: Henry Holt, 2000.

Calder, Nigel, *Einstein's Universe*. London: BBC Books, 1979.

——*The Comet Is Coming! The Feverish Legacy of Mr Halley*. London: BBC Books, 1980.

Canby, Courtlandt (ed.), *The Epic of Man*. New York: Time/Life, 1961.

Carey, John (ed.), *The Faber Book of Science*. London: Faber, 1995.

Chorlton, Windsor, *Ice Ages*. New York: Time-Life Books, 1983.

Christianson, Gale E., *In the Presence of the Creator: Isaac Newton and his Times*. New York: Free Press/Macmillan, 1984.

——*Edwin Hubble: Mariner of the Nebulae*. Bristol, England: Institute of Physics Publishing, 1995.

Clark, Ronald W., *The Huxleys*. London: Heinemann, 1968.

——*The Survival of Charles Darwin: A Biography of a Man and an Idea*. London: Daedalus Books, 1985.

——*Einstein: The Life and Times*. London: HarperCollins, 1971.

Cobb, Matthew, *The Egg and Sperm Race: The Seventeenth-Century Scientists Who Unravelled the Secrets of Sex, Life and Growth*. London: Free Press, 2006.

Coe, Michael, Dean Snow and Elizabeth Benson, *Atlas of Ancient America*. New York: Equinox/Facts on File, 1986.

BIBLIOGRAPHY

Colbert, Edwin H., *The Great Dinosaur Hunters and their Discoveries*. New York: Dover Publications, 1984.

Cole, K. C., *First You Build a Cloud: And Other Reflections on Physics as a Way of Life*. San Diego: Harvest/Harcourt Brace, 1999.

Conard, Henry S., *How to Know the Mosses and Liverworts*. Dubuque, Iowa: William C. Brown Co., 1956.

Conniff, Richard, *Spineless Wonders: Strange Tales from the Invertebrate World*. London and New York: Henry Holt, 1996.

Corfield, Richard, *Architects of Eternity: The New Science of Fossils*. London: Headline, 2001.

Coveney, Peter, and Roger Highfield, *The Arrow of Time: The Quest to Solve Science's Greatest Mystery*. London: Flamingo, 1991.

Cowles, Virginia, *The Rothschilds: A Family of Fortune*. London: Futura, 1975.

Crick, Francis, *Life Itself: Its Origin and Nature*. London: Macdonald, 1982.

——*What Mad Pursuit: A Personal View of Scientific Discovery*. London: Penguin Press, 1990.

Cropper, William H., *Great Physicists: The Life and Times of Leading Physicists from Galileo to Hawking*. Oxford: Oxford University Press, 2002.

Crowther, J. G., *Scientists of the Industrial Revolution*. London: Cresset, 1962.

Czerski, Helen, *Blue Machine: How the Ocean Shapes Our World*. London: Torva/Transworld, 2023.

Darwin, Charles, *On the Origin of Species by Means of Natural Selection, or the Preservation of Favoured Races in the Struggle for Life* (facsimile edn). London: AMSPR, 1972.

Davies, Paul, *The Fifth Miracle: The Search for the Origin of Life*. London: Penguin Books, 1999.

Dawkins, Richard, *The Blind Watchmaker*. London: Penguin Books, 1988.

——*River Out of Eden: A Darwinian View of Life*. London: Phoenix, 1996.

——*Climbing Mount Improbable*. London: Viking, 1996.

Dean, Dennis R., *James Hutton and the History of Geology*. Ithaca, NY: Cornell University Press, 1992.

de Duve, Christian, *A Guided Tour of the Living Cell*, 2 vols. New York: Scientific American/Rockefeller University Press, 1984.

Dennett, Daniel C., *Darwin's Dangerous Idea: Evolution and the Meanings of Life*. London: Penguin, 1996.

BIBLIOGRAPHY

Dennis, Jerry, *The Bird in the Waterfall: A Natural History of Oceans, Rivers and Lakes*. London and New York: HarperCollins, 1996.

Desmond, Adrian, and James Moore, *Darwin*. London: Penguin Books, 1992.

Dewar, Elaine, *Bones: Discovering the First Americans*. Toronto: Random House Canada, 2001.

Diamond, Jared, *Guns, Germs and Steel: The Fates of Human Societies*. New York: Norton, 1997.

Dickinson, Matt, *The Other Side of Everest: Climbing the North Face through the Killer Storm*. New York: Times Books, 1997.

Drury, Stephen, *Stepping Stones: The Making of our Home World*. Oxford: Oxford University Press, 1999.

Durant, Will and Ariel, *The Age of Louis XIV*. New York: Simon & Schuster, 1963.

Dyson, Freeman, *Disturbing the Universe*. London and New York: Harper & Row, 1979.

Easterbrook, Gregg, *A Moment on the Earth: The Coming Age of Environmental Optimism*. London: Penguin, 1995.

Ebbing, Darrell D., *General Chemistry*. Boston: Houghton Mifflin, 1996.

Elliott, Charles, *The Potting-Shed Papers: On Gardens, Gardeners and Garden History*. Guilford, Conn.: Lyons Press, 2001.

Engel, Leonard, *The Sea*. New York: Time-Life Books, 1969.

Erickson, Jon, *Plate Tectonics: Unravelling the Mysteries of the Earth*. London and New York: Facts on File, 1992.

Fagan, Brian M., *The Great Journey: The Peopling of Ancient America*. London: Thames & Hudson, 1987.

Falk, Dean, *The Fossil Chronicles: How Two Controversial Discoveries Changed Our View of Human Evolution*. Berkeley: University of California Press, 2022.

Fell, Barry, *America B.C.: Ancient Settlers in the New World*. London: Random House, 1976.

——*Bronze Age America*. London and Boston: Little, Brown, 1982.

Ferguson, Kitty, *Measuring the Universe: The Historical Quest to Quantify Space*. London: Headline, 1999.

Ferris, Timothy, *The Mind's Sky: Human Intelligence in a Cosmic Context*. New York: Bantam Books, 1992.

BIBLIOGRAPHY

———*The Whole Shebang: A State of the Universe(s) Report*. London: Phoenix, 1998.

———*Seeing in the Dark: How Backyard Stargazers Are Probing Deep Space and Guarding Earth from Interplanetary Peril*. New York: Simon & Schuster, 2002.

———*Coming of Age in the Milky Way*. London: HarperCollins, 2003.

Feynman, Richard P., *Six Easy Pieces*. London: Penguin Books, 1998.

Fisher, Richard V., Grant Heiken and Jeffrey B. Hulen, *Volcanoes: Crucibles of Change*. Princeton, NJ: Princeton University Press, 1997.

Flannery, Timothy, *The Future Eaters: An Ecological History of the Australasian Lands and People*. London: W. W. Norton, 1995.

———*The Eternal Frontier: An Ecological History of North America and its Peoples*. London: Heinemann, 2001.

———and Peter Schouten, *A Gap in Nature: Discovering the World's Extinct Animals*. London: Heinemann, 2001.

Fortey, Richard, *Life: An Unauthorised Biography*. London: Flamingo/HarperCollins, 1998.

———*Trilobite! Eyewitness to Evolution*. London: HarperCollins, 2000.

———*Dry Store Room No. 1: The Secret Life of the Natural History Museum*. London: HarperCollins, 2008.

Frayn, Michael, *Copenhagen*. London: Methuen, 1998; New York: Anchor Books, 2000.

Gamow, George, and Russell Stannard, *The New World of Mr Tompkins*. Cambridge: Cambridge University Press, 2001.

Gawande, Atul, *Complications: A Surgeon's Notes on an Imperfect Science*. New York: Metropolitan Books/Henry Holt, 2002.

Gee, Henry, *A (Very) Short History of Life on Earth*. London: Picador, 2021.

Giancola, Douglas C., *Physics: Principles with Applications*. London: Prentice-Hall, 1997.

Gjertsen, Derek, *The Classics of Science: A Study of Twelve Enduring Scientific Works*. New York: Lilian Barber Press, 1984.

Godfrey, Laurie R. (ed.), *Scientists Confront Creationism*. New York: W. W. Norton, 1983.

Goldsmith, Donald, *The Astronomers*. New York: St Martin's Press, 1991.

BIBLIOGRAPHY

'Gordon, Mrs', *The Life and Correspondence of William Buckland, D.D., F.R.S.* London: John Murray, 1894.

Gould, Stephen Jay, *Ever since Darwin: Reflections in Natural History.* London: Deutsch, 1978.

——*The Panda's Thumb: More Reflections in Natural History.* London and New York: W. W. Norton, 1980.

——*Hen's Teeth and Horse's Toes.* London: Penguin Books, 1984.

——*The Flamingo's Smile: Reflections in Natural History.* New York: W. W. Norton, 1985.

——*Wonderful Life: The Burgess Shale and the Nature of History.* London: Hutchinson Radius, 1990.

——*Bully for Brontosaurus: Reflections in Natural History*, London: Hutchinson Radius, 1991.

——*Time's Arrow, Time's Cycle: Myth and Metaphor in the Discovery of Geological Time.* Cambridge, Mass.: Harvard University Press, 1987.

——(ed.), *The Book of Life.* London: Ebury, 1993.

——*Eight Little Piggies: Reflections in Natural History.* London: Penguin, 1994.

——*Dinosaur in a Haystack: Reflections in Natural History.* London: Jonathan Cape, 1996.

——*Leonardo's Mountain of Clams and the Diet of Worms: Essays on Natural History.* London: Jonathan Cape, 1998.

——*The Lying Stones of Marrakech: Penultimate Reflections in Natural History.* London: Jonathan Cape, 2000.

Green, Bill, *Water, Ice and Stone: Science and Memory on the Antarctic Lakes.* New York: Harmony Books, 1995.

Gribbin, John, *In the Beginning: The Birth of the Living Universe.* London: Penguin, 1994.

——*Almost Everyone's Guide to Science: The Universe, Life and Everything.* London: Phoenix, 1998.

——and Mary Gribbin, *Being Human: Putting People in an Evolutionary Perspective.* London: Phoenix/Orion, 1993.

——*Fire on Earth: Doomsday, Dinosaurs and Humankind.* New York: St Martin's Press, 1996.

——*Ice Age.* London: Allen Lane, 2001.

BIBLIOGRAPHY

———and Jeremy Cherfas, *The First Chimpanzee: In Search of Human Origins*. London: Penguin Books, 2001.

Grinspoon, David Harry, *Venus Revealed: A New Look Below the Clouds of our Mysterious Twin Planet*. Reading, Mass.: Helix/Addison-Wesley, 1997.

Guth, Alan, *The Inflationary Universe: The Quest for a New Theory of Cosmic Origins*. London: Jonathan Cape, 1997.

Hafer, Abby, *The Not-So-Intelligent Designer: Why Evolution Explains the Human Body and Intelligent Design Does Not*. Eugene, Ore.: Cascade Books, 2015.

Haldane, J. B. S., *Adventures of a Biologist*. New York: Harper & Brothers, 1937.

———*What is Life?* New York: Boni & Gaer, 1947.

Hallam, A., *Great Geological Controversies*, 2nd edn. Oxford: Oxford University Press, 1989.

Hamblyn, Richard, *The Invention of Clouds: How an Amateur Meteorologist Forged the Language of the Skies*. London: Picador, 2001.

Hamilton-Paterson, James, *The Great Deep: The Sea and its Thresholds*. London: Random House, 1992.

Hapgood, Charles H., *Earth's Shifting Crust: A Key to Some Basic Problems of Earth Science*. New York: Pantheon Books, 1958.

Harrington, John W., *Dance of the Continents: Adventures with Rocks and Time*. Los Angeles: J. P. Tarcher, Inc., 1983.

Harrison, Edward, *Darkness at Night: A Riddle of the Universe*. Cambridge, Mass.: Harvard University Press, 1987.

Hartmann, William K., *The History of Earth: An Illustrated Chronicle of an Evolving Planet*. London: Workman Publishing, 1991.

Hawking, Stephen, *A Brief History of Time: From the Big Bang to Black Holes*. London: Bantam Books, 1988.

———*The Universe in a Nutshell*. London: Bantam Press, 2001.

Hazen, Rombert M., and James Trefil, *Science Matters: Achieving Scientific Literacy*. New York: Doubleday, 1991.

Heiserman, David L., *Exploring Chemical Elements and their Compounds*. Blue Ridge Summit, Pa.: TAB Books/McGraw Hill, 1992.

Hitchcock, A. S., *Manual of the Grasses of the United States*, 2nd edn. London: Peter Smith, 1971.

BIBLIOGRAPHY

Holmes, Hannah, *The Secret Life of Dust*. London: John Wiley & Sons, 2001.

Holmyard, E. J., *Makers of Chemistry*. Oxford: Clarendon Press, 1931.

Horwitz, Tony, *Blue Latitudes: Boldly Going Where Captain Cook Has Gone Before*. London: Bloomsbury, 2002.

Hough, Richard, *Captain James Cook*. London: Coronet, 1995.

Jardine, Lisa, *Ingenious Pursuits: Building the Scientific Revolution*. London: Little, Brown, 1999.

Johanson, Donald, and Blake Edgar, *From Lucy to Language*. London: Weidenfeld & Nicolson, 2001.

Jolly, Alison, *Lucy's Legacy: Sex and Intelligence in Human Evolution*. Cambridge, Mass.: Harvard University Press, 1999.

Jones, Steve, *Almost Like a Whale: The Origin of Species Updated*. London: Doubleday, 1999.

Judson, Horace Freeland, *The Eighth Day of Creation: Makers of the Revolution in Biology*. London: Penguin, 1995.

Junger, Sebastian, *The Perfect Storm: A True Story of Men Against the Sea*. London: Fourth Dimension, 1997.

Jungnickel, Christa, and Russell McCormmach, *Cavendish: The Experimental Life*. Bucknell, Pa: Bucknell Press, 1999.

Kaku, Michio, *Hyperspace: A Scientific Odyssey through Parallel Universes, Time Warps, and the Tenth Dimension*. Oxford: Oxford University Press, 1999.

Kastner, Joseph, *A Species of Eternity*. New York: Knopf, 1977.

Keller, Evelyn Fox, *The Century of the Gene*. Cambridge, Mass.: Harvard University Press, 2000.

Kemp, Peter, *The Oxford Companion to Ships and the Sea*. London: Oxford University Press, 1979.

Kevles, Daniel J., *The Physicists: The History of a Scientific Community in Modern America*. London: Random House, 1978.

Kitcher, Philip, *Abusing Science: The Case against Creationism*. Cambridge, Mass.: MIT Press, 1982.

Kolata, Gina, *Flu: The Story of the Great Influenza Pandemic of 1918 and the Search for the Virus that Caused It*. London: Pan, 2001.

Krebs, Robert E., *The History and Use of our Earth's Chemical Elements*. Westport, Conn.: Greenwood, 1998.

BIBLIOGRAPHY

Kunzig, Robert, *The Restless Sea: Exploring the World Beneath the Waves*. New York: W. W. Norton, 1999.

Kurlansky, Mark, *Cod: A Biography of the Fish That Changed the World*. London: Vintage, 1999.

Lane, Nick, *Power, Sex, Suicide: Mitochondria and the Meaning of Life*. Oxford: Oxford University Press, 2005.

——*Life Ascending: The Ten Great Inventions of Evolution*. London: Profile Books, 2009.

——*The Vital Question: Why Is Life the Way It Is?*. London: Profile Books, 2016.

Leakey, Richard, *The Origin of Humankind*. London: Phoenix, 1995.

——and Roger Lewin, *Origins*. New York: E. P. Dutton, 1977.

——*The Sixth Extinction: Biodiversity and its Survival*. London: Weidenfeld & Nicolson, 1996.

Leicester, Henry M., *The Historical Background of Chemistry*. New York: Dover, 1971.

Lemmon, Kenneth, *The Golden Age of Plant Hunters*. London: Phoenix House, 1968.

Lewin, Roger, *Bones of Contention: Controversies in the Search for Human Origins*, 2nd edn. Chicago: University of Chicago Press, 1997.

Lewis, Cherry, *The Dating Game: One Man's Search for the Age of the Earth*. Cambridge: Cambridge University Press, 2000.

Lewis, John S., *Rain of Iron and Ice: The Very Real Threat of Comet and Asteroid Bombardment*. Reading, Mass.: Addison-Wesley, 1996.

Lewontin, Richard, *It Ain't Necessarily So: The Dream of the Human Genome and Other Illusions*. London: Granta, 2001.

Little, Charles E., *The Dying of the Trees: The Pandemic in America's Forests*. New York: Viking, 1995.

Lynch, John, *The Weather*. Toronto: Firefly Books, 2002.

McGhee, Jr, George R., *The Late Devonian Mass Extinction: The Frasnian/Famennian Crisis*. New York: Columbia University Press, 1996.

McGrayne, Sharon Bertsch, *Prometheans in the Lab: Chemistry and the Making of the Modern World*. London: McGraw-Hill, 2002.

McGuire, Bill, *A Guide to the End of the World: Everything You Never Wanted to Know*. Oxford: Oxford University Press, 2002.

BIBLIOGRAPHY

McKibben, Bill, *The End of Nature*. London: Viking, 1990.

McPhee, John, *Basin and Range*. New York: Farrar, Straus & Giroux, 1980.

——*In Suspect Terrain*. New York: Noonday Press/Farrar, Straus & Giroux, 1983.

——*Rising from the Plains*. London: Farrar, Straus & Giroux, 1987.

——*Assembling California*. New York: Farrar, Straus & Giroux, 1993.

McSween, Harry Y., Jr, *Stardust to Planets: A Geological Tour of the Solar System*. New York: St Martin's Press, 1993.

Maddox, Brenda, *Rosalind Franklin: The Dark Lady of DNA*. London: HarperCollins, 2002.

Margulis, Lynn, and Dorion Sagan, *Microcosmos: Four Billion Years of Evolution from Our Microbial Ancestors*. London: HarperCollins, 2002.

Marshall, Nina L., *Mosses and Lichens*. New York: Doubleday, Page & Co., 1908.

Matthiessen, Peter, *Wildlife in America*. London: Penguin Books, 1995.

Moore, Patrick, *Fireside Astronomy: An Anecdotal Tour through the History and Lore of Astronomy*. Chichester: John Wiley & Sons, 1992.

Moorehead, Alan, *Darwin and the Beagle*. London: Hamish Hamilton, 1969.

Morowitz, Harold J., *The Thermodynamics of Pizza*. New Brunswick, NJ: Rutgers University Press, 1991.

Musgrave, Toby, Chris Gardner and Will Musgrave, *The Plant Hunters: Two Hundred Years of Adventure and Discovery around the World*. London: Ward Lock, 1999.

Norton, Trevor, *Stars Beneath the Sea: The Extraordinary Lives of the Pioneers of Diving*. London: Arrow Books, 2000.

Novacek, Michael, *Time Traveler: In Search of Dinosaurs and Other Fossils from Montana to Mongolia*. New York: Farrar, Straus & Giroux, 2001.

Nuland, Sherwin B., *How We Live: The Wisdom of the Body*. London: Vintage, 1998.

Nurse, Paul, *What Is Life?*. London: David Fickling, 2020.

Officer, Charles, and Jake Page, *Tales of the Earth: Paroxysms and Perturbations of the Blue Planet*. New York: Oxford University Press, 1993.

Oldroyd, David R., *Thinking about the Earth: A History of Ideas in Geology*. London: Athlone, 1996.

Oldstone, Michael B. A., *Viruses, Plagues and History*. New York: Oxford University Press, 1998.

Overbye, Dennis, *Lonely Hearts of the Cosmos: The Scientific Quest for the Secret of the Universe*. London: Macmillan, 1991.

Ozima, Minoru, *The Earth: Its Birth and Growth*. Cambridge: Cambridge University Press, 1981.

Pääbo, Svante, *Neanderthal Man: In Search of Lost Genomes*. New York: Basic Books, 2014.

Parker, Ronald B., *Inscrutable Earth: Explorations in the Science of Earth*. New York: Charles Scribner's Sons, 1984.

Pearson, John, *Serpents and Stags: The Story of the House of Cavendish and the Dukes of Devonshire*. London: Macmillan, 1983.

Peebles, Curtis, *Asteroids: A History*. Washington: Smithsonian Institution Press, 2000.

Plummer, Charles C., and David McGeary, *Physical Geology*. London: McGraw-Hill Education, 1997.

Pollack, Robert, *Signs of Life: The Language and Meanings of DNA*. London: Viking, 1995.

Powell, James Lawrence, *Night Comes to the Cretaceous: Dinosaur Extinction and the Transformation of Modern Geology*. New York: W. H. Freeman, 1998.

——*Mysteries of Terra Firma: The Age and Evolution of the Earth*. New York: Free Press/Simon & Schuster, 2001.

Psihoyos, Louie, with John Knoebber, *Hunting Dinosaurs*. London: Cassell Illustrated, 1995.

Putnam, William Lowell, *The Worst Weather on Earth*. London: Mountaineers Books, 1991.

Quammen, David, *The Song of the Dodo*. London: Hutchinson, 1996.

——*The Boilerplate Rhino: Nature in the Eye of the Beholder*. London: Touchstone, 2001.

——*Monster of God*. New York: W. W. Norton, 2003.

Rees, Martin, *Just Six Numbers: The Deep Forces that Shape the Universe*. London: Phoenix/Orion, 2000.

BIBLIOGRAPHY

———*On the Future: Prospects for Humanity*. Princeton, NJ: Princeton University Press, 2018.

Ridley, Matt, *The Red Queen: Sex and the Evolution of Human Nature*. London: Penguin, 1994.

———*Genome: The Autobiography of a Species*. London: Fourth Estate, 1999.

Ritchie, David, *Superquake! Why Earthquakes Occur and When the Big One Will Hit Southern California*. London: Random House, 1989.

Roberts, Callum, *Ocean of Life: How Our Seas Are Changing*. London: Allen Lane, 2012.

Rose, Steven, *Lifelines: Biology, Freedom, Determinism*. London: Penguin, 1997.

Rudwick, Martin J. S., *The Great Devonian Controversy: The Shaping of Scientific Knowledge among Gentlemanly Specialists*. Chicago: University of Chicago Press, 1985.

Rutherford, Adam, *Creation: The Origin of Life*. London: Viking, 2013.

———*A Brief History of Everyone Who Ever Lived: The Stories in Our Genes*. London: Weidenfeld & Nicolson, 2016.

Sacks, Oliver, *An Anthropologist on Mars: Seven Paradoxical Tales*. London: Picador, 1996.

———*Oaxaca Journal*. London: National Geographic, 2002.

Sagan, Carl, *Cosmos*. London: Random House, 1980.

———and Ann Druyan, *Comet*. London: Random House, 1985.

Sagan, Dorion, and Lynn Margulis, *Garden of Microbial Delights: A Practical Guide to the Subvisible World*. Boston: J. Harcourt Brace Jovanovich, 1988.

Sayre, Anne, *Rosalind Franklin and DNA*. London: W. W. Norton, 2002.

Schneer, Cecil J. (ed.), *Toward a History of Geology*. London: MIT Press, 1970.

Schopf, J. William, *Cradle of Life: The Discovery of Earth's Earliest Fossils*. Princeton, NJ: Princeton University Press, 1999.

Schultz, Gwen, *Ice Age Lost*. Garden City, NY: Anchor Press/Doubleday, 1974.

Schwartz, Jeffrey H., *Sudden Origins: Fossils, Genes and the Emergence of Species*. New York: John Wiley & Sons, 1999.

Semonin, Paul, *American Monster: How the Nation's First Prehistoric Creature Became a Symbol of National Identity*. New York: New York University Press, 2000.

Shore, William H. (ed.), *Mysteries of Life and the Universe*. San Diego: Harvest/Harcourt Brace & Co., 1992.

Silver, Brian, *The Ascent of Science*. New York: Solomon/Oxford University Press, 1998.

Simpson, George Gaylord, *Fossils and the History of Life*. New York: Scientific American, 1983.

Slimak, Ludovic, *The Naked Neanderthal*. London: Penguin Books, 2024.

Smith, Anthony, *The Weather: The Truth about the Health of our Planet*. London: Hutchinson, 2000.

Smith, Robert B., and Lee J. Siegel, *Windows into the Earth: The Geologic Story of Yellowstone and Grand Teton National Parks*. Oxford: Oxford University Press, 2002.

Snow, C. P., *Variety of Men*. London: Macmillan, 1967.

——*The Physicists*. London: House of Stratus, 1979.

Snyder, Carl H., *The Extraordinary Chemistry of Ordinary Things*. London: John Wiley & Sons, 1995.

Stalcup, Brenda (ed.), *Endangered Species: Opposing Viewpoints*. San Diego: Greenhaven, 1996.

Stanley, Steven M., *Extinction*. New York: *Scientific American*, 1987.

Stark, Peter, *Last Breath: Cautionary Tales from the Limits of Human Endurance*. New York: Ballantine Books, 2001.

Stephen, Sir Leslie, and Sir Sidney Lee (eds), *Dictionary of National Biography*. Oxford: Oxford University Press, 1973.

Stevens, William K., *The Change in the Weather: People, Weather, and the Science of Climate*. New York: Delacorte, 1999.

Stewart, Ian, *Nature's Numbers: Discovering Order and Pattern in the Universe*. London: Phoenix, 1995.

Strathern, Paul, *Mendeleyev's Dream: The Quest for the Elements*. London: Penguin Books, 2001.

Sullivan, Walter, *Landprints*. New York: Times Books, 1984.

Sulston, John, and Georgina Ferry, *The Common Thread: A Story of Science, Politics, Ethics and the Human Genome*. London: Bantam Press, 2002.

Swisher III, Carl C., Garniss H. Curtis and Roger Lewin, *Java Man: How Two Geologists' Dramatic Discoveries Changed our Understanding of the Evolutionary Path to Modern Humans*. London: Little, Brown, 2001.

BIBLIOGRAPHY

Sykes, Bryan, *The Seven Daughters of Eve*. London: Bantam Press, 2001.

Tattersall, Ian, *The Human Odyssey: Four Million Years of Human Evolution*. New York: Prentice-Hall, 1993.

———*The Monkey in the Mirror: Essays on the Science of What Makes Us Human*. Oxford: Oxford University Press, 2002.

———*The Strange Case of the Rickety Cossack*. London: Palgrave/Macmillan, 2015.

———and Jeffrey Schwartz, *Extinct Humans*. Boulder, Colo.: Westview/Perseus, 2001.

Thackray, John, and Bob Press, *The Natural History Museum: Nature's Treasurehouse*. London: Natural History Museum, 2001.

Thomas, Gordon, and Max Morgan Witts, *The San Francisco Earthquake*. London: Souvenir, 1971.

Thomas, Keith, *Man and the Natural World: Changing Attitudes in England, 1500–1800*. London: Penguin Books, 1984.

Thompson, Dick, *Volcano Cowboys: The Rocky Evolution of a Dangerous Science*. New York: St Martin's Press, 2000.

Thorne, Kip S., *Black Holes and Time Warps: Einstein's Outrageous Legacy*. London: Picador, 1994.

Tortora, Gerard J., and Sandra Reynolds Grabowski, *Principles of Anatomy and Physiology*. London: John Wiley & Sons, 1999.

Trefil, James, *The Unexpected Vista: A Physicist's View of Nature*. New York: Charles Scribner's Sons, 1983.

———*Meditations at Sunset: A Scientist Looks at the Sky*. New York: Charles Scribner's Sons, 1987.

———*Meditations at 10,000 Feet: A Scientist in the Mountains*. New York: Charles Scribner's Sons, 1987.

———*101 Things You Don't Know About Science and No One Else Does Either*. London: Cassell Illustrated, 1997.

Trinkaus, Erik, and Pat Shipman, *The Neandertals: Changing the Image of Mankind*. London: Pimlico, 1994.

Tudge, Colin, *The Time before History: Five Million Years of Human Impact*. New York: Touchstone/Simon & Schuster, 1996.

———*The Variety of Life: A Survey and a Celebration of All the Creatures that Have Ever Lived*. Oxford: Oxford University Press, 2002.

BIBLIOGRAPHY

Vernon, Ron, *Beneath our Feet: The Rocks of Planet Earth.* Cambridge: Cambridge University Press, 2000.

Vince, Gaia, *Adventures in the Anthropocene: A Journey to the Heart of the Planet We Made.* London: Chatto and Windus, 2014.

Vogel, Shawna, *Naked Earth: The New Geophysics.* New York: Dutton, 1995.

Walker, Alan, and Pat Shipman, *The Wisdom of the Bones: In Search of Human Origins.* London: Weidenfeld & Nicolson, 1996.

Wallace, Robert A., Jack L. King and Gerald P. Sanders, *Biology: The Science of Life*, 2nd edn. Glenview, Ill.: Scott, Foresman & Co., 1986.

Ward, Peter D., and Donald Brownlee, *Rare Earth: Why Complex Life Is Uncommon in the Universe.* New York: Copernicus, 1999.

Watson, James D., *The Double Helix: A Personal Account of the Discovery of the Structure of DNA.* London: Penguin Books, 1999.

Weinberg, Samantha, *A Fish Caught in Time: The Search for the Coelacanth.* London: Fourth Estate, 1999.

Weinberg, Steven, *The Discovery of Subatomic Particles.* London: W. H. Freeman, 1990.

——*Dreams of a Final Theory.* London: Vintage, 1993.

Whitaker, Richard (ed.), *Weather.* London: Warner Books, 1996.

White, Michael, *Isaac Newton: The Last Sorcerer.* London: Fourth Dimension, 1997.

——*Rivals: Conflict as the Fuel of Science.* London: Vintage, 2001.

Wilford, John Noble, *The Mapmakers.* London: Random House, 1981.

——*The Riddle of the Dinosaur.* London: Faber, 1986.

Williams, E. T., and C. S. Nicholls (eds), *Dictionary of National Biography, 1961–1970.* Oxford: Oxford University Press, 1981.

Williams, Stanley, and Fen Montaigne, *Surviving Galeras.* Boston: Houghton Mifflin, 2001.

Wilson, David, *Rutherford: Simple Genius.* London: Hodder, 1984.

Wilson, Edward O., *The Diversity of Life.* London: Allen Lane/Penguin Press, 1993.

Winchester, Simon, *The Map That Changed the World: The Tale of William Smith and the Birth of a Science.* London: Viking, 2001.

Woolfson, Adrian, *Life without Genes: The History and Future of Genomes.* London: Flamingo, 2000.

NOTES

Chapter 1: How to Build a Universe

p. 11   Protons are so small that: I am greatly obliged to Daniel Davis, Richard Craster and Cathal Hosty, all of Imperial College, London, for this calculation, which they stress is necessarily a rough approximation. I should note that a separate academic at Imperial, who perhaps slightly missed the point, insisted that such an exercise would never be possible because protons squeezed to such a density would turn into neutrons, changing the mathematics of the exercise utterly.

p. 11   Now pack into that tiny, tiny space: Guth, *The Inflationary Universe*, p. 254.

p. 12   using data from a NASA spacecraft with the uncompromisingly nerdy name of Wilkinson Microwave Anisotropy Probe: *New York Times*, 'Cosmos Sits for Early Portrait, Gives up Secrets', 12 Feb. 2003, p. 1.

p. 13   They climbed back into the dish with brooms and scrubbing brushes and carefully swept it clean: Lawrence M. Krauss, 'Rediscovering Creation', in Shore (ed.), *Mysteries of Life and the Universe*, p. 50.

p. 13   an instrument that might do the job: the Bell antenna: Overbye, *Lonely Hearts of the Cosmos*, p. 153, and *New York Times*, 'Where the Universe Began', 5 Sept. 2023.

NOTES

p. 14 They had found the edge of the universe: *Scientific American*, 'Echoes from the Big Bang', Jan. 2001, pp. 38–43.

p. 14 Penzias and Wilson's finding pushed our acquaintance with the visible: Guth, *The Inflationary Universe*, p. 101.

p. 15 about 1 per cent of the dancing static you see: Gribbin, *In the Beginning*, p. 18.

p. 15 'These are very close to religious questions': *New York Times*, 'Before the Big Bang, There Was . . . What?', 22 May 2001, p. F1.

p. 16 or one ten million trillion trillion trillionths: Alan Lightman, 'First Birth', in Shore (ed.), *Mysteries of Life and the Universe*, p. 13.

p. 16 He was thirty-two years old and, by his own admission, had never: Overbye, *Lonely Hearts of the Cosmos*, p. 216.

p. 16 The lecture inspired Guth to take an interest: Guth, *The Inflationary Universe*, p. 89.

p. 17 doubling in size every 10^{-34} seconds: Overbye, *Lonely Hearts of the Cosmos*, p. 242.

p. 17 it changed the universe from something you could hold in your hand to something at least 10,000,000,000,000,000,000,000,000 times bigger: *New Scientist*, 'The First Split Second', 31 March 2001, pp. 27–30.

p. 17 perfectly arrayed for the creation of stars, galaxies and other complex systems: *Scientific American*, 'The First Stars in the Universe', Dec. 2001, pp. 64–71; *New York Times*, 'Listen Closely: From Tiny Hum Came Big Bang', 30 April 2001, p. 1.

p. 17 A question that naturally arises is how the universe could expand faster than the speed of light: *BBC Sky at Night Magazine*, 'Does the Universe expand faster than light?', 26 Jan. 2024.

p. 18 'Tryon emphasized that no one had counted the failed attempts': quoted by Guth, *The Inflationary Universe*, p. 14.

p. 18 He makes an analogy with a very large clothing store: *Discover*, 'Why Is There Life?', Nov. 2000, p.66.

p. 19 with the slightest tweaking of the numbers the universe: Rees, *Just Six Numbers*, p. 147.

p. 19 In the long term, gravity may turn out to be a little too strong: *Financial Times*, 'Riddle of the Flat Universe', 1–2 July 2000; *Economist*, 'The World is Flat after All', 20 May 2000, p. 97.

NOTES

p. 20 the galaxies are rushing apart: Weinberg, *Dreams of a Final Theory*, p. 26.

p. 20 Scientists just assume that we can't really be the centre: Hawking, *A Brief History of Time*, p. 47.

p. 21 This visible universe – the universe we know and can talk about: Hawking, *A Brief History of Time*, p. 13.

p. 21 estimated to be about 94 billion light years across: NASA website, 'Age and Size of the Universe', undated.

p. 21 the number of light years to the edge of this larger, unseen universe: Rees, *Just Six Numbers*, p. 147, and interview with Martin Rees, Cambridge, 6 Aug. 2024.

Chapter 2: Welcome to the Solar System

p. 22 From the tiniest throbs and wobbles of distant stars: *New Yorker*, 'Among Planets', 9 Dec. 1996, p. 84.

p. 22 'less than the energy of a single snowflake striking the ground': Sagan, *Cosmos*, p. 261.

p. 22 In the summer of that year, a young astronomer named James Christy: US Naval Observatory press release, '20th Anniversary of the Discovery of Pluto's Moon Charon', 22 June 1998.

p. 23 Pluto was much smaller than anyone had supposed: *Atlantic Monthly*, 'When Is a Planet Not a Planet?', Feb. 1998, pp. 22–34.

p. 23 the International Astronomical Union recognized twelve new moons around Jupiter: *New York Times*, 12 May 2023.

p. 23 the total was increased again, by 128, to a truly spectacular 274: *Nature*, 'Saturn has a whopping 274 moons and scientists want to know why', 13 March 2025.

p. 23 Pluto, we now know, hasn't just one new moon, but *five*: *Scientific American*, 'Pluto's Secrets Revealed', 1 Dec. 2017.

p. 24 after a year's patient searching he somehow spotted Pluto: Tombaugh paper, 'The Struggles to Find the Ninth Planet', from NASA website.

p. 25 The Kuiper belt was actually theorized by an astronomer named F. C. Leonard in 1930: *Nature*, 'Almost Planet X', 24 May 2001, p. 423.

p. 25 Pluto doesn't act much like the other planets: *Scientific American*, 'Pluto's Secrets Revealed', 1 Dec. 2017.

NOTES

p. 25   Only on 11 Feb. 1999 did Pluto return to the outside lane: *Economist*, 'Pluto Out in the Cold', 6 Feb. 1999, p. 85.

p. 25   Matters came to a head of sorts in the summer of 2005: *London Review of Books*, 'When is a planet not a planet?', 18 Aug. 2005; *Nature*, 'The planet that never was', 3 Feb. 2011; *New York Review of Books*, 'In Search of Planet X', 24 Oct. 2019; *New York Times*, 'Do You Think Pluto Should Be a Planet?', 27 Oct. 2023.

p. 27   but all the visible stuff in it . . . fills less than a trillionth of the available space: *Natural History*, 'Between the Planets', Oct. 2001, p. 20.

p. 27   'have less memory than the key fob that opens your car door': quoted in *Scientific American*, 21 July 2022.

p. 28   Because no one has ever seen the Oort Cloud, we can't be exactly certain where it begins and ends: NASA website, 'Oort Cloud: Facts', undated.

p. 29   so the comets drift in a stately manner, moving at only about 220 miles an hour: Sagan and Druyan, *Comet*, p. 195.

p. 29   The most perfect vacuum ever created by humans is not as empty as the emptiness of interstellar space: Ball, H_2O, p. 15.

p. 30   Our nearest neighbour in the cosmos, Proxima Centauri: Guth, *The Inflationary Universe*, p. 1; Hawking, *A Brief History of Time*, p. 39.

p. 30   The average distance between stars: Dyson, *Disturbing the Universe*, p. 251.

p. 31   the number of advanced civilizations just in the Milky Way always works out to somewhere in the millions: *New Yorker*, 25 Jan. 2021, and *New York Times*, 5 Sept. 2022.

p. 31   Such is likely to be the case with a planet called K2-18b: *New York Times*, 'Astronomers Detect a Possible Signature of Life on a Distant Planet', 16 April 2025 and *Nature*, 'Signs of life on a distant planet? Not so fast, say these astronomers', 17 April 2025.

p. 32   'If we were randomly inserted into the universe,' Sagan wrote: Sagan, *Cosmos*, p. 5.

Chapter 3: The Reverend Evans's Universe

p. 34   releasing in an instant the energy of 100 billion suns: Ferris, *The Whole Shebang*, p. 37.

NOTES

p. 34 'It's like a trillion hydrogen bombs going off at once': Robert Evans, interviewed Hazelbrook, Australia, 2 Sept. 2001.

p. 34 devotes a passage to him in a chapter on autistic savants: Sacks, *An Anthropologist on Mars*, p. 189.

p. 35 'an irritating buffoon': Thorne, *Black Holes and Time Warps*, p. 164.

p. 35 refused to be left alone with him: Ferris, *The Whole Shebang*, p. 125.

p. 36 On at least one occasion Zwicky threatened to kill Baade: Overbye, *Lonely Hearts of the Cosmos*, p. 18, and *New York Review of Books*, 'The Power of Morphological Thinking', 16 Jan. 2020, and 'All Things Great and Small', 1 July 2021.

p. 36 Atoms would literally be crushed together: *Nature*, 'Twinkle, Twinkle, Neutron Star', 7 Nov. 2002, p. 31.

p. 36 enough to make the biggest bang in the universe: Thorne, *Black Holes and Time Warps*, p. 171.

p. 37 cosmic rays, though long since confirmed, still remain little understood: *University of Chicago News*, 'Cosmic rays, explained', 5 Dec. 2023.

p. 37 'one of the most prescient documents in the history of physics and astronomy': Thorne, *Black Holes and Time Warps*, p. 174.

p. 37 'he did not understand the laws of physics': Thorne, *Black Holes and Time Warps*, p. 175.

p. 37 wouldn't attract serious attention for nearly four decades: Overbye, *Lonely Hearts of the Cosmos*, p. 18.

p. 38 Only about 6,000 stars are visible to the naked eye: Harrison, *Darkness at Night*, p. 3.

p. 40 In 1987 Saul Perlmutter ... set out to find a more systematic method of searching for them: BBC *Horizon* documentary, 'From Here to Infinity', transcript of programme first broadcast 28 Feb. 1999.

p. 41 He had discovered forty-seven supernovae: *Australian Sky and Telescope*, Robert Evans obituary, 10 Nov. 2022.

p. 41 'The news of such an event travels out at the speed of light, but so does the destructiveness': interview with John Thorstensen, Hanover, NH, 5 Dec. 2001.

NOTES

p. 42 Only half a dozen times in recorded history have supernovae been close enough to be visible to the naked eye: note from Evans, 3 Dec. 2002.

p. 42 'cosmologist and controversialist': *Nature*, 'Fred Hoyle (1915–2001)', 17 Sept. 2001, p. 270.

p. 43 humans evolved projecting noses ... as a way of keeping cosmic pathogens from falling into them: Gribbin and Cherfas, *The First Chimpanzee*, p. 190.

p. 43 continually creating new matter as it went: Rees, *Just Six Numbers*, p. 75.

p. 43 100 million degrees or more: Bodanis, $E = mc^2$, p. 187.

p. 43 99.9 per cent of the mass of the solar system: Asimov, *Atom*, p. 294.

p. 44 In just 200 million years, possibly less: Stevens, *The Change in the Weather*, p. 6.

p. 44 Most of the lunar material, it is thought, came from the Earth's crust, not its core: *New Scientist* supplement, 'Firebirth', 7 Aug. 1999, n.p.

p. 44 in fact it was first proposed in the 1940s by Reginald Daly of Harvard: Powell, *Night Comes to the Cretaceous*, p. 38.

p. 44 the Earth might well have frozen over permanently: Drury, *Stepping Stones*, p. 144.

Chapter 4: The Measure of Things

p. 51 In the course of a long and productive career: Sagan, *Comet*, p. 52.

p. 51 'a very specific and precise curve': Feynman, *Six Easy Pieces*, p. 90.

p. 52 Hooke ... claimed that he had solved the problem already: Gjertsen, *The Classics of Science*, p. 219.

p. 52 and rubbed it around 'betwixt my eye and the bone': quoted by Ferris, *Coming of Age in the Milky Way*, p. 106.

p. 52 but then told no one about it for twenty-seven years: Durant, *The Age of Louis XIV*, p. 538.

p. 54 Even the great German mathematician Gottfried von Leibniz: Durant, *The Age of Louis XIV*, p. 546.

NOTES

p. 54 'one of the most inaccessible books ever written': Cropper, *Great Physicists*, p. 31.

p. 55 'proportional to the mass of each and varies inversely as the square of the distance between them': Feynman, *Six Easy Pieces*, p. 69.

p. 56 Newton, as was his custom, contributed nothing: Calder, *The Comet Is Coming!*, p. 39.

p. 56 He was to be paid instead in copies of *The History of Fishes*: Jardine, *Ingenious Pursuits*, p. 36.

p. 57 to 'within a scantling': Wilford, *The Mapmakers*, p. 98.

p. 59 The Earth was 43 kilometres stouter when measured equatorially than when measured from top to bottom around the poles: Asimov, *Exploring the Earth and the Cosmos*, p. 86.

p. 61 Unluckier still was Guillaume le Gentil: Ferris, *Coming of Age in the Milky Way*, p. 134.

p. 62 Mason and Dixon sent a note to the Royal Society: Jardine, *Ingenious Pursuits*, p. 141.

p. 64 'said to have been born in a coal mine': *Dictionary of National Biography*, vol. 12, p. 1302.

p. 64 We know that in 1772: *American Heritage*, 'Mason and Dixon: Their Line and its Legend', Feb. 1964, pp. 23–9.

p. 65 For convenience, Hutton had assumed: Jungnickel and McCormmach, *Cavendish*, p. 449.

p. 65 it was Michell to whom he turned for instruction in making telescopes: Calder, *The Comet Is Coming!*, p. 71.

p. 66 to a 'degree bordering on disease': Jungnickel and McCormmach, *Cavendish*, p. 306.

p. 66 'talk as it were into vacancy': Jungnickel and McCormmach, *Cavendish*, p. 305.

p. 68 he also foreshadowed 'the work of Kelvin and G. H. Darwin on the effect of tidal friction': Crowther, *Scientists of the Industrial Revolution*, pp. 214–15.

p. 68 At the heart of the machine were two 350-pound lead balls: *Dictionary of National Biography*, vol. 3, p. 1261.

p. 69 six billion trillion metric tons: *Economist*, 'G Whiz', 6 May 2000, p. 82.

NOTES

Chapter 5: The Stone-Breakers

p. 71 Hutton was by all accounts a man of the keenest insights and liveliest conversation: *Dictionary of National Biography*, vol. 10, pp. 354–6.

p. 71 'almost entirely innocent of rhetorical accomplishments': Dean, *James Hutton and the History of Geology*, p. 18.

p. 72 He became a leading member of a society called the Oyster Club: McPhee, *Basin and Range*, p. 99.

p. 74 quotations from French sources, still in the original French: Gould, *Time's Arrow*, p. 66.

p. 74 A third volume was so unenticing that it wasn't published until 1899: Oldroyd, *Thinking About the Earth*, pp. 96–7.

p. 74 Even Charles Lyell ... couldn't get through it: Schneer (ed.), *Toward a History of Geology*, p. 128.

p. 75 In the winter of 1807: Geological Society papers, *A Brief History of the Geological Society of London*.

p. 75 The members met twice a month from November until June: Rudwick, *The Great Devonian Controversy*, p. 25.

p. 76 As even a Murchison supporter conceded: Trinkaus and Shipman, *The Neandertals*, p. 28.

p. 76 In 1794 he was implicated in a faintly lunatic-sounding conspiracy: Cadbury, *Terrible Lizard*, p. 39.

p. 77 known ever since as Parkinson's disease: *Dictionary of National Biography*, vol. 15, pp. 314–15.

p. 77 because his mother was convinced that Scots were feckless drunks: Trinkaus and Shipman, *The Neandertals*, p. 26.

p. 78 Once Mrs Buckland found herself being shaken awake: Annan, *The Dons*, p. 27.

p. 78 His other slight peculiarity: Trinkaus and Shipman, *The Neandertals*, p. 30.

p. 78 Often when lost in thought: Desmond and Moore, *Darwin*, p. 202.

p. 79 but it was Lyell most people read: Schneer (ed.), *Toward a History of Geology*, p. 139.

p. 79 'and called for a new pack': Clark, *The Huxleys*, p. 48.

NOTES

p. 79 'Never was there a dogma more calculated to foster indolence': quoted in Gould, *Dinosaur in a Haystack*, p. 167.

p. 80 He failed to explain ... how mountain ranges were formed: Hallam, *Great Geological Controversies*, p. 135.

p. 80 'the refrigeration of the globe': Gould, *Ever Since Darwin*, p. 151.

p. 80 He rejected the notion that animals and plants suffered sudden annihilations: Stanley, *Extinction*, p. 5.

p. 80 'one yet saw it partially through his eyes': quoted in Schneer (ed.), *Toward a History of Geology*, p. 288.

p. 80 'De la Beche is a dirty dog': quoted in Rudwick, *The Great Devonian Controversy*, p.194.

p. 81 with the perky name of J. J. d'Omalius d'Halloy: McPhee, *In Suspect Terrain*, p. 190.

p. 82 Lyell originally intended to employ '-synchronous' for his endings: Gjertsen, *The Classics of Science*, p. 305.

p. 82 these number in the 'tens of dozens': McPhee, *In Suspect Terrain*, p. 50.

p. 83 In recent years, many observers have taken to using the term Anthropocene: *Nature*, 'Defining the Anthropocene', 12 March 2015, and 'The Meaning of the Anthropocene', 26 Aug. 2024.

p. 83 'I have seen grown men glow incandescent with rage': Fortey, *Trilobite!*, p. 238.

p. 83 When Buckland speculated: Cadbury, *Terrible Lizard*, p. 149.

p. 84 The best-known early attempt: Gould, *Eight Little Piggies*, p. 185.

p. 84 'most thinking people accepted the idea that the earth was young': cited in Gould, *Time's Arrow*, p. 114.

p. 84 'No geologist of any nationality whose work was taken seriously': Rudwick, *The Great Devonian Controversy*, p. 42.

p. 84 Even the Reverend Buckland: Cadbury, *Terrible Lizard*, p. 192.

p. 85 somewhere between 75,000 and 168,000 years old: Hallam, *Great Geological Controversies*, p.105 and Ferris, *Coming of Age in the Milky Way*, pp. 246–7.

p. 85 Darwin announced that the geological processes that created the Weald: Gjertsen, *The Classics of Science*, p. 335.

p. 86   The German scientist Hermann von Helmholtz: Cropper, *Great Physicists*, p. 78.

p. 86   and written (in French and English) a dozen papers in pure and applied mathematics of such dazzling originality that he had to publish them anonymously: Cropper, *Great Physicists*, p. 79.

p. 86   At the age of twenty-two he returned to Glasgow: *Dictionary of National Biography, Supplement 1901–1911*, p. 508.

Chapter 6: Science Red in Tooth and Claw

p. 89   who described it at a meeting of the American Philosophical Society: Colbert, *The Great Dinosaur Hunters and their Discoveries*, p. 4.

p. 89   The cause of this froth was a strange assertion by the great French naturalist the Comte de Buffon: Kastner, *A Species of Eternity*, p. 123.

p. 89   A Dutchman named Corneille de Pauw: Kastner, *A Species of Eternity*, p. 124.

p. 92   in 1796 Cuvier wrote a landmark paper, *Note on the Species of Living and Fossil Elephants*: Trinkaus and Shipman, *The Neandertals*, p. 15.

p. 92   Jefferson for one couldn't abide the thought that whole species would ever be permitted to vanish: Simpson, *Fossils and the History of Life*, p. 7.

p. 92   On the evening of 5 January 1796, he was sitting in a coaching inn in Somerset: Harrington, *Dance of the Continents*, p. 175.

p. 93   'The whys and wherefores cannot come within the Province of a Mineral Surveyor': Lewis, *The Dating Game*, pp. 17–18.

p. 93   Cuvier resolved the matter to his own satisfaction: Barber, *The Heyday of Natural History*, p. 217.

p. 94   In 1806 the Lewis and Clark expedition passed through the Hell Creek formation: Colbert, *The Great Dinosaur Hunters and their Discoveries*, p. 5.

p. 94   She is commonly held to be the source for the famous tongue-twister: Cadbury, *Terrible Lizard*, p. 3.

p. 95   The plesiosaur alone took her ten years of patient excavation: Barber, *The Heyday of Natural History*, p. 127.

NOTES

p. 95 Mantell could see at once it was a fossilized tooth: *New Zealand Geographic*, 'Holy Incisors! What a Treasure!', April–June 2000, p. 17.

p. 96 the name was actually suggested to Buckland by his friend Dr James Parkinson: Wilford, *The Riddle of the Dinosaur*, p. 31.

p. 97 Eventually he was forced to sell most of his collection to pay off his debts: Wilford, *The Riddle of the Dinosaur*, p. 34.

p. 97 the world's first theme park: Fortey, *Life*, p. 214.

p. 98 he sometimes illicitly borrowed limbs, organs and other parts: Cadbury, *Terrible Lizard*, p. 133.

p. 99 Once his wife returned home to find a freshly deceased rhinoceros filling the front hallway: Cadbury, *Terrible Lizard*, p. 200.

p. 99 some were no bigger than rabbits: Wilford, *The Riddle of the Dinosaur*, p. 5.

p. 99 the one thing they most emphatically were not was lizards: Bakker, *The Dinosaur Heresies*, p. 22.

p. 99 dinosaurs constitute not one but two orders of reptiles: Colbert, *The Great Dinosaur Hunters and their Discoveries*, p. 33.

p. 100 He was the only person Charles Darwin was ever known to hate: *Nature*, 'Owen's Parthian shot', 12 July 2001, p. 123.

p. 100 referred to his father's 'lamentable coldness of heart': Cadbury, *Terrible Lizard*, p. 321.

p. 100 Huxley was leafing through a new edition of *Churchill's Medical Directory*: Clark, *The Huxleys*, p. 45.

p. 101 His deformed spine was removed and sent to the Royal College of Surgeons: Cadbury, *Terrible Lizard*, p. 291.

p. 102 'not quite as original as it appeared': Cadbury, *Terrible Lizard*, pp. 261–2.

p. 102 he became the driving force behind the creation of London's Natural History Museum: Colbert, *The Great Dinosaur Hunters and their Discoveries*, p. 30.

p. 102 Before Owen, museums were designed primarily for ... the elite: Thackray and Press, *The Natural History Museum*, p. 24.

p. 103 He even proposed, very radically, to put informative labels on each display: Thackray and Press, *The Natural History Museum*, p. 98.

- p. 104 'lying everywhere like logs': Wilford, *The Riddle of the Dinosaur*, p. 97.
- p. 104 he managed to win them over by repeatedly taking out and replacing his false teeth: Wilford, *The Riddle of the Dinosaur*, p. 100.
- p. 104 it was an affront that he would never forget: Colbert, *The Great Dinosaur Hunters and their Discoveries*, p. 73.
- p. 105 increased the number of known dinosaur species in America from nine to almost 150: Colbert, *The Great Dinosaur Hunters and their Discoveries*, p. 93.
- p. 105 Nearly every dinosaur that the average person can name: Wilford, *The Riddle of the Dinosaur*, p. 90.
- p. 105 Between them they managed to 'discover' a species called *Uintatheres anceps* no fewer than twenty-two times: Psihoyos and Knoebber, *Hunting Dinosaurs*, p. 16.
- p. 106 mercifully obliterated by a German bomb in the Blitz: Cadbury, *Terrible Lizard*, p. 325.
- p. 106 much of it was taken to New Zealand by his son Walter: *Newsletter of the Geological Society of New Zealand*, 'Gideon Mantell – The New Zealand Connection', April 1992; *New Zealand Geographic*, 'Holy Incisors! What a Treasure!' April–June 2000, p. 17.
- p. 107 hence the name: Colbert, *The Great Dinosaur Hunters and their Discoveries*, p. 151.
- p. 107 He calculated that the Earth was 89 million years old: Lewis, *The Dating Game*, p. 37.
- p. 107 Such was the confusion: Hallam, *Great Geological Controversies*, p. 173.

Chapter 7: Elemental Matters
- p. 109 could make himself invisible: Ball, H_2O, p. 125.
- p. 110 An ounce of phosphorus retailed for 6 guineas: Durant, *Age of Louis XIV*, p. 516.
- p. 110 roughly £1,200 in today's money: Bank of England online inflation calculator.
- p. 110 and got credit for none of them: Strathern, *Mendeleyev's Dream*, p. 193.

NOTES

p. 111 which is why we ended up with two branches of chemistry: Davies, *The Fifth Miracle*, p. 14.

p. 112 perhaps £12 million in today's money: White, *Rivals*, p. 63.

p. 112 the fourteen-year-old daughter of one of his bosses: Brock, *The Norton History of Chemistry*, p. 92.

p. 112 *jour de bonheur* : Gould, *Bully for Brontosaurus*, p. 366.

p. 112 It was in such a capacity in 1780 that Lavoisier made some dismissive remarks: Brock, *The Norton History of Chemistry*, pp. 95–6.

p. 113 failed to uncover a single one: Strathern, *Mendeleyev's Dream*, p. 239.

p. 114 taken away and melted down for scrap: Brock, *The Norton History of Chemistry*, p. 124.

p. 114 'a highly pleasurable thrilling': Cropper, *Great Physicists*, p. 139.

p. 114 Theatres put on 'laughing gas evenings': Hamblyn, *The Invention of Clouds*, p. 76.

p. 115 What Brown noticed: Silver, *The Ascent of Science*, p. 201.

p. 116 'lukewarmness in the cause of liberty': *Dictionary of National Biography*, vol. 19, p. 686.

p. 117 a diameter of 0.00000001 centimetres: Asimov, *The History of Physics*, p. 501.

p. 119 Later, for no special reason: Ball, H_2O, p. 139.

p. 119 Luck was not always with the family: Brock, *The Norton History of Chemistry*, p. 312.

p. 120 There he was a competent but not terribly outstanding chemist: Brock, *The Norton History of Chemistry*, p. 111.

p. 120 this was an idea whose time had not quite yet come: Carey (ed.), *The Faber Book of Science*, p. 155.

p. 121 chemistry really is just a matter of counting: Ball, H_2O, p. 139.

p. 122 'the most elegant organizational chart ever devised': Krebs, *The History and Use of our Earth's Chemical Elements*, p. 23.

p. 122 Today we have 118 known elements: *Nature*, 'Extreme chemistry: experiments at the edge of the periodic table', 30 Jan. 2019.

p. 122 No one can say how high the numbers might go: *Chemistry World*, 'Moving beyond element 118', March 2016; *Nature*, 31 Jan.

2019; *Scientific American*, 'The Quest for Superheavy Elements and the Island of Stability', 1 March 2018.

p. 124 so it was big news, at least among chemists, in 2016 when IUPAC and IUPAP announced the names of four elements at once: *Chemistry World*, 'The periodic table name game', 5 Jan. 2016, and *Chemistry World*, 'Iupac announces new element names', July 2016.

p. 124 Marie Curie dubbed the effect 'radioactivity': Bodanis, $E = mc^2$, p. 75.

p. 126 He never accepted the revised figures: Lewis, *The Dating Game*, p. 55.

p. 127 'Appropriately . . . it is an unstable element': Strathern, *Mendeleyev's Dream*, p. 294.

p. 127 featured with pride the therapeutic effects of its 'Radio-active mineral springs': advertisement in *Time* magazine, 3 Jan. 1927, p. 24.

p. 127 It wasn't banned in consumer products until 1938: Biddle, *A Field Guide to the Invisible*, p. 133.

p. 128 Her lab books are kept in lead-lined boxes: *Science*, 'We Are Made of Starstuff', 4 May 2001, p. 863.

Chapter 8: Einstein's Universe

p. 132 his courses attracted an average of slightly over one student a semester: Cropper, *Great Physicists*, p. 106.

p. 132 a deck of cards: Ebbing, *General Chemistry*, p. 755.

p. 133 which dazzlingly elucidated the thermodynamic principles of, well, nearly everything: Cropper, *Great Physicists*, p. 109.

p. 133 In essence, what Gibbs did was show that thermodynamics didn't apply simply to heat and energy: Snow, *The Physicists*, p. 7.

p. 133 Gibbs's *Equilibrium* has been called 'the *Principia* of thermodynamics': Kevles, *The Physicists*, p. 33.

p. 134 he came to the United States with his family as an infant and grew up in a mining camp in California's gold rush country: Kevles, *The Physicists*, pp. 27–8.

p. 135 'The speed of light turned out to be the same in *all* directions and at *all* seasons': Thorne, *Black Holes and Time Warps*, p. 64.

NOTES

p. 135 'probably the most famous negative result in the history of physics': Cropper, *Great Physicists*, p. 208.

p. 135 Michelson counted himself among those who believed that the work of science was nearly at an end: *Nature*, 'Physics from the Inside', 12 July 2001, p. 121.

p. 136 of which three, according to C. P. Snow, 'were among the greatest in the history of physics': Snow, *The Physicists*, p. 101.

p. 137 His very first paper, on the physics of fluids in drinking straws: Bodanis, $E = mc^2$, p. 6.

p. 137 only to discover that the quietly productive J. Willard Gibbs in Connecticut had done that work as well: Boorse et al., *The Atomic Scientists*, p. 142.

p. 138 is one of the most extraordinary scientific papers ever published: Ferris, *Coming of Age in the Milky Way*, p. 193.

p. 138 It was . . . as if Einstein 'had reached the conclusions by pure thought, unaided': Snow, *The Physicists*, p. 101.

p. 139 you will contain within your modest frame no less than 7 3 10^{18} joules of potential energy: Thorne, *Black Holes and Time Warps*, p. 172.

p. 139 Even a uranium bomb . . . releases less than 1 per cent of the energy it could release: Bodanis, $E = mc^2$, p. 77.

p. 140 'Oh, that's not necessary,' he replied. 'It's so seldom I have one': *Nature*, 'In the Eye of the Beholder', 21 March 2002, p. 264.

p. 140 'it is undoubtedly the highest intellectual achievement of humanity': Boorse et al., *The Atomic Scientists*, p. 53.

p. 140 According to Einstein himself, he was simply sitting in a chair when the problem of gravity occurred to him: Bodanis, $E = mc^2$, p. 204.

p. 140 the publication in early 1917 of a paper entitled 'Cosmological Considerations on the General Theory of Relativity': Guth, *The Inflationary Universe*, p. 36.

p. 140 'Without it,' wrote Snow in 1979: Snow, *The Physicists*, p. 21.

p. 141 Crouch was hopelessly out of his depth, and got nearly everything wrong: Bodanis, $E = mc^2$, p. 215.

p. 141 'I am trying to think who the third person is': quoted in Hawking, *A Brief History of Time*, p. 91; Aczel, *God's Equation*, p. 146.

NOTES

p. 141 and the faster one moves the more pronounced these effects become: Guth, *The Inflationary Universe*, p. 37.

p. 142 a baseball thrown at 100 miles an hour will pick up 0.000000000002 grams of mass on its way to home plate: Brockman and Matson, *How Things Are*, p. 263.

p. 142 However, to turn to Bodanis again, we all commonly encounter other kinds of relativity: Bodanis, $E = mc^2$, p. 83.

p. 143 'the ultimate sagging mattress': Overbye, *Lonely Hearts of the Cosmos*, p. 55.

p. 143 'In some sense, gravity does not exist': Kaku, 'The Theory of the Universe?' in Shore (ed.), *Mysteries of Life and the Universe*, p. 161.

p. 145 and Edwin enjoyed a wealth of physical endowments, too: Cropper, *Great Physicists*, p. 423.

p. 146 At a single high-school track meeting: Christianson, *Edwin Hubble*, p. 33.

p. 148 One Harvard computer, Annie Jump Cannon, used her repetitive acquaintance with the stars to devise a system of stellar classifications: Ferris, *Coming of Age in the Milky Way*, p. 258.

p. 148 they are elderly stars that have moved past their 'main sequence phase': Ferguson, *Measuring the Universe*, pp. 166–7.

p. 148 They could be used as standard candles: Ferguson, *Measuring the Universe*, p. 166.

p. 149 was developing *his* seminal theory that dark patches on the Moon were caused by swarms of seasonally migrating insects: Moore, *Fireside Astronomy*, p. 63.

p. 149 In 1923 he showed that a puff of distant gossamer in the Andromeda constellation known as M31 wasn't a gas cloud at all: Overbye, *Lonely Hearts of the Cosmos*, p. 45; *Natural History*, 'Delusions of Centrality', Dec. 2002–Jan. 2003, pp. 28–32; *Scientific American*, 'How Astronomers Revolutionized Our View of the Cosmos', 1 Sept. 2020; *Nature*, 'The expanding universe – do ongoing tensions leave room for new physics?', 24 March 2025.

p. 150 The wonder, as Stephen Hawking has noted, is that no one had hit on the idea of the expanding universe before: Hawking, *The Universe in a Nutshell*, pp. 71–2.

NOTES

p. 151　In 1936 Hubble produced a popular book called *The Realm of the Nebulae*: Overbye, *Lonely Hearts of the Cosmos*, p. 13.

p. 151　More than seventy years later the whereabouts of the twentieth century's greatest astronomer remain unknown: Overbye, *Lonely Hearts of the Cosmos*, p. 28.

Chapter 9: The Mighty Atom

p. 152　'All things are made of atoms': Feynman, *Six Easy Pieces*, p. 4.

p. 152　45 billion billion molecules: Gribbin, *Almost Everyone's Guide to Science*, p. 250.

p. 153　up to a billion for each of us, it has been suggested: Davies, *The Fifth Miracle*, p. 127.

p. 153　Atoms themselves, however, go on practically for ever: Rees, *Just Six Numbers*, p. 96.

p. 154　If you wanted to see . . . a paramecium swimming in a drop of water: Feynman, *Six Easy Pieces*, pp. 4–5.

p. 155　'We might as well attempt to introduce a new planet into the solar system': Boorstin, *The Discoverers*, p. 679.

p. 155　In 1826, the French chemist P. J. Pelletier travelled to Manchester: Gjertsen, *The Classics of Science*, p. 260.

p. 156　a confused Pelletier, upon beholding the great man, stammered: Holmyard, *Makers of Chemistry*, p. 222.

p. 156　forty thousand people viewed the coffin and the funeral cortège stretched for 2 miles: *Dictionary of National Biography*, vol. 5, p. 433.

p. 156　For a century after Dalton made his proposal: von Baeyer, *Taming the Atom*, p. 17.

p. 156　it was said to have played a part in the suicide of . . . Ludwig Boltzmann: Weinberg, *The Discovery of Subatomic Particles*, p. 3.

p. 157　to raise a little flax and a lot of children: Weinberg, *The Discovery of Subatomic Particles*, p. 104.

p. 157　'Had she taken a bullfighter': quoted in Cropper, *Great Physicists*, p. 259.

p. 157　It was a feeling Rutherford would have understood: Cropper, *Great Physicists*, p. 317.

NOTES

p. 157 would give up halfway through and tell the students to work it out for themselves: Wilson, *Rutherford*, p. 174.

p. 157 'as far as he could see': Wilson, *Rutherford*, p. 208.

p. 158 He was one of the first . . . to see: Wilson, *Rutherford*, p. 208.

p. 158 'Why use radio?': quoted in Cropper, *Great Physicists*, p. 328.

p. 158 'Every day I grow in girth. And in mentality': Snow, *Variety of Men*, p. 47.

p. 159 gave it up when he was persuaded by a senior colleague that radio had little future: Cropper, *Great Physicists*, p. 94.

p. 159 Some physicists thought that atoms might be cube-shaped: Asimov, *The History of Physics*, p. 551.

p. 160 The number of protons is what gives an atom its chemical identity: Guth, *The Inflationary Universe*, p. 90.

p. 160 Add or subtract a neutron or two and you get an isotope: Atkins, *The Periodic Kingdom*, p. 106.

p. 160 only one-millionth of a billionth of the full volume of the atom: Gribbin, *Almost Everyone's Guide to Science*, p. 35.

p. 160 but a fly many thousands of times heavier than the cathedral: Cropper, *Great Physicists*, p. 245.

p. 161 'they could, like galaxies, pass right through each other unscathed': Ferris, *Coming of Age in the Milky Way*, p. 288.

p. 162 'Because atomic behaviour is so unlike ordinary experience': Feynman, *Six Easy Pieces*, p. 117.

p. 163 the delay in discovery was probably a very good thing: Boorse et al., *The Atomic Scientists*, p. 338.

p. 164 'I do not even know what a matrix *is*': Cropper, *Great Physicists*, p. 269.

p. 165 This isn't a matter of simply needing more precise instruments: Ferris, *Coming of Age in the Milky Way*, p. 288.

p. 165 until it is observed an electron must be regarded as being 'at once everywhere and nowhere': David H. Freedman, 'Quantum Liaisons', in Shore (ed.), *Mysteries of Life and the Universe*, p. 137.

p. 165 'a person who wasn't outraged on first hearing about quantum theory didn't understand what had been said': Overbye, *Lonely Hearts of the Cosmos*, p. 109.

NOTES

p. 165 'Don't try': von Baeyer, *Taming the Atom*, p. 43.

p. 165 The cloud itself is essentially just a zone of statistical probability: Ebbing, *General Chemistry*, p. 295.

p. 165 'an area of the universe that our brains just aren't wired to understand': Trefil, *101 Things You Don't Know About Science and No One Else Does Either*, p. 62.

p. 165 'things on a small scale behave *nothing* like things on a large scale': Feynman, *Six Easy Pieces*, p. 33.

p. 165 where matter could pop into existence from nothing at all: Alan Lightman, 'First Birth', in Shore (ed.), *Mysteries of Life and the Universe*, p. 13.

p. 166 It is as if . . . you had two identical billiard balls: Lawrence Joseph, 'Is Science Common Sense?', in Shore (ed.), *Mysteries of Life and the Universe*, pp. 42–3.

p. 166 Remarkably, the phenomenon was proved in 1997: *Christian Science Monitor*, 'Spooky Action at a Distance', 4 Oct. 2001.

p. 166 one cannot 'predict future events exactly': Hawking, *A Brief History of Time*, p. 61.

p. 167 Scientists have dealt with this problem . . . 'by not thinking about it': David H. Freedman, 'Quantum Liaisons', in Shore (ed.), *Mysteries of Life and the Universe*, p. 141.

p. 168 The weak nuclear force . . . is many billions of times stronger than gravity: Ferris, *The Whole Shebang*, p. 297.

p. 168 The grip of the strong force reaches out only to about one-hundred-thousandth of the diameter of an atom: Asimov, *Atom*, p. 258.

p. 168 'He devoted the rest of his life': Snow, *The Physicists*, p. 89.

Chapter 10: Getting the Lead Out

p. 170 Among the many symptoms associated with over-exposure are blindness, insomnia, kidney failure, hearing loss, cancer: McGrayne, *Prometheans in the Lab*, p. 88.

p. 170 the US Food and Drug Administration only got around to proposing limits on lead in baby food in 2023: *New York Times*, 24 Jan. 2023.

NOTES

p. 171 'These men probably went insane because they worked too hard': McGrayne, *Prometheans in the Lab*, p. 92.

p. 171 In fact, Midgley knew only too well the perils of lead poisoning: McGrayne, *Prometheans in the Lab*, p. 92.

p. 171 One leak from a refrigerator at a hospital in Cleveland, Ohio, in 1929 killed more than 100 people: McGrayne, *Prometheans in the Lab*, p. 96.

p. 172 A single kilogram of CFCs can capture and annihilate 70,000 kilograms of atmospheric ozone: Biddle, *A Field Guide to the Invisible*, p. 62.

p. 172 A single CFC molecule is about 10,000 times more efficient at exacerbating greenhouse effects than a molecule of carbon dioxide: *Science*, 'The Ascent of Atmospheric Sciences', 13 Oct. 2000, p. 299.

p. 172 His death was itself memorably unusual: *Nature*, 27 Sept. 2001, p. 364.

p. 173 Up to this time, the oldest reliable dates went back no further than the First Dynasty in Egypt: Willard Libby, 'Radiocarbon Dating', from Nobel Lecture, 12 Dec. 1960.

p. 173 After eight half-lives, only 0.39 per cent of the original radioactive carbon remains: Gribbin and Gribbin, *Ice Age*, p. 58.

p. 173 'every raw radiocarbon date you read today is given as too young by around 3 per cent': Flannery, *The Eternal Frontier*, p. 174.

p. 174 it is like miscounting by a dollar when counting to 1,000: Flannery, *The Future Eaters*, p. 151.

p. 174 Among the more dubious are dates just around the time that people first came to the Americas: Flannery, *The Eternal Frontier*, pp. 174–5.

p. 174 the long-running debate over whether syphilis originated in the New World or the Old: *Science*, 'Can Genes Solve the Syphilis Mystery?', 11 May 2001, p. 109.

p. 176 Unfortunately, he now met yet another formidable impediment to acceptance: Lewis, *The Dating Game*, p. 204.

p. 177 It was this that eventually led him to create a sterile laboratory: Powell, *Mysteries of Terra Firma*, p. 58.

NOTES

p. 177   'a figure that stands unchanged 50 years later': McGrayne, *Prometheans in the Lab*, p. 173.

p. 178   In one such study, a doctor who had no specialized training in chemical pathology: McGrayne, *Prometheans in the Lab*, p. 94.

p. 178   about 90 per cent of it appeared to come from car exhaust pipes: *Nation*, 'The Secret History of Lead', 20 March 2000.

p. 178   The notion became the foundation of ice core studies, on which much modern climatological work is based: Powell, *Mysteries of Terra Firma*, p. 60.

p. 179   Ethyl executives allegedly offered to endow a chair at Caltech 'if Patterson was sent packing': *Nation*, 'The Secret History of Lead', 20 March 2000.

p. 179   Almost immediately lead levels in the blood of Americans fell by 80 per cent: McGrayne, *Prometheans in the Lab*, p. 169.

p. 179   Americans alive today have several hundred times more lead in their blood than people did a century ago: *Nation*, 20 March 2000.

p. 179   '44 years after most of Europe': McGrayne, *Prometheans in the Lab*, p. 191.

p. 180   will almost certainly be around and devouring ozone long after you and I have shuffled off: Biddle, *A Field Guide to the Invisible*, pp. 110–11.

p. 180   illicit production is estimated to be equivalent to the annual carbon emissions of Switzerland: *New Scientist*, 3 April 2023.

p. 180   Two popular books on the history of the dating of the Earth actually manage to misspell his name: The books are *Mysteries of Terra Firma* and *The Dating Game*, both of which make his name 'Claire'. (Since this note first appeared, I have received a rather severe rebuke from the author of the latter book, Cherry Lewis, informing me that her choice of spelling was intentional and arose from correspondence she had had with Patterson's widow. Except for the other cited book, Lewis's choice of spelling accords with no other published sources I can find, including Patterson's many obituaries in leading journals – which were, after all, literally the last word on the man and his name. Nonetheless I am happy to accept that Ms Lewis's variant spelling of Patterson's name was done intentionally and I unreservedly apologize to her for any dismay caused.)

p. 180   made the additional, rather astounding error of thinking Patterson was a woman: *Nature*, 'The Rocky Road to Dating the Earth', 4 Jan. 2001, p. 20.

Chapter 11: Muster Mark's Quarks

p. 181   In 1911, a British scientist named C. T. R. Wilson: Cropper, *Great Physicists*, p. 325.

p. 182   'if I could remember the names of these particles, I would have been a botanist': quoted in Cropper, *Great Physicists*, p. 403.

p. 182   a single proton can do 11,000 laps . . . in under a second: *Scientific American*, 'Large Hadron Collider Seeks New Particles after Major Upgrade', 27 April 2022.

p. 182   Even the most sluggish: Guth, *The Inflationary Universe*, p. 121.

p. 183   Breaking up atoms . . . is easy: Trefil, *101 Things You Don't Know About Science and No One Else Does Either*, p. 48.

p. 183   took 10,000 scientists and engineers more than ten years and £5 billion to build: *London Review of Books*, 'At the Science Museum', 6 March 2014; *New York Review of Books*, 'The Crisis of Big Science', 10 May 2012

p. 185   'which are themselves universes at the next level and so on for ever': Sagan, *Cosmos*, pp. 265–6.

p. 185   'The charged pion and antipion decay respectively': Weinberg, *The Discovery of Subatomic Particles*, p. 163.

p. 185   'to restore some economy to the multitude of hadrons': Weinberg, *The Discovery of Subatomic Particles*, p. 165.

p. 185   wanted to call these new basic particles *partons*: von Baeyer, *Taming the Atom*, p. 17.

p. 186   Eventually out of all this emerged what is called the Standard Model: Institute of Physics, 'The Standard Model', undated; *London Review of Books*, 'Gremlin Fireworks', June 2009.

p. 186   bosons . . . are particles that produce and carry forces: *Scientific American*, 'Uncovering Supersymmetry', July 2002, p. 74.

p. 186   'It has too many arbitrary parameters': quoted on the PBS video *Creation of the Universe*, 1985; also quoted, with slightly different numbers, in Ferris, *Coming of Age in the Milky Way*, pp. 298–9.

NOTES

p. 186 the long-theorized Higgs particle: *Scientific American*, 'Beautiful Physics: The Search for New Particles at LHCb', 1 Nov. 2017, and *Nature*, 'Beyond the Higgs', 30 Aug. 2012.

p. 186 It was Peter Higgs of Edinburgh University whose name became indelibly attached to it: *Scientific American*, 'How the Higgs Boson Ruined Peter Higgs's Life', 24 June 2022.

p. 188 This postulates that all those fundamental entities like quarks and electrons: *Science News*, 22 Sept. 2001, p. 185; *Scientific American*, 'The Inner Life of Quarks', 21 May 2013.

p. 189 One puzzle with superstring theory was that it yielded five separate versions: *New York Times*, 'A Crisis at the Edge of Physics', 5 June 2015.

p. 189 Witten's new idea brought all the versions together into something called M theory: *Scientific American*, 'The Universe's Unseen Dimensions', Aug. 2000, pp. 62–9; *Science News*, 'When Branes Collide', 22 Sept. 2001, pp. 184–5; *New York Review of Books*, 'The Universe We Still Don't Know', 10 Feb. 2011; Centre for Theoretical Cosmology, 'M-theory, the theory formerly known as Strings', undated.

p. 189 'The ekpyrotic process begins far in the indefinite past': *New York Times*, 'Before the Big Bang, There Was . . . What?', 22 May 2001, p. F1.

p. 190 it is 'almost impossible for the non-scientist to discriminate between the legitimately weird and the outright crackpot': *Nature*, 27 Sept. 2001, p. 354.

p. 190 Karl Popper . . . once suggested that there may not in fact be an ultimate theory: Weinberg, *Dreams of a Final Theory*, p. 184.

p. 190 'we do not seem to be coming to the end of our intellectual resources': Weinberg, *Dreams of a Final Theory*, p. 187.

p. 190 Hubble calculated that the universe was about 2 billion years old: *US News and World Report*, 'How Old Is the Universe?', 25 Aug. 1997, p. 34.

p. 190 Almost the only thing constant about the Hubble constant has been the amount of disagreement over what value to give it: *Nature*, 'Mystery over Universe's expansion deepens with fresh data', 15 July 2020.

NOTES

p. 191   a new age for the universe of between 7 billion and 20 billion years: Trefil, *101 Things You Don't Know About Science and No One Else Does Either*, p. 91.

p. 191   In the years that followed there erupted a dispute that would run and run: Overbye, *Lonely Hearts of the Cosmos*, p. 268.

p. 191   In February 2003, a team from NASA: *New York Times*, 'Cosmos Sits for Early Portrait, Gives up Secrets', 12 Feb. 2003, p. 1.

p. 192   'a mountain of theory built on a molehill of evidence': *Economist*, 'Queerer than we can suppose', 5 Jan. 2002, p. 58.

p. 192   'may reflect the paucity of the data rather than the excellence of the theory': *National Geographic*, 'Unveiling the Universe', Oct. 1999, p. 25.

p. 193   Dark matter could hardly be more invisible: *Nature*, 'Dark matter and dark energy Q&A', 2 April 2009.

p. 193   Harvard professor Lisa Randall has suggested that it would be better called 'transparent matter': *New York Review of Books*, 'In Defense of "Dark Matter and the Dinosaurs"', 21 April 2016.

p. 193   'The universe does not give up its mysteries cheaply': interviews with Carlos Frenk, 6 Aug. and 30 Nov. 2024.

p. 195   The name axion, rather amazingly, is taken from a laundry product: *Symmetry*, 'The other dark matter candidate', 21 Jan. 2010; *Scientific American*, 'Searching for the Dark: The Hunt for Axions', 1 Jan. 2018.

p. 196   an international collaboration behind the Dark Energy Spectroscopic Instrument: *New York Times*, 'A Tantalizing "Hint" That Astronomers Got Dark Energy All Wrong', 4 April 2024; *Quanta Magazine*, 'Dark Energy May Be Weakening', 4 April 2024; *New York Times*, ' "More Than a Hint" That Dark Energy Isn't What Astronomers Thought', 19 March 2025.

Chapter 12: The Earth Moves

p. 198   In a tone that all but invited the reader to join him in a tolerant chuckle: Hapgood, *Earth's Shifting Crust*, p. 29.

p. 201   they posited ancient 'land bridges' wherever they were needed: Simpson, *Fossils and the History of Life*, p. 98.

NOTES

p. 201   Even land bridges couldn't explain some things: Gould, *Ever Since Darwin*, p. 163.

p. 201   full of 'numerous grave theoretical difficulties': *Encylopaedia Britannica*, vol. 6, p. 418.

p. 202   One reviewer there fretted . . . that students might actually come to believe them: Lewis, *The Dating Game*, p. 182.

p. 202   about half of those present now embraced the idea of continental drift: Hapgood, *Earth's Shifting Crust*, p. 31.

p. 202   'I feel the hypothesis is a fantastic one': Powell, *Mysteries of Terra Firma*, p. 147.

p. 202   Interestingly, oil company geologists had known for years: McPhee, *Basin and Range*, p. 175.

p. 203   Aboard this vessel was a fancy new depth sounder called a fathometer: McPhee, *Basin and Range*, p. 187.

p. 203   seamounts that he called guyots after an earlier Princeton geologist: Harrington, *Dance of the Continents*, p. 208.

p. 206   'probably the most significant paper in the earth sciences ever to be denied publication': Powell, *Mysteries of Terra Firma*, pp. 131–2.

p. 207   Well into the 1970s: Powell, *Mysteries of Terra Firma*, p. 141.

p. 207   one American geologist in eight still didn't believe in plate tectonics: McPhee, *Basin and Range*, p. 198.

p. 207   Today we know that the Earth's surface is made up of eight to twelve big plates: Simpson, *Fossils and the History of Life*, p. 113, and *Scientific American*, 'Geology's biggest mystery: when did plate tectonics start to reshape Earth?', 14 Aug. 2024.

p. 207   The connections . . . were found to be infinitely more complex than anyone had imagined: McPhee, *Assembling California*, pp. 202–8.

p. 208   at about the speed a fingernail grows: Vogel, *Naked Earth*, p. 19.

p. 208   one-tenth of 1 per cent of the Earth's history: Margulis and Sagan, *Microcosmos*, p. 44.

p. 208   It is thought . . . that tectonics is an important part of the planet's organic well-being: Trefil, *Meditations at 10,000 Feet*, p. 181.

p. 209   proposing that there may well be a relationship between the history of rocks and the history of life: *Science*, 'Inconstant Ancient Seas and Life's Path', 8 Nov. 2002, p. 1165.

NOTES

p. 209 'the whole earth suddenly made sense': McPhee, *Rising from the Plains*, p. 158.

p. 209 a habit of appearing inconveniently where they shouldn't and failing to be where they ought: Simpson, *Fossils and the History of Life*, p. 115.

p. 210 There are also many surface features that tectonics can't explain: *Scientific American*, 'Sculpting the Earth from Inside Out', March 2001.

p. 210 Alfred Wegener never lived to see his ideas vindicated: Kunzig, *The Restless Sea*, p. 51.

p. 211 One of his students was a bright young fellow named Walter Alvarez: Powell, *Night Comes to the Cretaceous*, p. 7.

Chapter 13: Bang!

p. 215 In 1905, a man drilling a well for the town water supply reported bringing up a lot of strangely deformed rock: Raymond R. Anderson, Geological Society of America GSA Special Paper 302, '*The Manson Impact Structure: A Late Cretaceous Meteor Crater in the Iowa Subsurface*', Spring 1996, and interview with Ray Anderson and Brian Witzke, Iowa City, Sept. 2024.

p. 216 Virtually the whole town turned out: *Des Moines Register*, 30 June 1979.

p. 217 'Very occasionally we get people coming in and asking where they should go to see the crater': Interview with Schlapkohl, Manson, Iowa, 18 June 2001.

p. 217 The leading early investigator, G. K. Gilbert of Columbia University: Lewis, *Rain of Iron and Ice*, p. 38.

p. 217 Gilbert conducted these experiments not in a laboratory at Columbia but in a hotel room: Powell, *Night Comes to the Cretaceous*, p. 37.

p. 218 'At the time we started, only slightly more than a dozen of these things had ever been discovered': transcript from BBC *Horizon* documentary, 'New Asteroid Danger', p. 4; programme first transmitted 18 March 1999.

p. 219 He called them asteroids – Greek for 'starlike': *Science News*, 'A Rocky Bicentennial', 28 July 2001, pp. 61–3.

NOTES

p. 219 it was finally tracked down in 2000 after being missing for eighty-nine years: Ferris, *Seeing in the Dark*, p. 150.

p. 220 down which we are cruising at over 100,000 kilometres an hour: Ferris, *Seeing in the Dark*, p. 147.

p. 220 'all of which are capable of colliding with the Earth and all of which are moving on slightly different courses through the sky at different rates': transcript from BBC *Horizon* documentary 'New Asteroid Danger', p. 5; first transmitted 18 March 1999.

p. 221 Altogether, as of 2024, the International Astronomical Union's Minor Planet Center had identified some 19,000 Near Earth Objects: *Nature Communications*, 1 March 2024; *Scientific American*, 'Are We Doing Enough to Protect Earth from Asteroids?', 1 June 2021.

p. 221 Easily the weirdest of all these unexpected visitors was a mysterious interstellar object called Oumuamua: *New Yorker*, 25 Jan. 2021, and *Science News*, 27 Feb. 2019.

p. 223 Every year the Earth accumulates some 30,000 tonnes of 'cosmic spherules': Vernon, *Beneath our Feet*, p. 191.

p. 224 'Well, they were very charming, very persuasive': telephone interview with Asaro, 10 March 2002.

p. 224 a Northwestern University astrophysicist named Ralph B. Baldwin had suggested such a possibility in an article in *Popular Astronomy* magazine: Powell, *Mysteries of Terra Firma*, p. 184.

p. 225 In 1956 a professor at Oregon State University, M. W. de Laubenfels: Peebles, *Asteroids: A History*, p. 170.

p. 225 may have been the cause of an earlier event known as the Frasnian extinction: Lewis, *Rain of Iron and Ice*, p. 107.

p. 226 'They're more like stamp collectors': quoted by Officer and Page, *Tales of the Earth*, p. 142.

p. 226 even while conceding in a newspaper interview that he had no actual evidence of it: *Boston Globe*, 'Dinosaur Extinction Theory Backed', 16 Dec. 1985.

p. 226 continued to believe that the extinction of the dinosaurs was in no way related to an asteroid or cometary impact: Peebles, *Asteroids: A History*, p. 175.

NOTES

p. 227 a big part of the work you do is to evaluate Manure Management Plans: Iowa Department of Natural Resources Publication, *Iowa Geology 1999*, Number 24, and *Iowa Capital Dispatch*, 'Report: Iowa produces the most factory farm waste in the country, report shows', 24 Sept. 2024.

p. 227 'Suddenly we were at the centre of things': interview with Anderson and Witzke, Iowa City, 15 June 2001.

p. 227 One of those moments came at the annual meeting of the American Geophysical Union in 1985: *Boston Globe*, 'Dinosaur Extinction Theory Backed', 16 Dec. 1985.

p. 230 The formation had been found by Pemex, the Mexican oil company, in 1952: Peebles, *Asteroids: A History*, pp. 177–8; *Washington Post*, 'Incoming', 19 April 1998.

p. 230 'I remember harboring some strong initial doubts about the efficacy of such an event': Gould, *Dinosaur in a Haystack*, p. 162.

p. 231 'Jupiter will swallow these comets up without so much as a burp': quoted by Peebles, *Asteroids: A History*, p. 196.

p. 231 One fragment, known as Nucleus G, struck with the force of about six million megatonnes: Peebles, *Asteroids: A History*, p. 202.

p. 231 Shoemaker was killed instantly, his wife injured: Peebles, *Asteroids: A History*, p. 204.

p. 233 nearly every standing thing would swiftly be flattened or on fire, and nearly every living thing would be dead: Anderson, Iowa Department of Natural Resources, *Iowa Geology 1999*, 'Iowa's Manson Impact Structure', and Iowa DNR publication *Iowa Outdoors*, 'The Day Iowa Instantly Ignited', July/Aug. 2008.

p. 234 fleeing would mean 'selecting a slow death over a quick one': Lewis, *Rain of Iron and Ice*, p. 209.

p. 234 analysed helium isotopes from sediments left from the later KT impact and concluded that it affected the Earth's climate for about 10,000 years: *Arizona Republic*, 'Impact Theory Gains New Supporters', 3 March 2001.

p. 234 First, as John S. Lewis has noted, our missiles are not designed for space work: Lewis, *Rain of Iron and Ice*, p. 215.

p. 234 NASA crashed a spacecraft called DART: *Nature Communications*, 'Planetary defense with the Double Asteroid Redirection Test

NOTES

(DART) mission and prospects', 1 March 2023, and *Scientific American*, 'NASA's Asteroid-Bashing DART Mission Was Wildly Successful', 2 March 2023.

p. 234   It has been suggested that even a year's warning would probably be insufficient: *New York Times* magazine, 'The Asteroids Are Coming! The Asteroids Are Coming!', 28 July 1996, pp. 17–19.

p. 235   Shoemaker–Levy 9 had been orbiting Jupiter in a fairly conspicuous manner since 1929, but it was over half a century before anyone noticed: Ferris, *Seeing in the Dark*, p. 168.

Chapter 14: The Fire Below

p. 237   'It was a dumb place to look for bones': interview with Mike Voorhies, Ashfall Fossil Beds State Park, Nebraska, 13 June 2001.

p. 237   At first they thought the animals were buried alive: *National Geographic*, 'Ancient Ashfall Creates Pompeii of Prehistoric Animals', Jan. 1981, p. 66.

p. 239   'far better than we understand the interior of the Earth': Feynman, *Six Easy Pieces*, p. 60.

p. 239   The distance from the surface of Earth to the middle is 6,370 kilometres: Williams and Montaigne, *Surviving Galeras*, p. 78.

p. 240   A modest fellow, he never referred to the scale by his own name: Ozima, *The Earth*, p. 49.

p. 240   It rises exponentially: Officer and Page, *Tales of the Earth*, p. 33.

p. 241   which was initially logged at 8.6 magnitude but later revised upwards by some authorities (including the US Geological Survey) to a truly grand-scale 9.5: *Nature*, 'The biggest one', 6 May 2010.

p. 242   'the city waiting to die': McGuire, *A Guide to the End of the World*, p. 21.

p. 242   Japanese seismologists have put the chances of another Kanto-sized earthquake in Tokyo by mid-century at 70 per cent: *Economist*, 'Japan is preparing for a massive earthquake', 31 Aug. 2023.

p. 243   'collapsed scaffolding erected around the Capitol Building': Trefil, *101 Things You Don't Know About Science and No One Else Does Either*, p. 158.

NOTES

p. 244   The project became known, all but inevitably, as the Mohole: Vogel, *Naked Earth*, p. 37.

p. 244   'like trying to drill a hole ... using a strand of spaghetti': *Valley News*, 'Drilling the Ocean Floor for Earth's Deep Secrets', 21 Aug. 1995.

p. 244   they had drilled to a depth of 12,262 metres: *Nature*, 'Journey to the Mantle of the Earth', 24 March 2011.

p. 244   the crust of the Earth represents only about 0.3 per cent of the planet's volume: Schopf, *Cradle of Life*, p. 73.

p. 245   We know a little bit about the mantle from what are known as kimberlite pipes: McPhee, *In Suspect Terrain*, pp. 16–18.

p. 245   Scientists are generally agreed: *Scientific American*, 'Sculpting the Earth from Inside Out', March 2001, pp. 40–7, and *New Scientist*, 'Journey to the Centre of the Earth', supplement, 14 Oct. 2000, p. 1.

p. 246   By all the laws of geophysics: *Earth*, 'Mystery in the High Sierra', June 1996, p. 16.

p. 247   The rocks are viscous ... in the same way that glass is: Vogel, *Naked Earth*, p. 31.

p. 247   The movements occur not just laterally: *Science*, 'Much About Motion in the Mantle', Feb. 2002, p. 982.

p. 247   an English vicar named Osmond Fisher presciently suggested: Tudge, *The Time Before History*, p. 43.

p. 248   'then had suddenly found out about wind': Vogel, *Naked Earth*, p. 53.

p. 248   'there are two sets of data, from two different disciplines': Trefil, *101 Things You Don't Know About Science and No One Else Does Either*, p. 146.

p. 248   82 per cent of the Earth's volume and 65 per cent of its mass: *Nature*, 'The Earth's Mantle', 2 Aug. 2001, pp. 501–6.

p. 248   something over 3 million times those found at the surface: Drury, *Stepping Stones*, p. 50.

p. 249   during the age of the dinosaurs, it was up to three times as strong as it is now: *New Scientist*, 'Dynamo Support', 10 March 2001, p. 27.

NOTES

p. 249 37 million years appears to be the longest stretch: *New Scientist*, 'Dynamo Support', 10 March 2001, p. 27.

p. 249 'the greatest unanswered question in the geological sciences': Trefil, *101 Things You Don't Know About Science and No One Else Does Either*, p. 150.

p. 250 'Geologists and geophysicists rarely go to the same meetings': Vogel, *Naked Earth*, p. 139.

p. 251 The seismologists resolutely based their conclusions on the behaviour of Hawaiian volcanoes: Fisher et al., *Volcanoes*, p. 24.

p. 251 It was the biggest landslide in human history: Thompson, *Volcano Cowboys*, p. 118.

p. 251 with the force of 25,000 Hiroshima-sized atomic bombs: website of the National Environmental Satellite, Data, and Information Service, part of the National Oceanic and Atmospheric Administration, undated.

p. 251 Fifty-seven people were killed: Fisher et al., *Volcanoes*, p. 12.

p. 252 An airliner . . . reported being pelted with rocks: Thompson, *Volcano Cowboys*, p. 123.

p. 253 Yet Yakima had no volcano emergency procedures: Fisher et al., *Volcanoes*, p. 16.

Chapter 15: Dangerous Beauty

p. 254 In 1943 at Paricutín in Mexico: Smith, *The Weather*, p. 112.

p. 256 a bang that reverberated around the world for nine days: Lewis, *Rain of Iron and Ice*, p. 152.

p. 257 The ash fall from the last Yellowstone eruption covered all or parts of nineteen western states: *Vox*, 'What would happen if the Yellowstone supervolcano actually erupted?', 15 Dec. 2014.

p. 257 The last supervolcano eruption on Earth was at Toba: McGuire, *A Guide to the End of the World*, p. 104.

p. 258 'It may not feel like it, but you're standing on the largest active volcano in the world': interview with Paul Doss, Yellowstone National Park, Wyoming, 16 June 2001, and telephone interviews and correspondence, 2024.

p. 261 'people spread misinformation': Yellowstone Forever website, undated.

NOTES

p. 261   the odds of a Yellowstone mega-eruption in any given year are just 0.00014 per cent: *Vox*, 14 Dec. 2014; *Nature*, 'The progression of basaltic-rhyolitic melt storage at Yellowstone Caldera', 1 Jan. 2025.

p. 261   A survey in 2022 using new techniques: *New York Times*, 'New Estimate Finds More Magma Under Yellowstone Supervolcano', 1 Dec. 2022.

p. 262   as was made devastatingly evident on the night of 17 August 1959 at a place called Hebgen Lake: Smith and Siegel, *Windows into the Earth*, pp. 5–6.

p. 262   just before 10 a.m. on a calm July morning, a massive explosion rent the air: *New York Times*, 'Hydrothermal Explosion at Yellowstone Sends Tourists Racing for Safety', 23 July 2024.

p. 264   only twenty-two people are known to have been killed by Yellowstone's boiling waters: *New York Times*, 'Yellowstone Visitor Sentenced to Jail for "Thermal Trespass"', 18 June 2024.

p. 265   as little as a single molecule in ideal conditions: Sykes, *The Seven Daughters of Eve*, p. 12.

p. 265   Meanwhile, scientists were finding even hardier microbes: Ashcroft, *Life at the Extremes*, p. 275.

Chapter 16: Lonely Planet

p. 269   no less than 99.5 per cent of the world's habitable space by volume: *New York Times Book Review*, 'Where Leviathan Lives', 20 April 1997, p. 9.

p. 269   water is about 1,300 times heavier than air: Ashcroft, *Life at the Extremes*, p. 51.

p. 270   your veins would collapse and your lungs would compress to the approximate dimensions of a Coke can: *New Scientist*, 'Into the Abyss', 31 March 2001.

p. 270   At least eleven people have died in attempts to break deep-dive records: *Science*, 'How Low Can You Go?', 13 July 2021.

p. 270   he blacked out from narcosis and suffered a series of paralysing strokes: DeeperBlue.com, 'Herbert Nitsch Talks About His Fateful Dive and Recovery', 6 June 2013.

NOTES

p. 270 Because we are made largely of water ourselves: Ashcroft, *Life at the Extremes*, p. 68.

p. 270 'humans may be more like whales and dolphins than had been expected': Ashcroft, *Life at the Extremes*, p. 69.

p. 271 'all that is left in the suit are his bones and some rags of flesh': Haldane, *What Is Life?*, p. 188.

p. 272 Ashcroft relates a story concerning the directors of a new tunnel under the Thames who held a celebratory banquet: Ashcroft, *Life at the Extremes*, p. 59.

p. 273 When roused, Haldane explained that he had found himself disrobing and assumed it was bedtime: Norton, *Stars Beneath the Sea*, p. 111.

p. 273 Haldane's gift to diving was to work out the rest intervals necessary to manage an ascent from the depths without getting the bends: Haldane, *What Is Life?*, p. 202.

p. 273 his blood saturation level had reached 56 per cent: Norton, *Stars Beneath the Sea*, p. 105.

p. 273 'But is it oxyhaemoglobin or carboxyhaemoglobin?': quoted in Norton, *Stars Beneath the Sea*, p. 121.

p. 274 called him 'the cleverest man I ever knew': Gould, *The Lying Stones of Marrakech*, p. 305.

p. 274 the younger Haldane found the First World War 'a very enjoyable experience': Norton, *Stars beneath the Sea*, p. 124.

p. 275 'Almost every experiment . . . ended with someone having a seizure, bleeding or vomiting': Norton, *Stars beneath the Sea*, p. 133.

p. 275 Perforated eardrums were quite common: Haldane, *What Is Life?*, p. 192.

p. 275 left Haldane without feeling in his buttocks and lower spine for six years: Haldane, *What Is Life?*, p. 202.

p. 275 It also produced wild mood swings: Ashcroft, *Life at the Extremes*, p. 78.

p. 276 'the tester was usually as intoxicated as the testee': Haldane, *What Is Life?*, p. 197.

p. 276 The cause of the inebriation is even now a mystery: Ashcroft, *Life at the Extremes*, p. 79.

NOTES

p. 276 Even in quite mild weather half the calories you burn go to keep your body warm: Attenborough, *The Living Planet*, p. 39.

p. 276 the portions of the Earth on which we are prepared or able to live are modest indeed: Smith, *The Weather*, p. 40.

p. 277 Had our sun been ten times as massive, it would have exhausted itself after 10 million years instead of 10 billion: Ferris, *The Whole Shebang*, p. 81.

p. 278 The Sun's warmth reaches it just two minutes before it touches us: Grinspoon, *Venus Revealed*, p. 9.

p. 278 It appears that during the early years of the solar system Venus was only slightly warmer than the Earth and probably had oceans: *National Geographic*, 'The Planets', Jan. 1985, p. 40.

p. 278 the atmospheric pressure at the surface is ninety times that of Earth: McSween, *Stardust to Planets*, p. 200.

p. 280 The Moon is slipping from our grasp at a rate of about 4 centimetres a year: Ward and Brownlee, *Rare Earth*, p. 33.

p. 281 That's the weight of a paper clip: Stockholm University Department of Physics website, 'The rarest element on earth is studied in detail', July 2024.

p. 282 discarded the state silver dinner service and replaced it with an aluminium one: Bodanis, *The Secret House*, p. 13.

p. 282 accounting for a very modest 0.048 per cent of the Earth's crust: Krebs, *The History and Use of our Earth's Chemical Elements*, p. 148.

p. 282 'If it wasn't for carbon, life as we know it would be impossible': Davies, *The Fifth Miracle*, p. 126.

p. 282 Of every 200 atoms in your body, 126 are hydrogen, 51 are oxygen, and just 19 are carbon: Snyder, *The Extraordinary Chemistry of Ordinary Things*, p. 24.

p. 283 The degree to which organisms require or tolerate certain elements is a relic of their evolution: Parker, *Inscrutable Earth*, p. 100.

p. 283 Drop a small lump of pure sodium into ordinary water and it will explode with enough force to kill: Snyder, *The Extraordinary Chemistry of Ordinary Things*, p. 42.

p. 284 The Romans also flavoured their wine with lead: Parker, *Inscrutable Earth*, p. 103.

NOTES

p. 285 The physicist Richard Feynman used to make a joke: Feynman, *Six Easy Pieces*, p. xix.

Chapter 17: Into the Troposphere

p. 286 Without it, Earth would be a lifeless ball of ice: Stevens, *The Change in the Weather*, p. 7.

p. 287 and was discovered in 1902 by a Frenchman in a balloon, Léon-Philippe Teisserenc de Bort: Stevens, *The Change in the Weather*, p. 56; *Nature*, '1902 and All That', 3 Jan. 2002, p. 15.

p. 287 it's from the same Greek root as *menopause*: Smith, *The Weather*, p. 52.

p. 287 would, at the very least, result in severe cerebral and pulmonary oedemas: Ashcroft, *Life at the Extremes*, p. 7.

p. 287 The temperature 10 kilometres up can be minus 57 degrees Celsius: Smith, *The Weather*, p. 25.

p. 288 about eight-millionths of a centimetre, to be precise: Allen, *Atmosphere*, p. 58.

p. 288 if an incoming vehicle hits the thermosphere at too shallow an angle, it could well bounce back into space: Allen, *Atmosphere*, p. 57.

p. 289 Dickinson records how Howard Somervell . . . 'found himself choking to death after a piece of infected flesh came loose and blocked his windpipe': Dickinson, *The Other Side of Everest*, p. 86.

p. 289 The absolute limit of human tolerance for continuous living appears to be about 5,500 metres: Ashcroft, *Life at the Extremes*, p. 8.

p. 289 above 5,500 metres even the most well-adapted women cannot provide a growing foetus with enough oxygen: Attenborough, *The Living Planet*, p. 18.

p. 290 'nearly half a ton has been quietly piled upon us during the night': quoted by Hamilton-Paterson, *The Great Deep*, p. 177.

p. 291 a typical weather front may consist of 750 million tonnes of cold air pinned beneath a billion tonnes of warmer air: Smith, *The Weather*, p. 50.

p. 291 an amount of energy equivalent to four days' use of electricity for the whole United States: Junger, *The Perfect Storm*, p. 102.

NOTES

p. 291   At any one moment 1,800 thunderstorms are in progress around the globe: Stevens, *The Change in the Weather*, p. 55.

p. 291   Much of our knowledge of what goes on up there is surprisingly recent: Biddle, *A Field Guide to the Invisible*, p. 161.

p. 293   a wind blowing at 300 miles an hour is not simply ten times stronger than a wind blowing at 30 miles an hour, but 100 times stronger: Bodanis, $E = mc^2$, p. 68.

p. 293   as much energy as a rich, medium-sized nation . . . uses in a year: Ball, H_2O, p. 51.

p. 293   The impulse of the atmosphere to seek equilibrium was first suspected by Edmond Halley: *Science*, 'The Ascent of Atmospheric Sciences', 13 Oct. 2000, p. 300.

p. 294   thus we call it the Coriolis effect: Roberts, *Oceans of Life*, p. 95.

p. 294   Coriolis's other distinction at the school was to introduce water coolers, which were once commonly known as Corios: Trefil, *The Unexpected Vista*, p. 24.

p. 294   gives weather systems their curl and sends hurricanes spinning off like tops: Drury, *Stepping Stones*, p. 25.

p. 295   Celsius made boiling point zero and freezing point 100: Trefil, *The Unexpected Vista*, p. 107.

p. 295   Howard is chiefly remembered now for giving cloud types their names in 1803: *Dictionary of National Biography*, vol. 10, pp. 51–2.

p. 296   Howard's system has been much added to over the years: Trefil, *Meditations at Sunset*, p. 62.

p. 296   That seems to have been the source of the expression 'to be on cloud nine': Hamblyn, *The Invention of Clouds*, p. 252.

p. 297   A fluffy summer cumulus several hundred metres to a side may contain no more than 100–150 litres of water: Trefil, *Meditations at Sunset*, p. 66.

p. 297   Only about 0.035 per cent of the Earth's fresh water is floating around above us at any moment: Ball, H_2O, p. 57.

p. 297   Depending on where it falls, the prognosis for a water molecule varies widely: Dennis, *The Bird in the Waterfall*, p. 8.

NOTES

p. 297 Even something as large as the Mediterranean would dry out in 1,000 years if it were not continually replenished: Gribbin and Gribbin, *Being Human*, p. 123.

p. 297 Such an event occurred a little under 6 million years ago: *New Scientist*, 'Vanished', 7 Aug. 1999.

p. 298 Every day, the Gulf Stream carries an amount of thermal energy: *Annual Review of Marine Science*, 'Overturning in the North Atlantic', Jan. 2012.

p. 298 For that reason there tends to be a lag in the official, astronomical start of a season and the actual feeling that that season has started: Stevens, *The Change in the Weather*, p. 111.

p. 299 As for the question of how anyone could possibly figure out how long it takes a drop of water to get from one ocean to another: *National Geographic*, 'New Eyes on the Oceans', Oct. 2000, p. 101.

p. 301 Every year now, the amount of carbon dioxide we expel into the atmosphere globally goes up by about 40 billion metric tons: MIT Climate Portal, undated.

p. 301 imagine a cube 3 miles across on each side filled to the very top with viscous black gunk: *New Yorker*, 'The Climate Fixers', 14 May 2012, p. 100.

Chapter 18: The Bounding Main

p. 303 Imagine trying to live in a world dominated by dihydrogen oxide: Margulis and Sagan, *Microcosmos*, p. 100.

p. 303 A potato is 80 per cent water, a cow 74 per cent, a bacterium 75 per cent: Schopf, *Cradle of Life*, p. 107.

p. 303 Almost nothing about it can be used to make reliable predictions about the properties of other liquids: Green, *Water, Ice and Stone*, p. 29; Gribbin, *In the Beginning*, p. 174.

p. 304 By the time it is solid, it is almost a tenth more voluminous than it was before: Trefil, *Meditations at 10,000 Feet*, p. 121.

p. 304 'an utterly bizarre property': Gribbin, *In the Beginning*, p. 174.

p. 304 like the ever-changing partners in a quadrille: Kunzig, *The Restless Sea*, p. 8.

NOTES

p. 304 At any given moment only 15 per cent of them are actually touching: Dennis, *The Bird in the Waterfall*, p. 152.

p. 305 Within days, the lips vanish 'as if amputated, the gums blacken, the nose withers to half its length': *Economist*, 13 May 2000, p. 4.

p. 305 A typical litre of sea water will contain only about 2.5 teaspoons of common salt: Dennis, *The Bird in the Waterfall*, p. 248.

p. 305 we sweat and cry sea water: Margulis and Sagan, *Microcosmos*, p. 184.

p. 306 There are 1.3 billion cubic kilometres of water on Earth and that is all we're ever going to get: Green, *Water, Ice and Stone*, p. 25.

p. 306 By 3.8 billion years ago, the oceans had (at least more or less) achieved their present volumes: Ward and Brownlee, *Rare Earth*, p. 360.

p. 306 Altogether the Pacific holds just over half of all the ocean water: Dennis, *The Bird in the Waterfall*, p. 226.

p. 306 we would better call our planet not Earth but Water: Ball, H_2O, p. 21.

p. 306 Of the 3 per cent of Earth's water that is fresh: Dennis, *The Bird in the Waterfall*, p. 6; *Scientific American*, 'On Thin Ice', Dec. 2002, pp. 100–5.

p. 306 Go to the South Pole and you will be standing on over 2 miles of ice, at the North Pole just 15 feet of it: Smith, *The Weather*, p. 62.

p. 306 enough to raise the oceans by a height of 200 feet if it all melted: Schultz, *Ice Age Lost*, p. 75.

p. 307 'driven to distraction by the mind-numbing routine of years of dredging': Weinberg, *A Fish Caught in Time*, p. 34.

p. 307 But they sailed across almost 70,000 nautical miles of sea: Hamilton-Paterson, *The Great Deep*, p. 178.

p. 308 female assistants whose jobs were inventively described as 'historian and technicist' or 'assistant in fish problems': Norton, *Stars Beneath the Sea*, p. 57.

p. 308 Soon afterwards he teamed up with Barton, who came from an even wealthier family: Ballard, *The Eternal Darkness*, pp. 14–15.

NOTES

p. 309   The sphere had no manoeuvrability ... and only the most primitive breathing system: Weinberg, *A Fish Caught in Time*, p. 158; Ballard, *The Eternal Darkness*, p. 17.

p. 310   Whatever it was, nothing like it has been seen by anyone since: Weinberg, *A Fish Caught in Time*, p. 159.

p. 310   In 1958, they did a deal with the US Navy: Broad, *The Universe Below*, p. 54.

p. 312   couldn't take even a single photograph: *National Geographic*, 'Unsinkable Don Walsh', 6 Dec. 2023, and *Scientific American*, 'Diving Deeper Than Any Human Ever Dove', 1 April 2014.

p. 312   The deepest fish ever found so far: *Scientific American*, 'Deepest Fish Discovered More Than Five Miles below the Sea Surface', 7 April 2023.

p. 313   'We didn't learn a hell of a lot from it, other than that we could do it': quoted in *Underwater* magazine, 'The Deepest Spot on Earth', Winter 1999.

p. 313   There was just one problem: the designers couldn't find anyone willing to build it: Broad, *The Universe Below*, p. 56.

p. 313   It has undertaken 5,200 missions: Woods Hole Oceanographic Institution, 'History of *Alvin*', 29 Dec. 2023.

p. 313   tube worms over 3 metres long, clams 30 centimetres wide, shrimps and mussels in profusion: Attenborough, *The Living Planet*, p. 30.

p. 314   Before this it had been thought that no complex organisms could survive in water warmer than about 54 degrees Celsius: *National Geographic*, 'Deep Sea Vents', Oct. 2000, p. 123.

p. 314   enough to bury every bit of land on the planet to a depth of about 150 metres: Dennis, *The Bird in the Waterfall*, p. 248.

p. 315   it can take up to 10 million years to clean an ocean: Vogel, *Naked Earth*, p. 182.

p. 315   Such ventures don't come cheap: *Forbes*, 'Victor Vescovo, A Walking Bucket List of Adventure', 9 June 2022.

p. 315   Famously, when Vescovo looked out his porthole he saw a plastic bag: *MIT Technology Review*, 'The deepest-ever dive to the bottom of the Mariana Trench found litter there', 14 May 2019.

NOTES

p. 315 As recently as 2017, just 6 per cent of the Earth's ocean floors had been mapped: *Scientific American*, 1 Aug. 2022.

p. 315 Soon afterwards, Woods Hole announced that the craft would not be replaced: *Nature*, 'Ocean-diving robot will not be replaced', 10 Dec. 2015.

p. 316 we have better maps of Mars than of our own sea floors: USGS, 'Why we have better maps of Mars than of the seafloor', 17 Nov. 2023.

p. 316 perhaps nothing speaks more clearly of our psychological distance from the ocean depths: Engel, *The Sea*, p. 183.

p. 316 When they failed to sink, which was usually, navy gunners riddled them with bullets to let water in: Kunzig, *The Restless Sea*, pp. 294–305.

p. 316 dumped seventeen 'entire nuclear reactors into the Arctic Sea': *New York Review of Books*, 'Where Wonders Await Us', 10 Dec. 2007.

p. 317 a layer of water called the mesopelagic: *Economist*, 'Mapping the Mesopelagic', 15 April 2017.

p. 317 an estimated 10 billion tons of them globally: *Nature*, 'The life of diatoms in the world's oceans', 14 May 2009, and Roberts, *Ocean of Life*, p. 155.

p. 317 It is the greatest daily migration on the planet: *Scientific American*, 'Greatest Migration on Earth Happens under Darkness Every Day', 1 Aug. 2022.

p. 317 they may be the most numerous vertebrates on the planet: *Economist*, 15 April 2017, p. 73.

p. 317 As to what else is down there, we have only barely begun to take note: *Nature*, 'Log of life beneath the waves', 19 Nov. 2009.

p. 318 there could be as many as 30 million species of animals living in the sea, most still undiscovered: *Time*, 'Call of the Sea', 5 Oct. 1998, p. 60.

p. 318 Blue whales will sometimes break off a song, then pick it up again at exactly the same spot six months later: Sagan, *Cosmos*, p. 271, and NOAA Fisheries website, Blue Whale, 28 Sept. 2023.

p. 318 Consider . . . the fabled giant squid: London Natural History Museum website, 'Giant and colossal squid: revealing the secrets of the largest invertebrates', undated.

NOTES

p. 319 Even at a depth of nearly 5 kilometres, they found some 3,700 creatures: Kunzig, *The Restless Sea*, pp. 104–5.

p. 320 Altogether less than a tenth of the ocean is considered naturally productive: *Economist* survey, 'The Sea', 23 May 1998.

p. 320 it doesn't even make it into the top fifty among fishing nations: Flannery, *The Future Eaters*, p. 104.

p. 321 Many fishermen 'fin' sharks: *Audubon*, May–June 1998, p. 54.

p. 321 and haul behind them nets big enough to hold a dozen jumbo jets: *Time*, 'The Fish Crisis', 11 Aug. 1997, p. 66, and *Proceedings of the Royal Society: Biological Sciences*, 'Effects of bottom trawling on fish foraging and feeding', 22 Jan. 2015.

p. 321 'We're still in the Dark Ages. We just drop a net down and see what comes up': *Economist*, 'Pollock Overboard', 6 Jan. 1996, p. 22.

p. 321 Perhaps as much as 40 million tonnes of such unwanted marine life is dumped back in the sea each year, mostly in the form of corpses: *Marine Policy*, 4 July 2009.

p. 322 yet still allows 60,000 pounds of halibut to be legally landed every year: NOAA website, undated.

p. 322 sailors scooped them up in baskets: *National Geographic*, Oct. 1993, p. 18.

p. 322 By 1990 this had crashed to a scant 22,000 tonnes: *Economist* survey, 'The Sea', 23 May 1998, p. 8.

p. 322 'Fishermen . . . had caught them all': Kurlansky, *Cod*, p. 186.

p. 322 In the end, Kurlansky noted drily, 'fish' is 'whatever is left': Kurlansky, *Cod*, p. 138.

p. 322 In 2008, the Canadian Department of Fisheries and Oceans: *Science*, 'No Recovery for Atlantic Cod Population', 25 Nov. 2008.

p. 323 a vast and little-explored sprawl of abyssal plains in the Pacific known as the Clarion-Clipperton Zone: Pew Charitable Trusts fact sheet, Dec. 2017, and *Scientific American*, 'Deep-Sea Mining Could Begin Soon, Regulated or Not', 1 Sept. 2023.

p. 324 Altogether, 90 per cent of all creatures in the zone were not known to science before 2023: BBC Radio 4 programme *Inside Science*, 28 Dec. 2023, and Natural History Museum press release, 25 May 2023.

NOTES

p. 324   In a scathing editorial: *Nature*, 13 July 2023.

p. 325   proposed that the nodules engage in chemical processes that produce significant amounts of oxygen: *Nature Geoscience*, 22 July 2024, and Scottish Association for Marine Science press release, 23 July 2024, and *Science*, 'Claim of "dark oxygen" on sea floor faces doubts', 18 Sept. 2024.

p. 325   as many as 15 million of them may live on the pack ice around Antarctica: BBC *Horizon* transcript, 'Antarctica: The Ice Melts', p. 16.

Chapter 19: The Rise of Life

p. 326   Scientists are now pretty certain that the early atmosphere was nothing like as primed for development as Miller and Urey's gaseous stew: *University of Chicago News*, 'The origin of life on Earth, explained', 19 Sept. 2022.

p. 327   Some 500 amino acids occur naturally on Earth: *Chemistry World*, 'Why does life use the same 20 amino acids?', Sept. 2019.

p. 328   the odds against all 200 coming up in a prescribed sequence are 1 in 10^{260}: Crick, *Life Itself*, p. 51.

p. 328   Haemoglobin is only 146 amino acids long, a runt by protein standards: Sulston and Ferry, *The Common Thread*, p. 14.

p. 328   DNA is a whiz at replicating – it can make a copy of itself in seconds: Margulis and Sagan, *Microcosmos*, p. 63.

p. 329   'If everything needs everything else, how did the community of molecules ever arise in the first place?': Davies, *The Fifth Miracle*, p. 71.

p. 329   there must have been some kind of cumulative selection process that allowed amino acids to assemble: Dawkins, *The Blind Watchmaker*, p. 45.

p. 329   Lots of molecules in nature get together to form long chains called polymers: Dawkins, *The Blind Watchmaker*, p. 115.

p. 330   'an obligatory manifestation of matter': quoted in Nuland, *How We Live*, p. 121.

p. 330   If you wished to create another living object . . . you would need really only four principal elements: Schopf, *Cradle of Life*, p. 107.

NOTES

p. 330 As one leading biology text puts it, with perhaps just a tiny hint of discomfort: Wallace et al., *Biology*, p. 428.

p. 331 Well into the 1950s, it was thought that life was less than 600 million years old: Margulis and Sagan, *Microcosmos*, p. 71.

p. 331 'We can only infer from this rapidity that it is not "difficult" for life of bacterial grade to evolve': *New York Times*, 'Life on Mars? So What?', 11 Aug. 1996.

p. 331 'life, arising as soon as it could, was chemically destined to be': Gould, *Eight Little Piggies*, p. 328.

p. 331 full of noxious vapours from hydrochloric and sulphuric acids powerful enough to eat through clothing and blister skin: Ferris, *Seeing in the Dark*, p. 200.

p. 335 There are many claimed examples of fossils from Canada, Greenland and elsewhere that go back almost 4 billion years: *New York Times*, 'World's Oldest Fossils Found in Greenland', 31 Aug. 2016, and *Scientific American*, 'The Rise of the First Animals', 1 June 2019.

p. 335 little more than 'bags of chemicals': Drury, *Stepping Stones*, p. 68.

p. 335 you could pack a billion into the space occupied by a grain of sand: Brown, *The Energy of Life*, p. 101.

p. 336 'the most important and dramatic event in the history of life': quoted in *New York Review of Books*, 'A Bird's-Eye View of Evolution', 27 June 2002.

p. 337 Unfortunately, that is stubbornly at odds with the fossil record: *Trends in Ecology & Evolution*, 'Fossilisation processes and our reading of animal antiquity', Nov. 2023.

p. 339 sometimes called the Great Oxidation Event: Gee, *A (Very) Short History of Life on Earth*, p. 9.

p. 339 'Animals could not summon up the energy to work': Fortey, *Life*, p. 89.

p. 339 the same stuff you find today ... on garden ponds: Gee, *A (Very) Short History of Life on Earth*, p. 8.

p. 339 Our white blood cells actually use oxygen to kill invading bacteria: note provided by Dr Laurence Smaje.

NOTES

Chapter 20: Small World

p. 341 Louis Pasteur, the great French chemist and bacteriologist, became so preoccupied with his that he took to peering critically with a magnifying glass at every dish placed before him: Biddle, *A Field Guide to the Invisible*, p. 16.

p. 341 If you are in good health and averagely diligent about hygiene, you will have a herd of about 1 trillion bacteria grazing on your fleshy plains: Ashcroft, *Life at the Extremes*, p. 248; Sagan and Margulis, *Garden of Microbial Delights*, p. 4.

p. 341 Your digestive system alone is host to at least 400 types of microbes: Biddle, *A Field Guide to the Invisible*, p. 57.

p. 342 A surprising number ... have no detectable function at all: *National Geographic*, 'Bacteria', Aug. 1993, p. 51.

p. 342 Every human body consists of about 37.2 trillion cells, but is host to up to 50 trillion bacterial cells: Annals of Human Biology, 'An Estimation of the Number of Cells in the Human Body', Nov.-Dec. 2013.

p. 342 We couldn't survive a day without them: *New York Times*, 'From Birth, Our Body Houses a Microbe Zoo', 15 Oct. 1996, p. C-3.

p. 343 *Clostridium perfringens*, the disagreeable little organism that causes gangrene, can reproduce in nine minutes: *Outside*, July 1999, p. 88.

p. 343 At such a rate, a single bacterium could theoretically produce more offspring in two days than there are protons in the universe: Margulis and Sagan, *Microcosmos*, p. 75.

p. 343 'a single bacterial cell can generate 280,000 billion individuals in a single day': de Duve, *A Guided Tour of the Living Cell*, vol. 2, p. 320.

p. 343 Some bacteria break down chemical materials from which, as far as we can tell, they gain no benefit at all: *National Geographic*, 'Bacteria', Aug. 1993, p. 39.

p. 344 Scientists in Australia found microbes known as *Thiobacillus concretivorans*: Davies, *The Fifth Miracle*, p. 145.

NOTES

p. 344 We now know that there are a lot of microbes living deep within the Earth: *New York Times*, 'The Mysterious, Deep-Dwelling Microbes That Sculpt Our Planet', 24 June 2024, and *New York Review of Books*, 'How You Consist of Trillions of Tiny Machines', 9 July 2015.

p. 344 It has even been suggested that their tireless nibblings created the Earth's crust: *New York Times*, 'Bugs Shape Landscape, Make Gold', 15 Oct. 1996, p. C-1.

p. 344 if you took all the bacteria out of the Earth's interior and dumped them on the surface, they would cover the planet to a depth of 15 metres: *Discover*, 'To Hell and Back', July 1999, p. 82.

p. 344 The liveliest of them may divide no more than once a century: *Scientific American*, 'Microbes Deep Inside the Earth', Oct. 1996, p. 71.

p. 345 'The key to long life, it seems, is not to do too much': *Economist*, 'Earth's Hidden Life', 21 Dec. 1996, p. 112.

p. 345 Other microorganisms have leaped back to life after being released from a 118-year-old can of meat and a 166-year-old bottle of beer: *Nature*, 'A Case of Bacterial Immortality?', 19 Oct. 2000, p. 844.

p. 345 Bacteria were usually lumped in with plants, too: Sagan and Margulis, *Garden of Microbial Delights*, p. 22.

p. 347 In 1969, in an attempt to bring some order to the growing inadequacies of classification: Sagan and Margulis, *Garden of Microbial Delights*, p. 23.

p. 347 By one calculation it contained as many as 200,000 species of organism: Sagan and Margulis, *Garden of Microbial Delights*, p. 24.

p. 347 At that time, according to Woese, only about 500 species of bacteria were formally recorded: *New York Times*, 'Microbial Life's Steadfast Champion', 15 Oct. 1996, p. C-3.

p. 348 Only about 1 per cent will grow in culture: *Science*, 'Microbiologists Explore Life's Rich, Hidden Kingdoms', 21 March 1997, p. 1740.

p. 348 'like learning about animals from visiting zoos': *New York Times*, 'Microbial Life's Steadfast Champion', 15 Oct. 1996, p. C-7.

p. 349 Woese ... 'felt bitterly disappointed': Ashcroft, *Life at the Extremes*, pp. 274–5.

NOTES

p. 349 'Biology, like physics before it . . . has moved to a level where the objects of interest and their interactions often cannot be perceived through direct observation': *Proceedings of the National Academy of Sciences*, 'Default Taxonomy: Ernst Mayr's View of the Microbial World', 15 Sept. 1998.

p. 350 'Woese was not trained as a biologist and quite naturally does not have an extensive familiarity with the principles of classification': *Proceedings of the National Academy of Sciences*, 'Two Empires or Three?', 18 Aug. 1998.

p. 351 Of the twenty-three main divisions of life only three . . . are large enough to be seen by the human eye: Schopf, *Cradle of Life*, p. 106.

p. 351 if you totalled up all the biomass of the planet . . . microbes would account for at least 80 per cent of all there is: *New York Times*, 'Microbial Life's Steadfast Champion', 15 Oct. 1996, p. C-7.

p. 351 the most rampantly infectious organism on Earth, a bacterium called *Wolbachia*: *Science*, '*Wolbachia*: a tale of sex and survival', 11 May 2001, p. 1093.

p. 351 only about one microbe in 1,000 is a pathogen for humans: *National Geographic*, 'Bacteria', Aug. 1993, p. 39.

p. 351 microbes are still the number-three killer in the Western world: *Outside*, July 1999, p. 88.

p. 352 History . . . is full of diseases that 'once caused terrifying epidemics and then disappeared as mysteriously as they had come': Diamond, *Guns, Germs and Steel*, p. 208.

p. 353 a disease called necrotizing fasciitis, in which bacteria essentially eat the victim from the inside out: Gawande, *Complications*, p. 234.

p. 354 'The time has come to close the book on infectious diseases': *New Yorker*, 'No Profit, No Cure', 5 Nov. 2001, p. 46.

p. 354 Even as he spoke, however, some 90 per cent of those strains were in the process of developing immunity to penicillin: *Economist*, 'Disease Fights Back', 20 May 1995, p. 15.

p. 355 in 1997 a hospital in Tokyo reported the appearance of a strain that could resist even that: *Boston Globe*, 'Microbe Is Feared to Be Winning Battle Against Antibiotics', 30 May 1997, p. A-7.

NOTES

p. 355   As James Surowiecki noted: *New Yorker*, 'No Profit, No Cure', 5 Nov. 2001, p. 46.

p. 355   America's National Institutes of Health ... didn't officially endorse the idea until 1994: *Economist*, 'Bugged by Disease', 21 March 1998, p. 93.

p. 355   'Hundreds, even thousands of people must have died from ulcers who wouldn't have': *Forbes*, 'Do Germs Cause Cancer?', 15 Nov. 1999, p. 195.

p. 355   Since then, further research has shown that there is or may well be a bacterial component in all kinds of other disorders: *Science*, 'Do Chronic Diseases Have an Infectious Root?', 14 Sept. 2001, pp. 1974–6.

p. 355   'a piece of nucleic acid surrounded by bad news': quoted in Oldstone, *Viruses, Plagues and History*, p. 8.

p. 356   Virus numbers are constantly changing as new species emerge: *Virology*, 'Human viruses: an ever-increasing list', March 2025.

p. 356   Smallpox in the twentieth century alone killed an estimated 300 million people: Oldstone, *Viruses, Plagues and History*, p. 1.

p. 356   In ten years the disease killed some 5 million people and then quietly went away: Kolata, *Flu*, p. 292.

p. 357   The First World War killed 21 million people in four years; swine flu did the same in its first four months: *American Heritage*, 'The Great Swine Flu Epidemic of 1918', June 1976, p. 82.

p. 357   In an attempt to devise a vaccine, medical authorities conducted experiments on volunteers at a military prison on Deer Island in Boston Harbor: *American Heritage*, 'The Great Swine Flu Epidemic of 1918', June 1976, p. 82.

p. 359   In 1969, a doctor at a Yale University lab in New Haven, Connecticut, who was studying Lassa fever: Oldstone, *Viruses, Plagues and History*, p. 126.

p. 359   In 1990, a Nigerian living in Chicago was exposed to Lassa fever on a visit to his homeland: Oldstone, *Viruses, Plagues and History*, p. 128.

NOTES

Chapter 21: Life Goes On

p. 361 The fate of nearly all living organisms: Schopf, *Cradle of Life*, p. 72.

p. 361 less than one species in 10,000 has made it into the fossil record: Trefil, *101 Things You Don't Know About Science and No One Else Does Either*, p. 280.

p. 362 statement ... that there are 250,000 species of creature in the fossil record: Leakey and Lewin, *The Sixth Extinction*, p. 45.

p. 362 About 95 per cent of all the fossils we possess are of animals that once lived under water: Leakey and Lewin, *The Sixth Extinction*, p. 45.

p. 362 'It seems like a big number,' he agreed: interview with Richard Fortey, Natural History Museum, London, 19 Feb. 2001.

p. 363 Humans ... have survived so far for one-half of 1 per cent as long: Fortey, *Trilobite!*, p. 24.

p. 363 'a whole *Profallotaspis* or *Elenellus* as big as a crab': Fortey, *Trilobite!*, p. 121.

p. 364 and built up a collection of sufficient distinction that it was bought by Louis Agassiz: 'From Farmer-Laborer to Famous Leader: Charles D. Walcott (1850–1927)', *GSA Today*, Jan. 1996.

p. 364 In 1879 Walcott took a job as a field researcher: Gould, *Wonderful Life*, pp. 242–3.

p. 364 'His books fill a library shelf': Fortey, *Trilobite!*, p. 53.

p. 365 'our sole vista upon the inception of modern life': Gould, *Wonderful Life*, p. 56.

p. 365 Gould, ever scrupulous, discovered: Gould, *Wonderful Life*, p. 71.

p. 366 140 species, by one count: Leakey and Lewin, *The Sixth Extinction*, p. 27.

p. 366 'a range of disparity ... never again equaled': Gould, *Wonderful Life*, p. 208.

p. 366 'Under such an interpretation,' Gould sighed: Gould, *Eight Little Piggies*, p. 225.

p. 366 Then, in 1973, a graduate student from Cambridge: *National Geographic*, 'Explosion of Life', Oct. 1993, p. 126.

NOTES

p. 367 There was so much unrecognized novelty . . . that at one point: Fortey, *Trilobite!*, p. 123.

p. 367 they all use architecture first created in the Cambrian party: *US News and World Report*, 'How Do Genes Switch On?', 18–25 Aug. 1997, p. 74.

p. 367 at least fifteen and perhaps as many as twenty: Gould, *Wonderful Life*, p. 25.

p. 368 'Wind back the tape of life': Gould, *Wonderful Life*, p. 14.

p. 368 In 1946 Sprigg, then just an assistant government geologist: Corfield, *Architects of Eternity*, p. 287.

p. 369 but it failed to find favour with the association's head: Corfield, *Architects of Eternity*, p. 287.

p. 369 Nine years later, in 1957: Fortey, *Life*, p. 85.

p. 371 'If only Stephen Gould could think as clearly as he writes!': Dawkins, *Sunday Telegraph*, 25 Feb. 1990.

p. 371 who startled many in the palaeontological community by rounding abruptly on Gould in a book of his own, *The Crucible of Creation*: *New York Times Book Review*, 'Rock of Ages', 10 May 1998, p. 15.

p. 371 'I have never encountered such spleen in a book by a professional': Fortey, *Trilobite!*, p. 138.

p. 371 Fortey gave as an example the idea of comparing a shrew and an elephant: Fortey, *Trilobite!*, p. 132.

p. 372 'None was as strange as a present day barnacle': Fortey, *Life*, p. 111.

p. 372 'no less interesting, or odd, just more explicable': Fortey, 'Shock Lobsters', *London Review of Books*, 1 Oct. 1998.

p. 372 In fact, it turns out the Cambrian explosion was not quite so explosive as once supposed: *Nature*, 'What sparked the Cambrian Explosion?', 18 Feb. 2016.

p. 373 It is one thing to have one well-formed creature like a trilobite burst forth in isolation: Fortey, *Trilobite!*, p. 137.

Chapter 22: Goodbye to All That

p. 375 In areas of Antarctica where virtually nothing else will grow: Attenborough, *The Living Planet*, p. 48.

NOTES

p. 375 'Spontaneously, inorganic stone becomes living plant!': Marshall, *Mosses and Lichens*, p. 22.

p. 376 The world has more than 20,000 species of lichens: Attenborough, *The Private Life of Plants*, p. 214.

p. 376 Those the size of dinner plates ... are therefore 'likely to be hundreds if not thousands of years old': Attenborough, *The Living Planet*, p. 42.

p. 376 if you imagine the 4,500 million years of Earth's history compressed into a normal earthly day: adapted from Schopf, *Cradle of Life*, p. 13.

p. 377 stretch your arms to their fullest extent and imagine that width as the entire history of the Earth: McPhee, *Basin and Range*, p. 126.

p. 379 Oxygen levels ... were as high as 35 per cent: Officer and Page, *Tales of the Earth*, p. 123.

p. 379 the isotopes accumulate at different rates depending on how much oxygen or carbon dioxide is in the atmosphere: Officer and Page, *Tales of the Earth*, p. 118.

p. 380 'The U.S. Air Force ... has put them in wind tunnels to see how they do it, and despaired': Conniff, *Spineless Wonders*, p. 84.

p. 380 In Carboniferous forests dragonflies grew as big as ravens: Fortey, *Life*, p. 201.

p. 381 Luckily the team found just such a creature: BBC *Horizon*, 'The Missing Link', first broadcast 1 Feb. 2001.

p. 382 The names simply refer to the number and location of small holes found in the sides of their owners' skulls: Tudge, *The Variety of Life*, p. 411.

p. 383 but the number has been put as high as 4,000 billion: Tudge, *The Variety of Life*, p. 9.

p. 383 'To a first approximation ... all species are extinct': quoted by Gould, *Eight Little Piggies*, p. 46.

p. 383 the average lifespan of a species is only about 4 million years: Leakey and Lewin, *The Sixth Extinction*, p. 38.

p. 383 'The alternative to extinction is stagnation': interview with Ian Tattersall, American Museum of Natural History, New York, 6 May 2002.

NOTES

p. 383 Crises in the Earth's history are invariably associated with dramatic leaps afterwards: Stanley, *Extinction*, p. 95; Stevens, *The Change in the Weather*, p. 12.

p. 384 In the Permian, at least 95 per cent of animals known from the fossil record checked out, never to return: *Harper's*, 'Planet of Weeds', Oct. 1998, p. 58.

p. 384 Even about a third of insect species went – the only occasion on which they were lost en masse: Stevens, *The Change in the Weather*, p. 12.

p. 384 'It was, truly, a mass extinction': Fortey, *Life*, p. 235.

p. 384 Estimates for the number of animal species alive at the end of the Permian range from as low as 45,000 to as high as 240,000: Gould, *Hen's Teeth and Horse's Toes*, p. 340.

p. 385 For individuals the death toll could be much higher – in many cases, practically total: Powell, *Night Comes to the Cretaceous*, p. 143.

p. 385 Grazing animals, including horses, were nearly wiped out in the Hemphillian event: Flannery, *The Eternal Frontier*, p. 100.

p. 385 At least two dozen potential culprits have been identified as causes or prime contributors: *Earth*, 'The Mystery of Selective Extinctions', Oct. 1996, p. 12.

p. 386 'tons of conjecture and very little evidence': *New Scientist*, 'Meltdown', 7 Aug. 1999.

p. 386 Such an outburst is not easily imagined: Powell, *Night Comes to the Cretaceous*, p. 19.

p. 387 The KT meteor had the additional advantage – advantage if you are a mammal, that is: Flannery, *The Eternal Frontier*, p. 17.

p. 387 'Why should these delicate creatures have emerged unscathed from such an unparalleled disaster?': Flannery, *The Eternal Frontier*, p. 43.

p. 387 In the seas it was much the same story: Gould, *Eight Little Piggies*, p. 304.

p. 387 'Somehow it does not seem satisfying just to call them "lucky ones" and leave it at that': Fortey, *Life*, p. 292.

p. 388 the period immediately after the dinosaur extinction could well be known as the Age of Turtles: Flannery, *The Eternal Frontier*, p. 39.

NOTES

p. 389   'Evolution may abhor a vacuum . . . but it often takes a long time to fill it': Stanley, *Extinction*, p. 92.

p. 389   For perhaps as many as 10 million years mammals remained cautiously small: Novacek, *Time Traveler*, p. 112.

p. 389   For a time, there were guinea pigs the size of rhinos and rhinos the size of a two-storey house: Dawkins, *The Blind Watchmaker*, p. 102.

p. 389   For millions of years, a gigantic, flightless, carnivorous bird called *Titanis* was possibly the most ferocious creature in North America: Flannery, *The Eternal Frontier*, p. 138.

p. 390   Until very recently, everything we know about the dinosaurs of this period came from only about 300 specimens: Powell, *Night Comes to the Cretaceous*, pp. 168–9.

p. 391   'Humans are here today because our particular line never fractured': Gould, *Eight Little Piggies*, p. 229.

Chapter 23: The Richness of Being

p. 393   The spirit room alone holds 15 miles of shelving: Thackray and Press, *The Natural History Museum*, p. 90.

p. 394   forty-four years after the expedition had concluded: Thackray and Press, *The Natural History Museum*, p. 74.

p. 394   published in 1956 and still to be found on some library shelves as almost the only attempt: Conard, *How to Know the Mosses and Liverworts*, p. 5.

p. 398   he endeavoured, unsuccessfully, to sneak a mistress onboard: *London Review of Books*, 'Keep him as a curiosity', 13 Aug. 2020.

p. 399   he sifted through a bale of fodder sent for the ship's livestock and made new discoveries: Barber, *The Heyday of Natural History: 1820–1870*, p. 17.

p. 401   According to Linnaeus's biographer Gunnar Broberg, as many as one in three of them never returned: *London Review of Books*, 'Unicorn or Narwhal?,' 22 Feb. 2024.

p. 401   To the parts of one species of clam he gave the names: Gould, *Leonardo's Mountain of Clams and the Diet of Worms*, p. 79.

NOTES

p. 401 'Love comes even to the plants. Males and females ... hold their nuptials': quoted by Gjertsen, *The Classics of Science*, p. 237, and at University of California/UCMP Berkeley website.

p. 401 Linnaeus lopped it back to *Physalis angulata*: Kastner, *A Species of Eternity*, p. 31.

p. 402 The first edition of his great *Systema Naturae*: Gjertsen, *The Classics of Science*, p. 223.

p. 402 John Ray's three-volume *Historia Generalis Plantarum* in England: Durant, *The Age of Louis XIV*, p. 519.

p. 402 just in time to make Linnaeus a kind of father figure to British naturalists: Thomas, *Man and the Natural World*, p. 65.

p. 403 gullibly accepted from seamen and other imaginative travellers, Schwartz, *Sudden Origins*, p. 59.

p. 403 he saw that whales belonged with cows, mice and other common terrestrial animals in the order quadrupedia (later changed to mammalia): Schwartz, *Sudden Origins*, p. 59.

p. 403 other names in everyday use included *mare's fart, naked ladies, twitch-ballock, hound's piss, open arse*, and *bum-towel*: Thomas, *Man and the Natural World*, pp. 82–5, and *Economist*, 'Namely offensive', 24 Sept. 2022.

p. 405 while Edward O. Wilson in *The Diversity of Life* puts the number at a surprisingly robust eighty-nine: Wilson, *The Diversity of Life*, p. 157.

p. 406 scientists have never been able to agree on what exactly a species is: *Economist*, 'On the origin of "species"', 28 Aug. 2021.

p. 406 A family of freshwater shrimp called *Gammarus fossarum* comprises thirty-two separate species or up to 152: *New York Times*, 'What Is a Species Anyway?', 19 Feb. 2024.

p. 407 were transferred, amid howls, to the genus *Pelargonium*: Elliott, *The Potting-Shed Papers*, p. 18.

p. 407 Estimates range from 3 million to 200 million: *Audubon*, 'Earth's Catalogue', Jan.-Feb. 2002; Wilson, *The Diversity of Life*, p. 132.

p. 408 he estimated the number of known species of all types – plants, insects, microbes, algae, everything – at 1.4 million: Wilson, *The Diversity of Life*, p. 133.

p. 408   Other authorities have put the number of known species slightly higher, at around 1.5 million to 1.8 million: *US News and World Report*, 18 Aug. 1997, p. 78.

p. 408   Its survey in 2023 concluded that there are 450,000 species of plant in the world: Kew Gardens publication, *State of the World's Plants and Fungi, 2023*.

p. 409   It put the number of known and catalogued species at 1.2 million: *PLoS Biology*, 'How Many Species Are There on Earth and in the Ocean?', 23 Aug. 2011.

p. 409   It took Groves four decades to untangle everything: *New Scientist*, 'Monkey Puzzle', 6 Oct. 2001, p. 54.

p. 409   estimated the global population of individual insects at 10 quintillion: *American Naturalist*, 'The Range and Pattern of Insect Abundance', March-April 1960.

p. 410   'It's not a biodiversity crisis, it's a taxonomist crisis!': interview with Koen Maes, National Museum, Nairobi, 2 Oct. 2002.

p. 410   launched an enterprise called the All Species Foundation: *The Times*, 'The List of Life on Earth', 30 July 2001.

p. 411   your mattress is home to perhaps 2 million microscopic mites: Bodanis, *The Secret House*, p. 16.

p. 411   to quote the man who did the measuring, Dr John Maunder of the British Medical Entomology Centre: *New Scientist*, 'Bugs Bite Back', 17 Feb. 2001, p. 48.

p. 411   These mites have been with us since time immemorial: Bodanis, *The Secret House*, p. 15.

p. 412   Your sample will also contain perhaps a million plump yeasts: *National Geographic*, 'Bacteria', Aug. 1993, p. 39.

p. 412   'If over 9,000 microbial types exist in two pinches of substrate from two localities in Norway': Wilson, *The Diversity of Life*, p. 144.

p. 412   and discovered 1,000 new species of flowering plant: Wilson, *The Diversity of Life*, p. 197.

p. 413   Overall, tropical rainforests cover only about 6 per cent of Earth's surface: Wilson, *The Diversity of Life*, p. 197.

NOTES

p. 413 'over three and a half billion years of evolution': *Economist*, 'Biotech's Secret Garden', 30 May 1998, p. 75.

p. 413 one ancient bacterium was found on the wall of a country pub: Fortey, *Life*, p. 75.

p. 414 about 450 species have been identified: Ridley, *The Red Queen*, p. 54.

p. 414 Gather together all the fungi found in a typical hectare of meadowland: Attenborough, *The Private Life of Plants*, p. 177.

p. 415 'We are a university with limited resources': *New York Times*, 'Duke Shuts Down Huge Plant Collection, Causing Scientific Uproar', 22 Feb. 2024.

p. 415 the horse was considered a rarity in the wider world: *New York Times*, 'A Stone-Age Horse Still Roams a Tibetan Plateau', 12 Nov. 1995.

p. 415 The large, flightless New Zealand bird called the takahe: Flannery and Schouten, *A Gap in Nature*, p. 2.

p. 416 animals that have been long feared extinct are sometimes rediscovered: *Scientific American*, 'Which Lost Species May be Found Again? Huge Study Reveals Clues', 17 Jan. 2024, and *New York Times*, 'Scientists Made a List of Lost Birds and Now They Want Us to Find Them', 23 Aug. 2024.

p. 416 A conservationist named Gonzalo Cardona nursed the birds back to a healthy population of nearly 3,000: *New York Times*, 23 Aug. 2024.

p. 417 A single line of text in a Crampton table: Gould, *Eight Little Piggies*, pp. 32–4.

p. 417 he hiked 4,000 kilometres to assemble a collection of 300,000 thousand wasps: Gould, *The Flamingo's Smile*, pp. 159–60.

Chapter 24: Cells

p. 420 you would have to miniaturize about the same number of components as are found in a Boeing 777 jetliner: *New Scientist*, 2 Dec. 2000, p. 37.

p. 420 we understand what no more than about 2 per cent of them do: Brown, *The Energy of Life*, p. 83.

NOTES

p. 421   Its purpose was at first a mystery, but then scientists began to find it all over the place: Brown, *The Energy of Life*, p. 229.

p. 421   It is converted into nitric oxide in the bloodstream, relaxing the muscle linings of vessels, allowing blood to flow more freely: Alberts, et al., *Essential Cell Biology*, p. 489.

p. 421   You possess 'some few hundred' different types of cell: de Duve, *A Guided Tour of the Living Cell*, vol. 1, p. 21, and *New York Review of Books*, 'How You Consist of Trillions of Tiny Machines', 9 July 2015.

p. 421   If you are an average-sized adult you are lugging around over 2 kilograms of dead skin: Bodanis, *The Secret Family*, p. 106.

p. 422   Liver cells can survive for years: de Duve, *A Guided Tour of the Living Cell*, vol. 1, p. 68.

p. 422   not so much as a stray molecule: Bodanis, *The Secret Family*, p. 81.

p. 422   spring-loaded boots that he said would allow the wearer to bound across the landscape with 12-foot strides: *London Review of Books*, 'Rough Trade', 6 March 2003.

p. 423   Hooke calculated that a one-inch square of cork would contain 1,259,712,000 of these tiny chambers: Nuland, *How We Live*, p. 100.

p. 424   Remarkably all were subsequently lost: *Federation of European Microscopical Societies Journal*, 6 April 2015.

p. 424   a Fellow of the society named Sir Everard Horne (1758–1832) borrowed them for private study: *Oxford Dictionary of National Biography*.

p. 424   the remaining two Leeuwenhoek originals similarly were borrowed: *Oxford Dictionary of National Biography*.

p. 425   After Leeuwenhoek reported finding 'animalcules' in a sample of pepper-water in 1676: Jardine, *Ingenious Pursuits*, p. 93.

p. 425   He calculated that there were 8,280,000 of these tiny beings in a single drop of water: Thomas, *Man and the Natural World*, p. 167.

p. 425   He called the little beings 'homunculi': Schwartz, *Sudden Origins*, p. 167.

p. 425   In one of his least successful experiments: Carey (ed.), *The Faber Book of Science*, p. 28.

p. 426   Only in 1839, however, did anyone realize that *all* living matter is cellular: Nuland, *How We Live*, p. 101.

NOTES

p. 426 The cell has been compared to many things: Trefil, *101 Things You Don't Know About Science and No One Else Does Either*, p. 33; Brown, *The Energy of Life*, p. 78.

p. 426 However, scale that up and it would translate as a jolt of 20 million volts per metre: Brown, *The Energy of Life*, p. 87.

p. 427 has the approximate consistency 'of a light grade of machine oil': Nuland, *How We Live*, p. 103.

p. 427 and flying into each other up to a billion times a second: Brown, *The Energy of Life*, p. 80.

p. 428 'the molecular world must necessarily remain entirely beyond the powers of our imagination': de Duve, *A Guided Tour of the Living Cell*, vol. 2, p. 293.

p. 428 'the total is still a very minimum of 100 million protein molecules in each cell': Nuland, *How We Live*, p. 157.

p. 429 At any given moment, a typical cell in your body will have about one billion ATP molecules in it: Alberts et al., *Essential Cell Biology*, p. 110.

p. 429 Every day you produce and use up a volume of ATP equivalent to about half your body weight: *Nature*, 'Darwin's Motors', 2 May 2002, p. 25.

p. 429 On average, humans suffer one fatal malignancy for each 100 million billion cell divisions: Ridley, *Genome*, p. 237.

p. 431 what has been called 'the single best idea that anyone has ever had': Dennett, *Darwin's Dangerous Idea*, p. 21.

Chapter 25: Darwin's Singular Notion

p. 432 'Everyone is interested in pigeons': quoted in Boorstin, *Cleopatra's Nose*, p. 176.

p. 433 'You care for nothing but shooting, dogs, and rat-catching': quoted in Boorstin, *The Discoverers*, p. 467.

p. 433 The experience of witnessing an operation on an understandably distressed child: Desmond and Moore, *Darwin*, p. 27.

p. 434 some 'bordering on insanity': Hamblyn, *The Invention of Clouds*, p. 199.

NOTES

p. 434   In five years . . . he had not once hinted at an attachment: Desmond and Moore, *Darwin*, p. 197.

p. 434   which suggested, not incidentally, that atolls could not form in less than a million years: Moorehead, *Darwin and the Beagle*, p. 191.

p. 435   It wasn't until the younger Darwin was back in England and read Thomas Malthus's *Essay on the Principle of Population* : Gould, *Ever Since Darwin*, p. 21.

p. 435   'How stupid of me not to have thought of it!': quoted in *Sunday Telegraph*, 'The Origin of Darwin's Genius', 8 Dec. 2002.

p. 436   It was his friend the ornithologist John Gould: Desmond and Moore, *Darwin*, p. 209.

p. 436   These he expanded into a 230-page 'sketch': *Dictionary of National Biography*, vol. 5, p. 526.

p. 436   'I hate a barnacle as no man ever did before': quoted in Ferris, *Coming of Age in the Milky Way*, p. 239.

p. 437   Some wondered if Darwin himself might be the author: Barber, *The Heyday of Natural History*, p. 214.

p. 438   'he could not have made a better short abstract': *Dictionary of National Biography*, vol. 5, p. 528.

p. 438   'This summer will make the 20th year (!) since I opened my first note-book': Desmond and Moore, *Darwin*, pp. 454–5.

p. 439   'whatever it may amount to, will be smashed': Desmond and Moore, *Darwin*, p. 469.

p. 439   'all that was new in them was false, and what was true was old': quoted by Gribbin and Cherfas, *The First Chimpanzee*, p. 150.

p. 439   Much less amenable to Darwin's claim of priority was a Scottish gardener named Patrick Matthew: Gould, *The Flamingo's Smile*, p. 336.

p. 440   He referred to himself as 'the Devil's Chaplain': Cadbury, *Terrible Lizard*, p. 305.

p. 440   felt 'like confessing a murder': quoted in Desmond and Moore, *Darwin*, p. xvi.

p. 441   'The case at present must remain inexplicable': quoted by Gould, *Wonderful Life*, p. 57.

p. 441   By way of explanation he speculated: Gould, *Ever Since Darwin*, p. 126.

NOTES

p. 441 'Darwin goes too far': quoted by McPhee, *In Suspect Terrain*, p. 190.

p. 441 Huxley was a saltationist: Schwartz, *Sudden Origins*, pp. 81–2.

p. 442 'The eye to this day gives me a cold shudder': quoted in Keller, *The Century of the Gene*, p. 97.

p. 442 it 'seems, I freely confess, absurd in the highest possible degree' that natural selection could produce such an instrument in gradual steps: Darwin, *On the Origin of Species* (facsimile ed.), p. 217.

p. 442 'Eventually . . . Darwin lost virtually all the support that still remained': Schwartz, *Sudden Origins*, p. 89.

p. 443 It had a library of 20,000 books: Lewontin, *It Ain't Necessarily So*, p. 91.

p. 445 And Darwin, for his part, is known to have studied Focke's influential paper: Ridley, *Genome*, p. 44.

p. 445 Huxley had been urged to attend by Robert Chambers: Trinkaus and Shipman, *The Neandertals*, p. 79.

p. 446 bravely slogged his way through two hours of introductory remarks: Clark, *The Survival of Charles Darwin*, p. 142.

p. 447 One of his experiments was to play the piano to them: Conniff, *Spineless Wonders*, p. 147.

p. 447 Having married his own cousin: Desmond and Moore, *Darwin*, p. 575.

p. 447 Darwin was often honoured in his lifetime, but never for *On the Origin of Species*: Clark, *The Survival of Charles Darwin*, p. 148.

p. 448 Darwin's theory didn't really gain widespread acceptance until the 1930s and 1940s: Tattersall and Schwartz, *Extinct Humans*, p. 45.

p. 448 seemed set to claim Mendel's insights as his own: Schwartz, *Sudden Origins*, p. 187.

Chapter 26: The Stuff of Life

p. 450 if you go back just 7,000 years, you reach a point of convergence where all our genealogical lines intersect: *New York Review of Books*, 'Our Twisted DNA', 7 March 2019.

p. 450 'roughly one nucleotide base in every thousand': Sulston and Ferry, *The Common Thread*, p. 198.

NOTES

p. 451 The exceptions are red blood cells, some immune system cells, and egg and sperm cells: Woolfson, *Life without Genes*, p. 12.

p. 451 'guaranteed to be unique against all conceivable odds': de Duve, *A Guided Tour of the Living Cell*, vol. 2, p. 314.

p. 451 'among the most nonreactive, chemically inert molecules': Lewontin, *It Ain't Necessarily So*, p. 142.

p. 452 It was discovered as far back as 1869: Ridley, *Genome*, p. 48.

p. 452 DNA didn't do anything at all, as far as anyone could tell: Wallace et al., *Biology*, p. 211.

p. 452 The necessary complexity, it was thought, had to exist in proteins in the nucleus: de Duve, *A Guided Tour of the Living Cell*, vol. 2, p. 295.

p. 454 Working out of a small lab (which became known, inevitably, as the Fly Room): Clark, *The Survival of Charles Darwin*, p. 259.

p. 455 no consensus 'as to what the genes are – whether they are real or purely fictitious': Keller, *The Century of the Gene*, p. 2.

p. 455 we are in much the same position today in respect of mental processes such as thought and memory: Wallace et al., *Biology*, p. 211.

p. 455 Chargaff . . . suggested that Avery's discovery was worth two Nobel Prizes: Maddox, *Rosalind Franklin*, p. 327.

p. 455 including, it has been said, lobbying the authorities . . . not to give Avery a Nobel Prize: White, *Rivals*, p. 251.

p. 456 a member of a highly popular radio programme called *The Quiz Kids*: Judson, *The Eighth Day of Creation*, p. 46.

p. 457 'it was my hope that the gene might be solved without my learning any chemistry': Watson, *The Double Helix*, p. 21.

p. 457 the results of which were obtained 'fortuitously': Jardine, *Ingenious Pursuits*, p. 356.

p. 457 In a severely unflattering portrait: Watson, *The Double Helix*, p. 17.

p. 458 'gratuitously hurtful': Jardine, *Ingenious Pursuits*, p. 354.

p. 459 To Wilkins's presumed dismay and embarrassment, in the summer of 1952 she posted a mock notice: White, *Rivals*, p. 257; Maddox, *Rosalind Franklin*, p. 185.

p. 459 'apparently without her knowledge or consent': PBS website, 'A Science Odyssey', n.d.

NOTES

p. 459   Years later, Watson conceded that it 'was the key event . . . it mobilized us': quoted in Maddox, *Rosalind Franklin*, p. 317.

p. 460   a 900-word article by Watson and Crick titled 'A Structure for Deoxyribose Nucleic Acid': de Duve, *A Guided Tour of the Living Cell*, vol. 2, p. 290.

p. 460   It received a small mention in the *News Chronicle* and was ignored elsewhere: Ridley, *Genome*, p. 50.

p. 460   Franklin rarely wore a lead apron and often stepped carelessly in front of a beam: Maddox, *Rosalind Franklin*, p. 144.

p. 460   'It took over twenty-five years for our model of DNA to go from being only rather plausible, to being very plausible': Crick, *What Mad Pursuit*, pp. 73–4.

p. 461   by 1968 the journal *Science* could run an article entitled 'That Was the Molecular Biology That Was': Keller, *The Century of the Gene*, p. 25.

p. 461   A good 98 per cent of your DNA consists of nothing but long stretches of seemingly meaningless garble: *Scientific American*, 'Mapping the "Unknome" May Reveal Critical Genes Scientists Have Ignored', 1 Nov. 2023, and *Scientific American*, 'What Makes Us Different?', 1 Nov. 2012.

p. 461   In this sense, they are rather like the keys of a piano, each playing a single note and nothing else: *National Geographic*, 'Secrets of the Gene', Oct. 1995, p. 55.

p. 461   Guanine, for instance, is the same stuff that abounds in, and gives its name to, guano: Pollack, *Signs of Life*, pp. 22–3.

p. 463   The 0.1 per cent difference between your genes and mine is accounted for by our SNPs: *Methods of Molecular Biology*, 'SNPs: impact on gene function and phenotype', 2009.

p. 463   'you could say all humans share nothing, and that would be correct, too': *Discover*, 'Bad Genes, Good Drugs', April 2002, p. 54.

p. 464   'exist for the pure and simple reason that they are good at getting themselves duplicated': Ridley, *Genome*, p. 127.

p. 464   'Empires fall, ids explode, great symphonies are written, and behind all of it is a single instinct that demands satisfaction': Nuland, *How We Live*, p. 158.

NOTES

p. 464   Non-coding DNA is employed in DNA fingerprinting: *National Geographic*, 'The New Science of Identity', May 1992, p. 118.

p. 465   Here were two creatures that hadn't shared a common ancestor for 500 million years: BBC *Horizon*, 'Hopeful Monsters', first transmitted 1998.

p. 465   At least 90 per cent correlate at some level with those found in mice: *Nature*, 'Sorry, dogs – man's got a new best friend', 19–26 Dec. 2002, p. 734.

p. 465   We even have the same genes for making a tail, if only they would switch on: *Los Angeles Times* (reprinted in *Valley News*), 9 Dec. 2002.

p. 465   dubbed homeotic (from a Greek word meaning 'similar') or hox genes: BBC *Horizon*, 'Hopeful Monsters', first transmitted 1998.

p. 465   We have forty-six chromosomes, but some ferns have more than 600: Gribbin and Cherfas, *The First Chimpanzee*, p. 53.

p. 465   The lungfish, one of the least evolved of all complex animals, has forty times as much DNA as we have: Schopf, *Cradle of Life*, p. 240.

p. 466   Initially it was thought that humans had at least 100,000 genes: *Chemistry World*, 'A splice of life', May 2020.

p. 466   Perhaps the apogee (or nadir) of this faith in biodeterminism was a study published in the journal *Science* in 1980: Lewontin, *It Ain't Necessarily So*, p. 215.

p. 466   retinitis pigmentosa, a common cause of blindness, has been associated with 3,000 rogue mutations involving seventy different genes: *New York Review of Books*, 'Seeing the Power in Blindness', 18 April 2024.

p. 467   How fast a man's beard grows ... is partly a function of how much he thinks about sex: *Wall Street Journal*, 'What Distinguishes Us from the Chimps? Actually, Not Much', 12 April 2002, p. 1.

p. 468   Depending on mood and metabolic circumstance, they will allow themselves to be phosphorylated, glycosylated, acetylated, ubiquitinated: *The Bulletin*, 'The Human Enigma Code', 21 Aug. 2001, p. 32.

p. 468   Drink a glass of wine ... and you materially alter the number and types of proteins at large in your system: *Scientific American*, 'Move Over, Human Genome', April 2002, pp. 44–5.

NOTES

p. 468 'Anything that is true of *E. coli* must be true of elephants, except more so': *Nature*, 'From *E. coli* to Elephants', 2 May 2002, p. 22.

Chapter 27: Ice Time

p. 471 In London, *The Times* ran a small story: Williams, *Surviving Galeras*, p. 198.

p. 472 Spring never came and summer never warmed: Officer and Page, *Tales of the Earth*, pp. 3–6.

p. 472 One French naturalist named de Luc: Hallam, *Great Geological Controversies*, p. 89.

p. 473 and the other abundant clues that point to passing ice sheets: Hallam, *Great Geological Controversies*, p. 90.

p. 473 The naturalist Jean de Charpentier told the story: Hallam, *Great Geological Controversies*, p. 90.

p. 474 He lent Agassiz his notes: Hallam, *Great Geological Controversies*, pp. 92–3.

p. 474 Humboldt . . . observed that there are three stages in scientific discovery: Ferris, *The Whole Shebang*, p. 173.

p. 474 In his quest to understand the dynamics of glaciation, he went everywhere: McPhee, *In Suspect Terrain*, p. 182.

p. 474 William Hopkins, a Cambridge professor and leading member of the Geological Society: Hallam, *Great Geological Controversies*, p. 98.

p. 476 He began to find evidence for glaciers practically everywhere: Hallam, *Great Geological Controversies*, p. 99.

p. 476 Eventually he became convinced that ice had once covered the whole Earth: Gould, *Time's Arrow*, p. 115.

p. 477 When he died in 1873 Harvard felt it necessary to appoint three professors to take his place: McPhee, *In Suspect Terrain*, p. 197.

p. 477 Less than a decade after his death: McPhee, *In Suspect Terrain*, p. 197.

p. 478 For the next twenty years, even while on holiday: Gribbin and Gribbin, *Ice Age*, p. 51.

p. 478 The cause of ice ages, Köppen decided, is to be found in cool summers, not brutal winters: Chorlton, *Ice Ages*, p. 101.

NOTES

p. 479   'It is not necessarily the *amount* of snow that causes ice sheets but the fact that snow, however little, lasts': Schultz, *Ice Age Lost*, p. 72.

p. 479   'The process is self-enlarging': McPhee, *In Suspect Terrain*, p. 205.

p. 479   'you would have been hard pressed to find a geologist or meteorologist who regarded the model as being anything more than an historical curiosity': Gribbin and Gribbin, *Ice Age*, p. 60.

p. 480   The fact is, we are still very much in an ice age: Schultz, *Ice Age Lost*, p. 5.

p. 480   a situation that may be unique in the Earth's history: Gribbin and Gribbin, *Fire on Earth*, p. 147.

p. 480   it appears that we have had at least seventeen severe glacial episodes in the last 2.5 million years: Flannery, *The Eternal Frontier*, p. 148.

p. 481   Before 50 million years ago the Earth had no regular ice ages: Stevens, *The Change in the Weather*, p. 10.

p. 481   the Cryogenian, or super ice age: McGuire, *A Guide to the End of the World*, p. 69.

p. 481   The entire surface of the planet may have frozen solid: *Valley News* (from *Washington Post*), 'The Snowball Theory', 19 June 2000, p. C1; *New York Times*, 'How Earth Might Have Turned Into a Snowball', 7 Feb. 2024.

p. 482   the wildest weather it has ever experienced: BBC *Horizon* transcript, 'Snowball Earth', broadcast 22 Feb. 2001, p. 7.

p. 483   in an event known to science as the Younger Dryas: Stevens, *The Change in the Weather*, p. 34.

p. 484   'the last thing you'd want to do is conduct a vast unsupervised experiment on it': *New Yorker*, 'Ice Memory', 7 Jan. 2002, p. 36.

p. 484   The idea is that a slight warming would enhance evaporation rates: Schultz, *Ice Age Lost*, p. 72.

p. 485   No less intriguing are the known ranges of some late dinosaurs: Drury, *Stepping Stones*, p. 268.

p. 485   'Did you have a good ice age?': Flannery, *The Eternal Frontier*, p. 267.

NOTES

Chapter 28: The Mysterious Biped

p. 486 Just before Christmas 1887: *National Geographic*, May 1997, p. 87.

p. 487 found by railway workers in a cave at a cliff called Cro-Magnon: Tattersall and Schwartz, *Extinct Humans*, p. 149.

p. 487 So instead the name and credit for the discovery of the first early humans went to the Neander valley in Germany: Trinkaus and Shipman, *The Neandertals*, pp. 3–6.

p. 488 though just a footnote to history, he was an extraordinary figure: *American Anthropologist*, 'We Are Not Alone: William King and the Naming of the Neanderthals', Dec. 2021.

p. 489 Hearing of this, T. H. Huxley in England drily observed: Trinkaus and Shipman, *The Neandertals*, p. 59.

p. 489 He did no digging himself, but instead used fifty convicts lent by the Dutch authorities: Gould, *Eight Little Piggies*, pp. 126–7.

p. 489 In fact, many anthropologists think it *is* modern, and has nothing to do with Java Man: Walker and Shipman, *The Wisdom of the Bones*, p. 39.

p. 490 If it is an *erectus* bone, it is unlike any other found since: Trinkaus and Shipman, *The Neandertals*, p. 144.

p. 490 He also produced, with nothing but a scrap of cranium and one tooth, a model of the complete skull, which also proved uncannily accurate: Trinkaus and Shipman, *The Neandertals*, p. 154.

p. 490 To Dubois' dismay, Schwalbe thereupon produced a monograph: Walker and Shipman, *The Wisdom of the Bones*, p. 42.

p. 491 He realized that it was unlike anything ever seen before: Walker and Shipman, *The Wisdom of the Bones*, p. 74.

p. 491 Dart spent five years working up a monograph, but could find no one to publish it: Lewin, *Bones of Contention*, p. 82.

p. 491 For years, the skull . . . sat as a paperweight on a colleague's desk: Walker and Shipman, *The Wisdom of the Bones*, p. 93.

p. 492 found a single fossilized molar and on the basis of that alone quite brilliantly announced the discovery of *Sinanthropus pekinensis*: Swisher et al., *Java Man*, p. 75.

p. 493 then discovered to his horror that they had been enthusiastically smashing large pieces into small ones: Swisher et al., *Java Man*, p. 77.

NOTES

p. 493 The Solo People were known variously as *Homo soloensis*, *Homo primigenius asiaticus*: Swisher et al., *Java Man*, p. 211.

p. 493 in 1960 F. Clark Howell of the University of Chicago, following the suggestions of Ernst Mayr and others the previous decade: Trinkaus and Shipman, *The Neandertals*, pp. 267–8.

p. 493 After about a decade of comparative calm, palaeoanthropology embarked on another period of swift and numerous discovery: *Nature*, 'Fifty years after *Homo habilis*', 3 April 2014.

p. 494 the whole of our understanding of human prehistory is based on the remains of just 6,000 or so individuals: *Washington Post*, 'Skull Raises Doubts about our Ancestry', 22 March 2001. Smithsonian Museum website, undated.

p. 494 'You could fit it all into the back of a pickup truck if you didn't mind how much you jumbled everything up': interview with Ian Tattersall, American Museum of Natural History, New York, 6 May 2002.

p. 495 For the first 99.99999 per cent of our history as organisms, we were in the same ancestral line as chimpanzees: Tattersall, *The Human Odyssey*, p. 60.

p. 495 Then, about 7 million years ago, something major happened: *Chemistry World*, 'Dating the Age of Humans', Dec. 2015.

p. 495 'She is our earliest ancestor, the missing link between ape and human': PBS *Nova*, 'In Search of Human Origins', first broadcast Aug. 1999.

p. 496 Johanson breezily replied that he had discounted the 106 bones of the hands and feet: Walker and Shipman, *The Wisdom of the Bones*, p. 147.

p. 496 when you are a small, vulnerable australopithecine, with a brain about the size of an orange, the risk must have been enormous: Stevens, *The Change in the Weather*, p. 3; Drury, *Stepping Stones*, pp. 335–6.

p. 497 'but that the forests left them': Gribbin and Gribbin, *Being Human*, p. 135.

p. 497 For some 2 million years, Lucy and her fellow australopithecines scarcely changed at all: Correspondence with Chris Stringer, 2025.

NOTES

p. 497   by the beginning of the present century, some twenty types of hominid were recognized: *Nature*, 'On the origin of our species', 8 June 2017; *New York Times*, 'How Did We Get to Be Human', 19 Nov. 2018.

p. 498   Nothing, it turns out, was nearly as linear or straightforward as once thought: *Quaternary Science Reviews*, 'Close encounters vs missed connections?', Sept. 2023; *Nature*, 'Human evolution: The Neanderthal in the Family', 26 March 2014; *New York Times*, 'Early Humans Left Africa Much Earlier Than Previously Thought', 11 July 2024.

p. 498   The discovery was a tiny fragment of finger bone: *Nature*, 'Fossil finger points to new human species', 24 March 2010; *New Scientist*, 'Mystery Relations', 5 April 2014.

p. 499   it was from a child who was neither Neanderthal nor *Homo sapiens*: *New York Times*, 'A Blended Family: Her Mother Was Neanderthal, Her Father Was Something Else Entirely', 22 Aug. 2018.

p. 499   the comings and goings of the three groups over thousands of years was anything but straightforward: *Nature*, 28 Feb. 2019, p. 445; *Nature Genetics*, 'A history of multiple Denisovan introgression events in modern humans', 5 Nov. 2024.

p. 500   a jawbone dredged from the sea floor off the coast of Taiwan: *Nature*, 'Mysterious human fossil found in Taiwan was a Denisovan', 11 April 2025.

p. 500   a skull found years earlier in Harbin, China: *Science*, 'The proteome of the Late Middle Pleistocene individual', 18 June 2025, and *New York Times*, 'Mysterious Ancient Humans Now Have a Face', 18 June 2025.

p. 500   Their DNA has been found in modern populations thousands of miles away: *Nature*, 'Finding our inner Denisovan', 23 March 2016.

p. 500   Virtually everything about them is unexpected: *Scientific American*, 'Rethinking "Hobbits": What They Mean for Human Evolution', 1 Nov. 2009; *Quaternary Science Reviews*, 'The origins and persistence of *Homo floresiensis* on Flores', 1 May 2013; *Nature Communications*, 'Early evolution of small body size in *Homo floresiensis*', 6 Aug. 2024.

p. 501   'No hominid fossil find . . . had ever come from farther out in left field': Tattersall, *The Strange Case of the Rickety Cossack*, p. 192.

NOTES

p. 501 'it is hard to see how reducing body size could have been advantageous': Tattersall, *The Strange Case of the Rickety Cossack*, p. 193.

p. 501 All this raises the possibility that some primitive, small-brained hominids left Africa well over a million years ago: *Nature*, 'Hominins on Flores, Indonesia, by one million years ago', 17 March 2010.

p. 501 'Regardless of the route taken, transit is thought to have been accidental': *Proceedings of the National Academy of Sciences*, 'When did *Homo sapiens* first reach Southeast Asia and Sahul?', 21 Aug. 2018.

p. 502 In 2019 the puzzle grew even more complicated: *Nature*, 'Unknown human species found in Asia', 11 April 2019.

p. 502 It's a real challenge': telephone interview with Matt Tocheri, Aug. 2024.

p. 503 The bones came from at least fifteen individuals of a species never before seen: *Scientific American*, 'Controversy and Excitement Swirl around New Human Species', 1 March 2016; *New York Times*, 'Fossils Are Filling Out the Human Family Tree', 10 April 2019.

p. 503 we *Homo sapiens* may once almost have gone extinct ourselves: *New York Times*, 31 Aug. 2023.

Chapter 29: The Restless Ape

p. 505 Here, along the baking valley floor, is an ancient site called Olorgesailie: Just for the record, the name is also commonly spelled Olorgasailie, including in some official Kenyan materials.

p. 507 'the most recent major event in human evolution – the emergence of our own species – is perhaps the most obscure of all': Tattersall, *The Human Odyssey*, p. 150.

p. 507 it was widely believed that *Homo sapiens* arose as a distinct species only about 40,000 years ago: conversation with Professor John Gowlett of the University of Liverpool, 25 Nov. 2024.

p. 508 fossil bones of *Homo sapiens* found at Jebel Irhoud in Morocco were about 300,000 years old: *Scientific American*, Sept. 2018.

p. 509 'Modern humans neutralized this advantage . . . with better clothing, better fires and better shelter': Stevens, *The Change in the Weather*, p. 30.

NOTES

p. 510 1.8 litres for Neanderthals versus 1.4 for modern people: Flannery, *The Future Eaters*, p. 301.

p. 513 'These were not unsophisticated people': *Nature*, 'How Neanderthal minds took flight', 7 Feb. 2019.

p. 514 'they were selecting golden eagles for their feathers': *Nature*, 'Trail of feathers to the Neanderthal mind', 4 Feb. 2019.

p. 515 you will have somewhere between 1 and 4 per cent Neanderthal DNA within you: *Science*, 'Recurrent gene flow between Neanderthals and modern humans over the past 200,000 years', 12 July 2024.

p. 516 they may well have been the last surviving archaic humans: *Smithsonian Magazine*, 'The Rock of Gibraltar: Neanderthals' Last Refuge', 19 Sept. 2012.

Chapter 30: Goodbye

p. 518 'unscientific voyagers, three or four oil paintings, and a few scattered osseous fragments': quoted in Gould, *Leonardo's Mountain of Clams and the Diet of Worms*, p. 237.

p. 519 Australia ... lost no less than 95 per cent: Flannery and Schouten, *A Gap in Nature*, p. xv.

p. 519 'There's no material benefit to hunting dangerous animals more often than you need to – there are only so many mammoth steaks you can eat': *New Scientist*, 'Mammoth Mystery', 5 May 2001, p. 34.

p. 520 only four types of really hefty ... land animals: Flannery, *The Eternal Frontier*, p. 195.

p. 520 human-caused extinction now may be running at as much as 120,000 times that level: Leakey and Lewin, *The Sixth Extinction*, p. 241.

p. 522 He set off at once for the island, but by the time he got there the cat had killed them all: Flannery, *The Future Eaters*, pp. 62–3.

p. 522 'At each successive discharge': quoted in Matthiessen, *Wildlife in America*, pp. 114–15.

p. 523 The zoo lost it: Flannery and Schouten, *A Gap in Nature*, p. 125.

p. 523 Rothschild was a strange and reclusive fellow: *New York Review of Books*, 'A Bird's-Eye View of Evolution', June 27, 2002.

p. 524 Hugh Cuming, who became so preoccupied with accumulating objects that he built a large ocean-going ship and employed a crew to sail the world full-time picking up whatever they could find: Desmond and Moore, *Darwin*, p. 342.

p. 524 Millions of years of isolation had allowed Hawaii: *National Geographic*, 'On the Brink: Hawaii's Vanishing Species', Sept. 1995, pp. 2–37.

p. 524 The greater koa finch, an innocuous member of the honeycreeper family: Flannery and Schouten, *A Gap in Nature*, p. 84.

p. 524 a bird so sublimely rare that only one has ever been seen: Flannery and Schouten, *A Gap in Nature*, p. 76.

p. 527 'One planet, one experiment': Wilson, *Diversity of Life*, p. 182.

INDEX

Acheulean tools 505–6
Agassiz, Louis 80, 364, 473–6
Aharanov, Yakir 167
AIDS 353, 356
air pressure 290, 292–3
alchemy 53, 108, 125
algae: blue-green 339; calcareous 300; classification 345; lichens 375–6
All Souls, University of Oxford 332–3
All Species Foundation 410
Alpha Centauri 30, 41
altitude 288–90
aluminium 116, 282
Alvarez, Luis 223–6, 231
Alvarez, Walter 211, 222–6
Alvin, submersible 313–15
amino acids 326, 327, 329
ammonites 387
amoebas 345
amphibians 379, 387
anapsids 382
ancestors 449–50
Anderson, Ray 227–30, 231–2, 235
Anderson, Ross 333–5, 337–9
Anning, Mary 94–5

Antarctica: ecosystem 325; ice 306, 485; ice-free 485; lichens 375; seals 325
Anthropocene period 83
antibiotics 354–5
archaea 348
Archaean period 84, 335
archaeopteryx 43, 99, 441n
Armitage, Eleanora 397
Armstrong, Richard 246–7
arsenic 283
Asaro, Frank 223–4, 225
Ashcroft, Frances 270, 272, 289, 349
Ashfall Fossil Beds 236–8
Asimov, Isaac 161
Askesian Society 114, 295–6
astatine 281
asteroids 218–22, 232
Atlantic floor 202
atmosphere 286–9
atomic bomb 163, 168
atoms: atomic number 121; atomic weight 121, 155; behaviour 159–60, 161–3; cells and 328–9; durability 153; evidence of existence 156; idea of 155; image 161, 165; life of 1–2; molecules and 152–3;

atoms – *cont.*
   nucleus 160–1, 165, 168; shape 159; shell 165; sizes 153–4; structure 159–62, 163
ATP (adenosine triphosphate) 429
Attenborough, David 318, 376
Audubon, John James 243
Australia, sea life 320
australopithecines 496–7, 501
Australopithecus 491, 493, 495
Avery, Oswald 455–6, 460
Avogadro, Lorenzo 117–18
Avogadro's Number 118n
Avogadro's Principle 117–18, 120
axions 195

Baade, Walter 35–6
Bachman's warbler 525
backbones 368, 378
bacteria: cyanobacteria 339–40, 343, 481; discovery 426; first life on Earth 335; hot-water 265–6, 344; in human body 341–2; mutation 343; reproduction 343; survival 343–5, 413; types 412
bacteriophages 355
Baldwin, Ralph B. 224
Ball, Philip 121, 306
balloon ascents 287, 289–90
Banks, Sir Joseph 66, 393, 398–9, 403
barnacles 436, 524
Barringer, Daniel M. 217
Barton, Otis 307–11
Bastin, Edson 344
bathyscaphe 310–11
bathysphere 308–10
bdelloid rotifers 414
Beagle voyage 433–5, 436
Becher, Johann 109
Becker, George 107
Becquerel, Henri 124, 158
Beebe, Charles William 307–10
belemnites 102
Bell, Alexander Graham 135

bends, the 271–2
Bennett, Victoria 344
Berzelius, J.J. 119
Besso, Michele 138
Betelgeuse 42
Big Bang 11, 13–18, 21, 43, 150
biodiversity 318, 526
birds 100, 389, 416, 514, 517–19, 522–5
Black, Davidson 492
black mamos 525
Blackett, Patrick 205–6
blood cells 419n, 463
blue whale 318
Bodanis, David 141, 142
Bohr, Niels 162–3, 166
Boltzmann, Ludwig 156
Bonnichsen, Bill 238
Bose, S.N. 186
bosons 186
Bouguer, Pierre 49–50, 59–60
Boyle, Robert 109
Bradshaw, Corey J. A. 502
brain size 497, 501, 510
Braithwaite, Robert 396
Brand, Hennig 109
Briggs, Derek 367, 371
bristlemouths 317
British Museum 102, 307
Broad, William J. 313
Broberg, Gunnar 401
Brock, Thomas and Louise 264–5
Broglie, Prince Louis-Victor de 164
Brown, Guy 426
Brown, Harrison 176, 177
Brown, Robert 115, 398
Brownian motion 115, 137, 156, 398
Brunhes, Bernard 205
Bryan, Alanson 525
Bryce, David 414
bryophytes 394–7
Buckland, William 76, 77–8, 83–4, 96, 106
Buffon, Comte de 85, 89–90, 91, 402
Bullard, E.C. 249

INDEX

Burgess Shale, fossils 365–8, 370–2, 377
Burrell, William Holmes 397
Byron, Lord 471

Cabot, John 322
Cadbury, Deborah 102
calcium 116, 330
calculus 52, 54
Cambrian: explosion 363, 365, 367, 372–3, 383, 441, 482; period 80–1
Cameron, James 315
cancer 429–30
Cannon, Annie Jump 149
carbon: atoms 282; composition of living objects 282, 330; formation 21; long-term cycle 300–1, 324, 379–80
carbon-14 160, 173–4
carbon dioxide 44, 301
Carboniferous period 379, 381–2
Carolina parakeet 522–3
Carr, Geoffrey 192
Cassini, Giovanni and Jacques 59
Caster, K.E. 198
catastrophism 79, 225–6
cattle 283
Cavendish, Henry 66–9, 154, 158n, 517
Cavendish Laboratory 157, 158–9, 163, 181, 456
cells: actions 430; composition 419–20, 426–8; death 429; discovery 422–3; eukaryotes 333, 335; human 419–20, 421–2; life 421–2; membrane 427; mitochondria 335–6, 428–9; nucleus 426; splitting 419; types 421; yeast 420
Celsius, Anders 295
Cepheids 148, 149, 191
CERN 161, 182, 183, 187–8
CFCs (chlorofluorocarbons) 171–2, 179–80
Chadwick, James 36, 157, 163

Challenger, HMS 307
Chambers, Robert 437, 445
Chappe, Jean 61
Chargaff, Erwin 455
Charpentier, Jean de 473–4, 475
chemical abbreviations 118–19
chemosynthesis 314
Chicxulub crater 230, 231
chimpanzees 495
chlorine 110, 111, 283–4
Christiansen, Bob 254, 255, 259
Christy, James 22–3
chromosomes 451, 453
chrysanthemums 406–7
Clarion-Clipperton Zone 323
Clark, William 92, 94
classification: Linnaean system 400–3; of organisms 345–51; of stars 148; universal code 405
climate 471–2, 481–7, 516
cloud forests 412–13
clouds 296–7, 306
cobalt 282
cod 322
colossal squid 318–19
Columbia, space shuttle 288
comets: entering Earth's atmosphere 232–5; Halley's 51; long-period 29; Shoemaker-Levy 230–1, 235
Conard, Henry S. 394
continental drift 199–202, 206–7
convection process 247–8, 292–3
Conway Morris, Simon 366–7, 371
Cook, James 63, 398–9
Cope, Edward Drinker 103–6
copper 283
coral reefs 320, 388
Coriolis effect 294–5, 306
Coriolis, Gustave-Gaspard de 294
cosmic background radiation 14–15, 150
cosmic rays 37, 250
cosmological constant 144
Covid-19 358–9
Cox, David 463

621

INDEX

Crab Nebula 42
Crampton, Henry Edward 417
Cretaceous period 81, 384, 390–1, 485
Crick, Francis 158, 456–60
Crisp, Frank 425
Cro-Magnons 487, 489, 509, 510
crocodiles 387, 389
Croll, James 475–7
Cropper, William H. 133, 135, 146, 160
Crouch, Henry 141
Crowther, J.G. 68
crustaceans 379
Cryogenian age 481–2
Crystal Palace 97–8
crystallography 458
crystals 329
Cuming, Hugh 524
Curie, Marie and Pierre 124–5, 127, 281
Cuvier, Georges 91–2, 96, 99, 473
cyanobacteria 339–40, 343, 481
cyclotron 181

Dalton, John 117, 154–6
Daly, Reginald 44, 202
dark matter 37–8, 193–6
Dart, Raymond 491, 495
Darwin, Charles: on age of Earth 85–6; on barnacles 436, 524; *Beagle* voyage 393, 433–5, 436; birth 432–3, 449; *Descent of Man* 446–7; early life 433; on evolution 364, 436; illness 436–7; on Lyell 78, 80; *Origin of Species* 364, 432, 434, 440–1, 447; on Owen 100; responses to his work 440–2, 445–6; statue 103; tomb 448; worms 447
Darwin, Erasmus 435
Davies, Paul 190, 282, 329
Davis, Marc 194
Davy, Humphry 111, 116–17, 154, 282

Dawkins, Richard 371
de Bort, Léon-Philippe Teisserenc 287
de Duve, Christian 330, 343, 421, 428, 451
de Vries, Hugo 448
Deane, Charles 271
Deccan Traps 226, 256
deep-sea vents 313–14, 315
DeMoivre, Abraham 53
Denosovians 499–500
Denver, height 210
Descartes, René 134
DESI (Dark Energy Spectroscopic Instrument) 196–7
Devonian: extinction 384, 386; period 80–1, 379, 381
Diamond, Jared 352
diamonds 245
diapsids 382–3
diatoms 387
Dicke, Robert 13–15, 17
Dickinson, Matt 289
Dickson, Tony 209
Dimetrodon 382
dinosaurs: age of 363, 377, 382, 390–1; discoveries 89, 94–7, 104, 107; extinction 222, 225, 228, 234, 280, 386; life-sized models 97–8; museum displays 390; name 99–100; number of species 105, 390–1; ranges 485
diseases 352–9
diving 57, 270–3, 274–5, 308–15
Dixon, Hugh 397
Dixon, Jeremiah 62–3, 64
DNA: damage 427; discovery of structure 456–60; experiments 455; fingerprinting 464n; mitochondrial 336; molecular clock 336–7; Neanderthals 515–16; proteins and 328, 453, 461; replication 462–3; role 431, 451–3, 461, 463–7; shape 461–2
dodo 99, 517–19

INDEX

Doppler shift 145
Doss, Paul 258–63
Douglas, David 399
Dover, White Cliffs 300
dragonflies 380
Drake, Frank 30
Draper, John William 446
Drury, Stephen 297, 335, 485
Dubois, Eugène 486–7, 489–90, 502
Duke University, herbarium 415
dwarf planets 26

Earth: age 69–70, 84–6, 87–8, 107–8, 126, 169, 176–8; atmosphere 286–9; climate 471–2, 481–5, 516; cores 239–40, 248–9; crust 245–7; dawn of life 330–2, 333–4, 337; distance from Sun 63, 277–8; formation 43–5; history 376–7; human living conditions 276–7, 280–1; interior 239–40, 245–8; layers 246n; magnetic field 205–6, 249–50; mantle 248; mass 64–5, 66, 68–70; Moon's influence 279–80; orbit 220; revolution 294, 307; shape 56–7, 59; size 57–9; temperature at core 249
earthquakes 240–4, 260, 262
earthworms 408
Ebbings, Darrell 132n
Ebola virus 356, 359
Eddington, Sir Arthur 141
Ediacaran Hills, fossils 368–70, 373–4, 376
Einstein, Albert: on Brownian motion 137, 156; career 136–8, 139, 150, 168; on cosmological constant 144; death 211; Hapgood foreword 198, 211; on mass and energy 124, 138–40; Nobel Prize 137; on quantum theory 136, 167–8; relativity theory 20, 137–44, 150, 156; reputation 141, 150; works 136–7

electromagnetism 17
electrons: discovery 158; formation 17; Mendeleyev's opinion 127; negative charge 159; number in atom 160; orbits 162–3, 165; Uncertainty Principle 164–5; wave or particle 163–6
elements 120–4, 281–4
Elliott, Charles 407
Ellis, Len 394–6, 398
Elwin, Whitwell 432
Emerald Pool 264–5
Endeavour voyage 398–9
Englert, François 188
entropy 132
enzymes 427–8
Eris 26
erosion 73
ether, the 133–5
Ethyl Corporation 170–1, 179
eukaryotes 333, 335, 348, 350
euryapsids 382
Evans, Matthew 144
Evans, Robert 33–5, 38–41
evaporation 297
evolution 274, 436, 438–41
Exclusion Principle 166
extinction 91–2, 383–91, 518–26
extremophiles 278n

Fahrenheit, Daniel Gabriel 295
Falconer, Hugh 100, 487
Fallarone Islands 316
Feldhofer fossil 488
Fermi, Enrico 182, 186
fermions 186
Ferris, Timothy 61, 161
Feynman, Richard: on atoms 152, 154, 162, 165; on elements 122; on interior of Earth 239; on particle physics 184; on planetary orbits 51; on reasoning 285
finches: Galápagos 436; koa 524
Finlayson, Clive 511–16
Finlayson, Stewart 511–13

INDEX

fish 320–2, 378–9
Fisher, Osmond 247
fishing 320–2
FitzRoy, Robert 433–4, 446
Flannery, Tim: on Australian coastline 320; on extinction 388, 520–2; on ice ages 485; on KT impact 387; on radioactive dumping 316; on radiocarbon dating 173–4
Flores people 500–2
flu epidemic (1918) 357–8
Focke, Wilhelm Olbers 445
foraminiferans 300, 387
Forbes, Edward 307
Fortey, Richard: on bacteria 413; on dinosaur models 97; on early life forms 339, 370, 373; on extinction 384, 387–8; on geological disputes 83, 371–2; on specialists 393–4, 418; on trilobites 362–3; on Walcott 364
fossil fuels 301
fossils: Anning's work 94–5; anomalies 199; Ashfall Fossil Beds 236–8; Burgess Shale 365–8, 370–2, 377; dating rocks 93; dinosaurs 89, 94–5, 96, 103–7, 390; Ediacaran Hills 368–70, 373–4, 377; evidence 88; extinction theories 92–4; formation 334–5, 337, 361–2; Hell Creek Formation 94, 391; location 72–3, 199, 200, 337–8; mammoth 91–2; Mantell's work 95–7, 100–1; record 361–2, 389–90, 441
Fowler, W.A. 43
Franklin, Benjamin 67
Franklin, Rosalind 456, 457–60
Fraser, John 399
Frenk, Carlos 193–7
fruit flies 454, 465
Fuhlrott, Johann Karl 487–8
fungi 345–6, 375–6, 408, 414

Gabr, Ahmed 270
Galápagos Islands 308, 313, 436
galaxies 147, 149, 194
Galeras, eruption 252
Gamow, George 14
Gardiner, William 397
Gauteng caves 503
Geiger, Hans 159
Gell-Mann, Murray 185
genes 450–1, 453, 454–5, 461
genetics 274, 444, 460–1
genome 450–1, 461, 467
Geological Society 75, 77, 118, 175
geology: eras and epochs 80–4; rock dating 174–8
geysers 260, 262–3
giant squid 318–19
Gibbs, J. Willard 132–3, 137
Gibraltar 511–12, 516
Gilbert, G.K. 217–18
glaciation 472–80, 483
Glicken, Harry 252
global warming 484–5
Glossopteris 210
gluons 186
Goethe, Johann Wolfgang von 296
Goksøyr, Jostein 412
Gold, Thomas 344
Goldilocks effect 19
Gondwana 209–10
Gotham 333
Gould, John 436
Gould, Stephen Jay: on Burgess Shale 365–8, 370–2; on Crampton 417; on emergence of life 331; on human history 391; on Lyell 84; on meteor impact 230
Grant, Robert 100
Grant, Ulysses S. 134
grasses 405–6
gravitational waves 40, 144
gravity: Einstein's work 140, 144; formation 17; gravitational constant 60, 68; Newton's

INDEX

work 55, 60; spacetime 143; Standard Model omission 188–9; strength 19
Great Rift Valley 505
greenhouse effect 44–5
Greenland ice sheet 300, 306, 483–4
Greer, Frank 344
Gregory, Bruce 147
Gregory, J.W. 505
Gribbin, John 304, 497
Groves, Colin 409
Gulf Stream 298, 484
Gutenberg, Beno 240
Guth, Alan 14, 17–18
Guyot, Arnold 203

Habeler, Peter 289
Hadley cells 294
Hadley, George 294
hadrons 185
Haeckel, Ernst 345, 404
Haldane, J.B.S. 20, 271, 272–6
Haldane, John Scott 272–3
half-life 125–6
halibut 322
Halley, Edmond: on atmosphere 293–4; career 51; on dating the Earth 85, 107; Newton visit 53–4; planetary orbit wager 51–2, 517; production of *Principia* 55–6; on transits of Venus 60–1
Halley's comet 51
Halloy, J.J. d'Omalius d' 81
Hamilton, William Phillips 397
hand-axes 506–7
Hapgood, Charles 198, 202, 211
Harrington, Robert 22
Hart, Michael 277
Hartsoecker, Nicolaus 425
Haughton, Samuel 107
Hawaii, birds 524
Hawking, Stephen 143, 150, 166
Hebgen Lake, earthquake 262
Heisenberg, Werner 164–5
Helin, Eleanor 218

helium 12, 19, 21, 121–2
Hell Creek Formation, fossils 94, 391
Helmholtz, Hermann von 86
Hemphillian extinction 385
herbariums 415
Herschel, William 65, 219
Hess, Harry 203–5, 211, 311n
Hessler, Robert 319
Higgs boson 186–7, 188, 196
Higgs, Peter 186–7, 188
Hildebrand, Alan 230
HIV 353, 356
Hogg, John 347
Holmes, Arthur 175–6, 202
Holmyard, E.J. 156
Holocene period 484
hominids: australopithecines 496–7; bipedalism 496–7; brain size 497, 501; classification 491n, 493; remains 493–5; Sumatran fossils 486–7
Homo erectus: brain size 507, 510; classification 493, 494; dates 493; Dubois' discovery 489–90; remains 494; Solo People 493; territory 494
Homo floresiensis 500–2
Homo habilis 492
Homo luzonensis 502
Homo naledi 503
Homo sapiens: almost extinction 503–4; appearance 507–8; brain size 508–9, 510; classification 493; interbreeding 499, 515–16; territory 508–9, 515; type specimen 105–6
Hooke, Robert 51–2, 55–6, 422–3, 517
Hooker, Joseph 439
Hopkins, William 474–5
hormones 430
Horne, Everard 424
horses 385, 415
Howard, Luke 295–6

INDEX

Howell, F. Clark 493
hox genes 465
Hoyle, Fred 42–3, 328
Hubble, Edwin 146–7, 149–51, 190
Hubble Space Telescope 151, 191
Human Genome Project 462
Humboldt, Alexander von 393, 474
Hunter, John 99, 424
Hunterian Museum 96, 101, 106, 487
Hutton, Charles 64–5
Hutton, James 71–5, 77, 200, 473
Huxley, Aldous 273, 274
Huxley, T.H.: on catastrophism 79; on Darwin's work 435, 441–2; on Neanderthal bones 489; Owen's career 100, 102, 103; saltationist 441–2; on *Vestiges* 437; Wilberforce encounter 446
Hyde, Jack 251
hydrocyanic acid 110
hydrogen: Cavendish's work 67; combustion 283; commonest element 121; composition of living objects 330; conversion to helium 19; formation 12, 21; Lavoisier's work 113; Rozier's work 67; spectrum 162, 163; water 283, 304
hydrosphere 306
hydrothermal explosions 262–3
hyperthermophiles 265–6, 313–14

ice 299, 303, 304, 306
ice age(s): Agassiz's theory 474–6; causes 478–9; Croll's work 475–7; Cryogenian 481–2; current 480–1, 484–5; cycles 477–9; Milankovitch's work 477–9; Schimper's work 474; term 474
ichthyosaurus 94
iguanodon 96, 97, 106
illness 352–9
Indian Ocean earthquake 241
inflation theory 16–17

insects 380, 384, 387–8, 408, 409–10, 411
International Astronomical Union 23, 26, 221
International Commission on Stratigraphy 83
International Seabed Authority 324
International Union of Pure and Applied Chemistry 122–4
inverse square law 54, 55–6, 422
ionosphere 286
iridium 224, 226
iron 282, 330, 339, 340
Isaacs, John 319
isotopes 160, 379–80
Izett, Glenn 228

Jardine, Lisa 457
Jarvik, Erik 381–2
Java Man 489–90, 493
Jefferson, Thomas 90, 92
Jeffreys, Harold 207
Jenkin, Fleeming 442
jet streams 291
Johanson, Donald 495, 496
Johnston, David 252
Joly, John 107, 126
Joseph, Lawrence 166
Jupiter 23, 28
Jurassoc period 81

K2-18b 31
Kaku, Michio 143, 189
Kelly, Kevin 410
Kelvin, William Thomson, Lord 86–8, 107, 126, 127, 440
Kew Gardens 408
Keynes, John Maynard 53
kimberlite pipes 245
King, William 488
Kinsey, Alfred C. 417
koa finch 524
Koenigswald, Ralph von 492–3
Kola hole 244–5
Kolbert, Elizabeth 484

INDEX

Köppen, Wladimir 478
Krakatau, eruption 256–7
Krebs, Robert E. 122
KT: boundary 222; impact 234, 386–9, 391
Kuiper belt 25–6, 219
Kuiper, Gerard 219
Kunzig, Robert 304
Kurlansky, Mark 322

La Condamine, Charles Marie de 49–50, 59–60
Lakes, Arthur 104
Lalande, Joseph 63
Lamont Doherty Laboratory 222
land bridges 201
Lander, Eric 467
Large Magellanic Cloud 42
Lascaux caves 173
Lassa fever 359–60
Laubenfels, M.W. de 225
Lavoisier, Antoine-Laurent 111–14, 154
Lavoisier, Mme 112–13, 116
Lawrence Berkeley Laboratory 40, 223
Lawrence, Ernest 181
le Gentil, Guillaume 61–2
lead 169–71, 175–6, 178–9, 284
Leakey, Louis 506
Leakey, Mary 506
Leakey, Richard 362, 520
Leavitt, Henrietta Swan 148, 149
Leclerc, Georges-Louis, Comte de Buffon 85, 89–90, 91, 402
Lederman, Leon 186
Leeuwenhoek, Antoni van 423–6
Lehmann, Inge 240
Leibniz, Gottfried von 54
Lemaître, Georges 13, 150
Leonard, F.C. 25
leptons 186
Lever, Sir Ashton 77
Levy, David 218, 230
Lewin, Roger 362, 520

Lewis, John S. 234
Lewis, Meriwether 92, 94
Lewontin, Richard 451
Libby, Willard 172–3
lichens 375–6, 394–5
life in the universe, chances of 30–1
light, speed of 17n, 134–5, 138, 139, 150
Lightman, Alan 166
lightning 291
LIGO (Laser Interferometer Gravitational-Wave Observatory) 144
limestone 300, 379
Lincoln, Abraham 432n, 449
Linde, Andrei 15
Linnaean Society 295, 403, 439
Linnaeus, Carolus 400–3
lithium 12, 21
lithosphere 247
lizards 388
Loeb, Avi 221–2
Lorentz transformations 141
Lowell Observatory 22, 24, 145
Lowell, Percival 24–5
Lucy, australopithecine 495–7
Luzon 502
Lyell, Charles: career 77–9, 222; Darwin friendship 439, 441; geological epochs 81–2; on Hutton's work 73–4, 79; on ice age 475; influence 79–80, 84; lectures 76; *Principles of Geology* 79, 80, 81–2
Lyon, John 399–400
lystrosaurus 210

M theory 189
Mach, Ernst 156
MacPhee, Ross 519
Maddox, Brenda 460
Maes, Koen 410, 412–13
magnesium 116
magnetosphere 41
Malthus, Thomas 435

INDEX

mammals 389
mammoths 519–20
Manning, Aubrey 300
Manson crater: extinction question 228, 235, 386; Shoemaker's studies 226, 227–8; size 215–17, 231, 377
Mantell, Gideon Algernon 95–7, 100–1, 106
Marat, Jean-Paul 112, 113
Marcy, Geoffrey 10
Margulis, Lynn 305, 342
Mariana Trench 266, 311, 312n, 315
Marić, Mileva 138
Mars: canal theory 24; distance from Sun 279
Marsh, Bryan 354
Marsh, Othniel Charles 103–6
Marshall, Barry 355
Maskelyne, Nevil 60, 62–3, 64
Mason, Charles 60, 62–3, 64
Mason, Roger 369
mass 68n
mastodon 91–2
matrix mathematics 164
Matthew, Patrick 439–40
Matthews, Drummond 206
Maunder, John 411
Maxwell, James Clerk 67
Mayer, August 489
Mayr, Ernst 336, 349–50, 493
McDonald Observatory 219
McGrayne, Sharon Bertsch 170–1, 178, 179
McGuire, Bill 242
McLaren, Dewey J. 225
McPhee, John: on geological timespans 82, 377; on ice ages 479; kimberlite pipes 245; on plate tectonics 207
Medawar, Peter 274, 355
megadynasties 382–3
megalosaurus 96
megatherium 434
Meinertzhagen, Richard 393

Mendel, Gregor 274, 443–5, 448
Mendeleyev, Dmitri Ivanovich 119–22, 127
Mendes, J.C. 198
meningitis 354
mesosphere 286, 287
Messinian Salinity Crisis 297–8
Meteor Crater, Arizona 217, 218
meteorites 177, 217–18, 232–4, 377
meteorology: Howard's work 296; long-term carbon cycle 300–1; oceans and atmosphere 298–300; term 295
methane 44
Michel, Helen 224
Michell, John 65–6, 68
Michelson, Albert 133–5, 146, 150
microbes: hot-water 265–6; role 342–3
microscopes 423–4
Midgley, Thomas Junior 169, 171–2, 176, 179
Miescher, Johann Friedrich 452
Milankovitch, Milutin 477–9, 485
Milky Way 29–30, 147, 149
Miller, Stanley 326, 329
Millikan, Robert 37n
millipedes 380, 413
mining, oceans 323–5
Miocene epoch 82, 236
Mirsky, Alfred 455
mites 411–12
mitochondria 335–6, 428–9
Modern Synthesis 274, 448
Moho discontinuity 239–40, 243–4
Mohole 244
Mohorovičić, Andrija 239
molecules 152–3, 329, 427
Monod, Jacques 468
Moody, Plinus 94
Moon: craters 217–18; formation 44, 202; influence on Earth 279–80; journey to 313; size 280
moons elsewhere in the solar system 22–3

INDEX

Morgan, Thomas Hunt 453–5
Morley, Edward 133–4, 135
Morley, Lawrence 206
mosses 394–7
motion, laws of 55
Mount St Helens, eruption 250–3
Mount Unzen, eruption 252
Mount Wilson Observatory 147, 149, 150, 191
mountain lion 525
Mullis, Kary B. 265
Murchison, Roderick 76, 80–1, 474
Murray, John 440
mutation 343, 454
myxomycetes 346–7

Nagaoka, Hantaro 161
Nägeli, Karl-Wilhelm von 444
NASA (National Aeronautics and Space Agency) 191, 220–1, 234, 255, 364
Natural History Museum: ancient marine reptiles 95; author's visit 362, 392–4, 417; collection 43, 392–8; creation 102–3; dinosaur displays 390; ocean species survey 324
nautiloids 387
Neanderthals 447, 452, 487–8, 499, 509–10, 512–16
nebulae 149
necrotizing fasciitis 353–4
Negrín, Juan 275
Neptune, orbit 25, 27–8
Neptunists 73, 79
neutrinos 137n, 182–3, 194–5
neutron stars 36–7
neutrons: in atomic nucleus 160; discovery 158, 163; formation 17; Zwicky's theories 36
New Horizons spacecraft 26
New Zealand: fishing 320–1
Newlands, John 120

Newton, Isaac: alchemy 53, 108; character 52; on ether 134; on gravity 55, 60; Halley's visit 53–4; inverse square law 55–6, 422; laws of motion 55, 56, 135–6; *Principia* 54–6, 60, 154, 517; tomb 448; on weight of Earth 69; work 52–3
Ngalli, Jillani 506, 507
nitric oxide 421
nitrogen: composition of living objects 330; Earth's atmosphere 44, 285; formation 21; intoxication 275–6
nitrous oxide 114–15, 117
Nitsch, Herbert 270
Nobel Prize: Alvarez 223; Becquerel 125; Bohr 163; Chadwick 163; Crick 457, 460; Curies 125; Einstein 137; Englert 188; Fowler 43; Higgs 188; Libby 173; Medawar 355; Michelson 135; Morgan 454–5; Mullis 265; Pääbo 498; Pauling 456; Reines 137n; Ruska 137n; Rutherford 157, 159; Sulston 450; Urey 326; Watson 457, 460; Weinberg 20
North Star 33n, 148
Norton, Trevor 273, 275
Norwood, Richard 57–8
nuclear forces, strong and weak 17, 167–8
nucleus of a cell 426
Nuland, Sherwin B. 427, 428, 464
numbers, very large 16n, 118n, 191n
Nuttall, Thomas 94, 399

ocean: chemistry 209; deep-sea vents 313–14, 315; floor 203–5, 307, 312, 315–16, 324; influence on climate 298–300; life 318–20; mesopelagic layer 317; mining 323–5; 'road kill' 323; saltiness 314–15, 484; sea level 306–7; share of Earth's water 306

629

INDEX

oceanography 307–8, 314–15
Officer, Charles 226
oganesson 122, 124
okapi 415
Oldham, R.D. 239
Olorgesailie axes 506–7
Oort cloud 28–9
Oort, Jan 28n
Öpik, Ernst 28n, 224
Oppenheimer, Robert 37
Ordovician: extinction 383, 384, 386; period 81
organelles 335
Ostro, Steven 220
Oumuamua 221–2
Overbye, Dennis 14–15, 143, 165
Owen, Richard 98–104, 446
oxygen: at altitude 288, 289; bacteria and 339; breathing 269, 285, 288, 378, 379; Cavendish's work 67; cell requirements 428; combustion 283; commonest element 281; composition of living objects 330; discovery 110, 111; formation 21; Lavoisier's work 113; levels 379–80; ocean floor nodules 325; supply 343; water 283, 304
Oyster Club 72
ozone 172, 180, 287

Pääbo, Svante 498–9, 516
palaeogenomics 498
Paley, William 442
Palgrave, William 397
Palmer, Harry 524
Palomar Observatory 218
Pangaea 199, 209, 378
Parkinson, James 76–7, 96
particle detector 181
particles 159, 166, 181–8, 291
Pasteur, Louis 341, 426
Patterson, Clair 169, 176–80
Pauli, Wolfgang 157, 166
Pauling, Linus 456, 458, 459

Pauw, Corneille de 90
Peabody, George 104
Peabody Museum 94
Peale, Charles Willson 522–3
Pearce, Chaning 102
Peebles, Curtis 230
Peking Man 492, 493
Pelletier, P.J. 155–6
penguins 325
penicillin 354
Penzias, Arno 13–15, 150
peridotite 245, 248
Periodic Table 120–4
Perlmutter, Saul 40
Permian: extinction 363, 384–5, 386; period 81
Perutz, Max 328
phosphorus 109–10, 330
photoelectric effect 136
photons 17, 166, 187
photosynthesis 314, 339, 345, 481
phyla 405
phytoplankton 317, 325
Piazzi, Giuseppe 219
Picard, Jean 58–9
Piccard, Auguste and Jacques 310–12
Pickering, William H. 149
Pillmore, C.L. 228
pitchblende 126
Planck, Max 131–3, 136, 137
planets: definition 26; number in universe 32
plankton 379, 387
plants 378, 405–6
plate tectonics 74, 177, 206–8
Playfair, John 71, 75
plesiosaurus 94–5
Pluto: discovery 24–5; distance from Earth 25, 28, 29; dwarf planet 26; moons 22–4; orbit 25
Plutonists 73, 79
Polaris 33n, 148
polonium 125
Popper, Karl 190
potassium 116, 282

INDEX

Powell, James Lawrence 386
Precambrian era 83, 377, 441
Priestley, Joseph 111
Princeton University 14–15
prokaryotes 339, 350
proteins 327–8, 427–8, 453, 461, 467
proteome 467
protons 11, 17, 121, 160
protozoa 347, 425
Proxima Centauri 28, 30
pterodactyls 94

quantum leap 163
quantum theory 136, 165, 167–8
quarks 185–6

Rabone, Muriel 324
radiation: cosmic background 14–15, 150; Einstein's work 139; sickness 127–8
radio waves 158–9
radioactive waste 316
radioactivity 124–8, 159
radiocarbon dating 172–4
radium 125
rainforests 412–13
Raup, David 383, 520
Ray, John 402
red giants 148
Rees, Martin 18–19, 21, 31, 153, 192
refrigerators 171
Reines, Frederick 137n
relativity theory: bending of universe 20, 143–4; General Theory 140–2, 143–4, 145, 150; gravity 168; Special Theory 137, 138–41; understanding 141–4
reptiles 387
Richter, Charles 240
Richter scale 240–1
Rickover, Hyman G. 312–13
Ridley, Matt 464
RNA 453, 468
Roentgen, Wilhelm 158

Rothschild, Lionel Walter, Baron 523–4
roughy, orange 321
Royal Institution 116, 126
Royal Society: biodiversity 526; *Challenger* expedition 307; Dalton's work 156; Leeuwenhoek's work 423–5; Mantell's paper 96; Mason's work 60; Owen's career 101–2; publications 56; status 75; symposium 206
Rozier, Pilatre de 67
Rubin, Vera 193
Rudwick, Martin 81, 84
Rumford, Benjamin Thompson, Count von 115–16, 126, 247, 299
Runcorn, S.K. 205–6
Ruska, Ernst 137n
Russell, Bertrand 142
Rutherford, Ernest: on age of Earth 108, 125–6; career 156–8, 159; character 157–8; on heat reserves 200; work on atoms 158–9, 161–3, 175
rutherfordium 122

Sacks, Oliver 34
Sagan, Carl 22, 32, 184–5, 382
Sagan, Dorion 305, 342
saltationists 441–2
salts 305–6, 314–15
San Andreas Fault 242
San Francisco earthquake 241
Sandage, Allan 191
Sandler, Howard 319
Saturn 23
Scheele, Karl 110–11
Schiehallion, mountain 64–5, 396
Schimper, Karl 474, 475
Schlapkohl, Ann 217
Schouten, Peter 520–2
Schrödinger, Erwin 110, 164, 166
Schultz, Gwen 479
Schwalbe, Gustav 490

INDEX

Schwann, Theodor 426
Schwartz, Jeffrey 442
scientific notation 16n
sea cow 521
sea level 306–7
seaborgium 124
seals 322, 325
Sedgwick, Adam 80, 441
selenium 283
sharks 320
Sheehan, Peter 391
sheep 283
Shelley, Mary 111
Shoemaker, Carolyn 218
Shoemaker, Eugene: comet 230–1; death 231; Manson crater work 226, 227–8; Meteor Crater work 217, 218, 222, 230
Shoemaker–Levy comet 230–1, 235
silicon 281
Silurian period 80, 81
Simpson, George Gaylord 209
Sinanthropus pekinensis 492
singularity 11–12, 15
Sirius 30
sleeping sickness 355–7
slime moulds 346–7
SLiME (sub-surface lithoautotrophic microbial ecosystems) 344
Slipher, Vesto 145, 149
Smith, Anthony 291
Smith, William 92–3
snakes 387
Snow, C.P. 136, 138, 140, 158, 168
Snowball Earth 481
SNP (single nucleotide polymorphism) 462
Soddy, Frederick 125, 200
sodium 116, 282, 283–4
solar flares 385–6
solar system: birth 43–4; size 26–9
Sollas, William 490
Solo People 493
Somervell, Howard 289

space: curvature of 20; distances between stars 30
spaceships 288
spacetime 144
speech 508
Spencer, Herbert 435
spin 166
sponges 430–1
Sprigg, Reginald 368–9, 373, 376
squeeze, the 271
squid, colossal 318–19
squid, giant 318–19
Standard Model 186, 188
Stanley, Steven M. 389
stars: classification 148; neutron 36–7; visible 38
Steller, Georg 521
Stewart, William 354
stratosphere 286, 287
Strickland, H.E. 518
Stricklandian Code 404
stromatolites 335
strontium 116
Suess, Eduard 200
Sullivan, John 90
sulphur 44, 330
Sulston, John 450
Sun: appearance from Pluto 28; formation 43; helium 121–2; mass 65; solar flares 385–6
supernovae 33–42
superplumes 255–6
superstring theory 188–9
Surowiecki, James 355
Swift, Jonathan 400
synapsids 382
syphilis 106, 174

tachyons 184
takahe 416
Tambora eruption 471–2
Tasmanian tiger 525–6
Tattersall, Ian: on extinction 383; on fossil record 494; on *Homo floresiensis* 501; on *Homo sapiens* 507

INDEX

Taung child 491–2, 495
taxonomy 401–2, 405–8, 410
Taylor, Frank Bursley 199
tetraethyl lead 170–1, 178
tetrapods 381
therapsids 382
thermodynamics 87, 132–3
Thermodynamics, Second Law of 87n
thermohaline circulation 299
thermometers 295
thermosphere 287–8
Thomson, J.J. 134, 158, 163
Thomson, William. *see* Kelvin
Thomson, Wyville 290
Thorne, Kip S. 37, 135
Thorstensen, John 41–2
thunderstorms 291
Toba eruption 257–8
Tocheri, Matt 502
Tokyo earthquake 242
Tombaugh, Clyde 24–5
tools 505–7
Torsvik, Vigdis 412
trawlers 321
trees 380–1
Trefil, James: on atoms 165, 183; on cells 426; on clouds 297; on convection process 248; on tectonic plates 208–9
triangulation 49n, 58–9
Triassic period 83, 384
Trieste, bathyscaphe 311–12
trilobites 200, 362–4, 377, 384
Trinil skullcap 489
troposphere 286–7
Tryon, Edward P. 18
Tsurutani, Bruce 386
turbulence 292
Turner, J.M.W. 471
Turner, Michael 196
turtles 382, 388
tyrannosaurus rex 105n, 485

Uncertainty Principle 164–5

uniformitarianism 79
universe: age 12, 190–2; beginning 12–13, 15–18, 150, 194; composition 192–3, 197; edge 19–22; expansion 17–18, 19–20, 149–50, 194, 196; size 20–1; superstring theory 189
universes, numbers of 15, 18–19
uranium 124–6, 139, 175–6
Uranus 65n
Urey, Harold 224, 326, 329
Ussher, James 84

Valéry, Paul 140–1
Van Allen belts 250
Vaucouleurs, Gérard de 191
Venus: distance from Sun 278–9; transit of 61–3
Vescovo, Victor 315
Vine, Allyn C. 313
Vine, Fred 206
Virdee, Tejinder 187
viruses 355–9
Vogel, Shawna 248, 250
volcanoes: Galeras eruption 252; Krakatau eruption 256–7; Mount St Helens eruption 250–3; Mount Unzen 252; Tambora eruption 471–2; Toba eruption 257–8; Yellowstone 254–7, 258–65
Voorhies, Mike 236–8, 257
Voyager expeditions 27

Walcott, Charles Doolittle 364–6, 372
Walker, Alan 496
Wallace, Alfred Russel 438–9, 440
Walsh, Don 311–12
water: chemical formula 304; pressure 269–70; properties 303–5; salts 305–6
Watson, James 158, 456–60
Wedlake, George 338
Wegener, Alfred 199–202, 210–11, 478

INDEX

weight 68n
Weinberg, Samantha 307
Weinberg, Steven 20, 157, 185, 190
Weitz, Charles 463
Weryk, Robert 221
whales 318, 322
Whewell, William 82
white blood cells 352–3
White, Nathaniel 58
Whittaker, R.H. 347
Whittington, Harry 367
Wilberforce, Samuel 446
Wilbraham, Jo 397
Wilkins, Maurice 456, 457–60
Wilkinson Microwave Anisotropy Probe (WMAP) 12, 192
Williams, C. B. 408–9
Williams, Stanley 252
Wilson, C.T.R. 158, 181
Wilson, Edward O. 405, 407–8, 412, 526–7
Wilson, Robert 13–15, 150
WIMPs (Weakly Interacting Massive Particles) 195
Winchester, Simon 93
wind speeds 293
Wistar, Caspar 89, 92, 93–4, 106, 399
Witten, Edward 189
Witzke, Brian 227–30, 231–2, 235
Woese, Carl 347–50, 351
Wren, Sir Christopher 51, 517

Xena 25–6

Yellowstone National Park 238, 253, 254–7, 258–65
Younger Dryas 484

zinc 282
zooplankton 317
Zwicky, Fritz 35–8, 42, 193

ABOUT THE AUTHOR

Bill Bryson was born in Des Moines, Iowa, in 1951. He is the author of eighteen books and holds the record of having the most bestsellers of any author on the *Sunday Times* bestseller list in the last fifty years. *A Short History of Nearly Everything*, first published in 2003, spent 106 weeks in the chart, won both the Aventis Prize and the Descartes Prize and is the biggest-selling non-fiction book of the twenty-first century.

Bill Bryson is a former Chancellor of Durham University and is an Honorary Fellow of the Royal Society. He lives in England.

BILL BRYSON

The Body, A Guide for Occupants

SUNDAY TIMES SCIENCE BOOK OF THE YEAR 2019

'So packed with arresting facts (you eat 60 tons of food in a lifetime) and unlikely anecdotes (such as Isambard Kingdom Brunel's six weeks with a half sovereign lodged in his throat) that you barely notice the sheer volume of anatomical knowledge you're digesting . . . Makes complex subjects simple and eminently entertaining.'
Sunday Times

'We spend our whole lives in one body and yet most of us have practically no idea how it works and what goes on inside it. The idea of the book is simply to try to understand the extraordinary contraption that is us. What I learned is that we are infinitely more complex and wondrous, and often more mysterious, than I had ever suspected. There really is no story more amazing than the story of us.'
Bill Bryson

A wonderful successor to *A Short History of Nearly Everything*, this book will have you marvelling at the form you occupy and celebrating the genius of your existence.